An Amateur's Guide to Particle Physics

A Primer for the Lay Person

BY

RICHARD Z. ZIMMERMANN

DORRANCE PUBLISHING CO., INC.
PITTSBURGH, PENNSYLVANIA 15222

ISBN # 0–8059–6061–9
Printed in the United States of America

First Printing

For information or to order additional books, please write:
Dorrance Publishing Co., Inc.
701 Smithfield Street
Third Floor
Pittsburgh, Pennsylvania 15222
U.S.A.
1-800-788-7654
Or visit our web site and on–line catalog at www.dorrancepublishing.com

I wish to express my gratitude to my wife, Helen, for her patience and tolerance during the many hours in which this book preoccupied—a gratitude which is second only to my appreciation for her loyalty and faithfulness through many years of marriage, which is the most precious of many gifts that a woman may bring to the fortunate ordinary husband.

List of Chapters

Preface

This modest monograph is not intended as a scientific treatise or comprehensive textbook. Rather, it is intended for the interest and entertainment of the intelligent lay person, whether undergraduate student or not, who has an active curiosity about recent trends of thought by the world's science community on the nature of matter as the twenty-first century begins. In particular, as the title suggests, it attempts to deal in a simple way with our newly acquired knowledge of the atom and its parts. We do not concern ourselves with molecular reactions, which effectively eliminates almost the entire subject of chemistry and biochemistry. Moreover, the author feels unqualified even to attempt any serious discussion of the recent and more abstract and metaphysical "Theories of Everything," such as the Grand Unified Theory {GUT} or Superstrings. These have all multiplied in the last three decades and fascinate enough people to have acquired a devoted cult and respectable status, but experimental verification in the laboratory still remains a long way off.

The assumption is made here that there are many people who are not active participants in the fraternity of scholars and who are otherwise busily engaged in diverse occupations; but who, nevertheless, maintain an alert and continuous sensitivity to the efforts of thoughtful human beings over the centuries to better comprehend how the universe works and how we all got here in the first place. Merely having been born into the post-industrial revolution era of history should stimulate an intuitive realization of the prime importance of science in our technologically dependent society of nations. Yet there is far more to the story than that. The intellectual challenge, the excitement of the search for truth, mental encounters with many personalities of brilliant minds, the intrigue of decoding the secret construct of Nature and, ultimately, a sense of wonder and awe. Many of the prevailing theories today stretch one's imagination to the limit and physicists routinely discuss ideas among themselves which exceed even the fantasies of the science fiction magazines of our childhood. How did the stars and galaxies come into being? What makes the stars burn? What really is "space?" Is there undetected "dark matter" in space and, if so, of what is it composed? Is it possible that there are other worlds or universes outside our own space-time continuum which we cannot perceive? Why are we all locked into a one–way time frame with relentless entropy? How is it that the matter of our reality exceeds the anti-matter in the universe as we know it? To what extent is probability or chance built into the behavior of the universe? If we reverse all the plus and minus signs in our mathematical equations for the "state" of an atom, do all the objects and the physics of the atoms on the other side remain exactly the same except for being opposite hand? If mathematics leads the way for our understanding, then who wrote the blueprint anyway?

Clearly, mankind does not know all the answers yet, and there are those who contend that the brain is utterly inadequate for the task anyhow. For the theist, we are probing the mind of God. Notwithstanding all this, the mere fact that men and women presume to persist in the effort to comprehend is a form of evidence of some spark of divinity in a creature still evolving from the animal brute of evolution. This reader is thus encouraged to share the excitement of the search and to appreciate the brilliance of both the theoretical and experimental physicists who conduct it. Where this path of discovery may lead one cannot be predicted here; for all of us, savant or dilettante, are ultimately confronted with metaphysics and the unknown. A learned humility may not be all that bad, for it is the beginning of real wisdom. Nature continues to present us with many mysteries; the scientists, of course, already know this and admit it among themselves.

The reading public is most fortunate that a new development in the publishing business has taken place in the last two decades or so. Many excellent books of a far more profound nature than this necessarily limited summary have come into print from scientists of unimpeachable credentials. Although this new trend might not have happened if the book publishers had not perceived a viable market, the phenomenon, nevertheless, has emerged for a secondary reason. The younger generation of scientists have become less reticent to venture into popular print; and, in fact, many are even eager to explain their work in understandable terms for a wider audience than the college lecture hall. This was not always the case, especially in Europe. Few professors of earlier generations would scarcely dare to write for the general public, considering it either not worth the effort or not even respectable. Technical journals of their discipline were usually the only appropriately decorous means of written expression for stature and dignity. Today scientists of acknowledged achievement, including some Nobel prize winners, are writing their memoirs. Thus a bibliographic treasure trove of information and history has become readily available to those who desire to expand their learning and share the enthusiasm of the scientific fraternity.

There happens to be a fringe benefit to this trend for the scientific community itself, however. The informed or curious population of the world is fully entitled to an understanding of their work. Much of the experimental research in laboratories today involve complicated and hugely expensive apparatus and corps of highly educated technicians, a fact which makes a very significant proportion of the profession dependent upon college or government grants and contracts ultimately derived from taxes by public consent. People, as events have testified, have begun to wonder what exactly all those brainy people with their atom smashers and super-colliders are trying to accomplish which is worthwhile.

Anyone so presumptuous as to undertake a coherent and comprehensible picture of the nuclear particle specialty is immediately confronted with two significant literary difficulties for achieving clarity. The first is chronological in nature. Constructing the story on a very strict format of historical sequence for discoveries may not do the reader any great service for aiding his or her full understanding of the overall harmony of the developed theories. The path to those theories and a consensus of agreement was always a long and tortuous one filled with false leads and ruthless criticism. To follow rigidly the chronological evolution of thought on the subject would be similar to compiling a list of pieces in a large jig saw puzzle in the exact order in which the participants endeavoring to complete the puzzle picked up each piece of the game and tried to discover if it interlocked with another. No picture of the entire puzzle picture could ever appear to this reader until close to the end of the book; and at that time he would be obliged to reassemble all the facts in some form of organization mentally. For this reason this writer chose to adopt a little flexibility in the arrangement of accepted ideas, if not discoveries, without strict regard for the historical sequence of dates, although the real progression is not ignored entirely.

The other difficulty to which we alluded is that the second language of scientists the world over is really mathematics. Worse yet, it is mathematics of a very advanced and intricate order. Most of us ordinary citizens do not possess adequate training with these tools of symbolic logic. The mathematics undergirding the theoretical physics of the atom in these times is truly formidable, involving wave mechanics, the Fourier series, differential calculus in several ascensions, matrix algebra, and relativity. Superimposed are Laplace transformations, group and gauge theory, scaling and Monte Carlo probabilities, to mention a few. Consequently, all but the simplest algebra and trigonometry familiar in a good High School has been assumed as obvious. To the average nuclear physicist a popular survey of their field may seem like attempting to describe a chicken salad sandwich without knowing the attributes of any of the ingredients. It may even earn the scorn of some. Notwithstanding, many of the concepts may be expressed in ordinary language to the satisfaction of most people not involved in the field professionally.

The writer first became fascinated with particle physics as a late teenager attending science lectures at the Franklin Institute. At one lecture by Enrico Fermi, who had then just emigrated to the United States, the speaker happened to mention that nuclear fission could be accompanied by positrons.I had the temerity of youth to go down to his desk in the auditorium and ask, if an electron and a positron both came into close proximity, why would they not annihilate each other? His exasperated reply was "Of course, you idiot, what else would you expect!"Little did I know that he was implying the feasibility of an atomic bomb to more knowledgeable members of the audience!

Partially through the direction of my High School Counselor and partially influenced by close family relatives, but mostly because of the need to select an occupation, I happened to select a career as a Mechanical and Electrical engineer where the mathematical foundations appealed.However, three other facts contributed to the writer's interest.First was my later attendance as an engineer at lectures on nuclear reactors and atomic power sponsored by the Philadelphia Electric Company.Second was the impact of the two atomic bombs dropped on Japan which ended the war in the Pacific and probably saved my life as a serviceman in the Army Air Force in the Phillipine Islands. The third catalyst was simply the mystery and enigma of quantum mechanics itself and what it means for reality.The author hopes to persuade the casual reader that exploring the subject as an amateur is fascinating and fun.This book is an invitation to start your own journey.More than that, it is peering into the future of human knowledge. Professor John Wheeler of the University of Texas and Princeton University, who was originally educated as an engineer as a young man before turning to physics and the design of accelerators, plutonium reactor piles and atomic bombs during World War II, put the reason for his fascination succinctly: "I continue to say that the quantum is the crack in the armor that covers the secret of existence."[1] That is certainly not the statement of a mere dreamer.

[1] Jeremy Bernstein, *Quantum Profiles* (Princeton University Press, 1991).

Introduction

> In experimental philosophy we are to look upon propositions inferred by general information from phenomena as accurately or very nearly true, notwithstanding any hypotheses that may be imagined, until such time as other phenomena occur by which they may be made more accurate, or liable to exceptions.
>
> —Isaac Newton of London in 1868

The world of the atom is a very strange place. When we venture inside it mentally, we all take with us many of our "common sense" notions and preconceptions derived from experiences in the much larger macroscopic world of matter outside the atom. Most of these ideas are essentially Newtonian or, as the scientists say, "classical" in essence. Moreover, most of them involve a conciousness of local time and a confidence in the permanence or immutability of most of matter, however those words might be defined. Yet exploring intellectually into the sub-microscopic world of individual atoms violates many of our familiar expectations and confounds our innate sense of reality. Things are not always what they seem by our "common sense" intuitions; transformations from matter to energy and back again take place in fractions of time too small for our honest comprehension. Some sub-particles appear instantaneously with velocities near the ultimate speed of light in a world of Einsteinian relativity. Our time is warped by these extreme velocities in a moving system.

Considering how small the diameter of even an atom of heavy gold actually is in special dimensions (about 2.6×10^{-8} centimeter or .00000003 centimeter), it is extraordinary how much the scientists have been able to learn about the atom thus far. Equally remarkable is the fact that there are times when the behavior of atomic components may only be expressed by mathematical equations of a very high order. These are often beyond the capacity of the human imagination to fully picture in terms which correlate with our experience. This mysterious connection between mathematics and Nature has always been a source of wonder and delight to all scientists and philosophers. It led Sir James Jeans back in 1930 to write in his book, *The Mysterious Universe*, that "the universe begins to look more like a great thought than like a great machine." Romantic as that observation may strike the reader, it is echoed today by John D. Barrow in his current book, titled *Pi in The Sky*.

When dealing with physics, a problem with some words will inevitably be encountered by most of us. This difficulty is not merely the result of the fact that the discipline has accumulated, over time, many terms which refer to a most precisely defined phenomenon, quantity or magnitude in the natural world. The problem is that there is a shortage of universally accepted and commonly understood words in our everyday English language for many of the involved and abstract ideas encountered in the study of the mechanics of the interior of the atom. Physicists are human beings also, after all. They are just as guilty as us of carrying over words belonging to comfortable old associations into their conversations about more arcane matters as symbols or pictures for phenomena which take place in the atom which they are attempting to describe. Often these words, which they have been accustomed to using, have been arbitrarily assigned to designate certain attributes or behavior of forces or particles which have little correlation with their meaning in the ordinary language of the man in the street. Worse yet, structural visualizations are often given to energy transformations for the purpose of convenient mental imagery which are, in fact, mere crutches to aid our limited earthly imaginations and may be entirely fictitious. Thus we are dealing with a new world which is strange indeed and often materially insubstantial. The emotional reaction of the layman is apt to be instinctive skepticism, if not outright incredulity. We must learn to comprehend concepts which are frequently described with words having an entirely unfamiliar connotation or which stand merely as symbols having a different meaning than that of Webster's dictionary. The vocabulary may be easily learned; but the theoretical concepts are more subtle.

It therefore happens that the reader must be warned at the outset that even the term "particle" should be viewed with suspicion, since some have the provisional appellation of "virtual" and others may exist only to decay almost instantaneously into relatively more permanent particles capable of observation. Alas, even the electron, which is the fundamental source of energy for all our electrically powered machines and the more recently developed electronics industry, is not exempt from this ambiguity of words. Once upon a time it was regarded as an extremely light and tiny, but substantial, particle temporarily locked into a tight orbit around the nucleus of a host atom. Today it is more fashionable to think of it as a point source for a cloud of electromagnetic negative charge which may temporarily be found in one of several possible elliptical orbits at an energy level location subject to statistical probabilities which may be calculated only by means of the new science of Quantum Mechanics. Such language is apt to change the mentally assumed "ground under our feet," figuratively speaking, into a quicksand of speculation. The explorer of the atom must be a diligent soul capable of an active imagination and resilient mind, mixed perhaps with a resolute dedication to the proposition that the truth may ultimately be discovered.

As anyone familiar with the subject of science knows, the entire body of scientific knowledge to which we all have access in this early twenty-first century represents a gradual accumulation of many discrete contributions by thinking individuals preoccupied with a greater understanding of the natural world. All the textbook explanations were built methodically upon the work of past investigators with "the bricks" of this intellectual edifice going back to four thousand years before Christ. The consequence is that no matter how appealing and exciting a new theoretical construction may sometimes appear, the responsible scientist is always loath to casual discard the teachings of the past. This conservatism has proven desirable and provides time for a more critical review by peers. One should many times regard the older concepts as more often corrected in fine particulars instead of totally discarded.

These previous comments are intended as a warning to the reader, since he or she will encounter many new and currently prevailing ideas on the working of the atom. Considering the popular enthusiasm of our own technologically oriented society, one has no difficulty in finding numerous publications on the latest "discoveries," whether in book, magazine or newspaper form. North Americans are generally not bashful about striving to be the first or the newest. This becomes evident in the universal bantering about in print of the word "quantum," which may be applied to the brain, art or the stock market. It is also fashionable to accept the attitude that classical or Newtonian physics, meaning all physics before 1920, is hopelessly antiquated and inadequate. People talk knowingly about the revolutionary new ways of thinking ushered in by Special and General Relativity and Quantum Mechanics. In truth, there are very important philosophical implications to these new developments in physics which do affect the individual's thinking about himself or herself and the universe around all of us. Nevertheless, such grand eloquence can be a misleading generality. Our own minds and those of our scientists belonging to the so-called Western Civilization have not really thrown off all the acquired habits of thought from previous centuries. One reason is that the old "classical" physics has continued to work fairly well for our normal macroscopic problems, though not always with absolute precision. Nevertheless, our own individual intuitions are thus satisfied. After all, the entire space shuttle and satellite probe program of the United States for the exploration of the planetary system is built upon Newtonian physics.

Sir Ernest Rutherford used Newtonian physics to calculate particle momenta from measurements made in his laboratory while observing their scattering trajectories which resulted in his discovery of the existence of atoms only 10^{-14} centimeters across. Although his calculations were not totally accurate by today's standards, the correction for Special Relativity was sufficiently small as to be insignificant for affecting his essentially correct conclusions.

Nevertheless, it is entirely possible that the study of particle physics has reached an intellectual threshold where a new and, perhaps, revolutionary concept may be required and is yet to be enunciated. The reader should not be surprised by this statement and should keep it in mind. Some of the physicists' theories today may very well be incomplete; although the scientists are loath to believe they are totally erroneous. Calculated predictions match observed results too closely not to believe they are on the correct track. Although change or revisions may be psychologically unsettling to those people who need didactic certainty, the prospect must be conceded. In nuclear physics today, imaginative theories abound; but only time will tell us which ones will eventually stand recognized as truth.

Placing such skepticism aside, there are certain definite verities and traditional intellectual tools which all scientists employ in their work and which we, as intelligent readers, must accept. The concept of particle *mass* is certainly one of them; and standard units of measurement are another.

Mass is an attribute, characteristic or quality of all matter. So, then, what can the matter be? If one takes the trouble to look up the word "matter" in the large Webster's New International Dictionary, one may be surprised to discover those savants expended almost an entire page on that one subject in order to sort out and define the various popular connotations of the word so casually used in common parlance. The fact is that our generally acquired thinking on the word is astonishingly vague and imprecise. Speaking for the philosophers, the dictionary states that matter is "the indeterminate subject of reality; the wholly or virtually passive element in the universe; the unorganized basis or stuff of experience which, when combined with form, gives phenomena or real objects." Speaking for the physics profession, on the other hand, they say that matter is "whatever occupies space; that which with energy forms the basis of objective phenomena. The nature of matter is unknown." And you thought this study would be easy!

Objects in our world of daily experience all tend to fall down towards the center of the earth unless otherwise restrained and, if they are large enough, mutually attract one another with a force which we call gravity. Since we measure the inertial mass of any physical object by weighing or measuring the magnitude of this mysterious force field, we might well be tempted to define mass in this manner. Gravitational mass is that attribute of substantial matter which feels or experiences mutual attraction to other separate masses by Newton's laws established for this gravitational force. If we demur from the venerable name of Newton and the classical past and wish to strive to be more up to date, we might revise our arbitrary definition to include the General Theory of Relativity. That might entail adding some abstruse language such as stating that any mass of significant magnitude so warps the geodesic world lines of the space-time continuum surrounding it as to create a vectorial acceleration of any other adjacent inertial mass system in its direction.

As the reader undoubtedly suspects by this time, this suggested approach above is absurdly impractical for nuclear particles or even whole atoms. It takes millions upon millions of elemental atoms of carbon, hydrogen and oxygen in order to make a teaspoonful of sugar; so it is easy to perceive that gravitational mass of individual atoms becomes virtually meaningless for all practical purposes. More than that, the gravitational forces on an atom are so much weaker than any other of the forces working within the atom that they become negligible when compared to the electrostatic or electromagnetic. Nuclear physicists soon recognized this and very seldom include gravity into their calculations for particle behavior, which is mildly embarrassing to them if challenged.

There is, however, a second property of mass which the physicists prefer to work with as a matter of routine; and that is momentum. Now you understand why we keep talking about inertial mass. A body at rest tends to remain at rest unless acted upon by an external force. A body in motion in a straight line tends to remain in the same motion in the same direction unless acted upon by an external force, to paraphrase Isaac Newton. Surely those sentences ring a bell in the memory of the reader from their school days. There are at least two big reasons why scientists find momentum so useful in particle physics: (1) The Conservation of Momentum Law means that when two bodies or atomic particles smash into one another, some energy is lost in heat or radiation usually, but the total momenta of the smaller particles after the collision must remain very closely the same as that which existed before impact; and (2) one can exert a small force on a moving particle, such as by a magnetic field, and then measure the velocity of the particle with timing devices so as to be able to calculate its mass. Sir Joseph J. Thompson in 1897, did exactly this by measuring the curved deflection of moving electrons in a magnetic field of known strength and came up with a value of 9.108×10^{-31} kilogram for the mass of the electron. Now you can appreciate the reason that he was knighted for his cleverness.

As the exponents of an entirely new branch of science, the nuclear physicists appropriately took a novel approach to measuring or evaluating mass. The reader may remember, energy represents the ability to do work or produce some action. Work can be measured in the laboratory. It is perfectly true that they all could have struggled along with the old conventional units of Joules or Ergs (meaning dyne centimeters) for the magnitudes of their energies; but such a system was awkward. Not only were the units of measurement too large; but they were quite arbitrary. The scientists were looking for a simple relationship between mass and energy and, moreover, something which could be used as a single universal constant applicable to the entire universe. It happened that Albert Einstein helped to point the way in 1905.

Einstein almost single-handedly revolutionized the field of physics; and it all began quite innocuously. He had been conspicuously good in mathematics in school and was a moderately good student in other subjects; but he also possessed a keen intelligence and an original, imaginative, and perceptive mind. He was not interested in gymnastics or athletics and did much of his studying outside of the classroom. He hated memorization and learning by rote, as many talented people still do; and he sometimes succeeded in arousing the hostility of his teachers by arguing with their statements or smiling condescendingly at their errors. All this is not unfamiliar to good teachers today; but in the late years of the nineteenth century in Austria and Switzerland the school system was autocratic and not inclined towards tolerant leniency. Despite his quite superior record of grades, Einstein had difficulty in securing his first permanent job after graduating from the Federal Institute of Technology in Zurich in 1900. Eventually he was accepted by the Swiss patent office in Bern and started work there as a clerk in June of 1902. He stayed there for nine years and the tranquil routine that he enjoyed there allowed him the time and energy to pursue his own ideas on physics and to write papers for publication. His first dealt principally with statistical mechanics and the kinetics of molecules involved with the thermodynamics of gases. Einstein was fascinated with the well-known Brownian movement or erratic jiggling of very tiny dust particles in warm water or other liquid while viewed under a microscope. His many technical articles (forty or more), published over a duration of twenty-two years revealed his total acceptance of molecules and particles as a physical reality of our world. Thus it became quite natural for him to eventually conclude that light rays were composed of discrete energy quanta or corpuscular "photons," as they were later named. Isaac Newton, over two hundred years earlier, had toyed with this same idea, which testifies to the prescience of that genius.

The year of 1905 was an extraordinarily productive one for Einstein. By that time he was a married man with one infant son and had been accepted as a member of the permanent staff of technical experts at the Patent Office. All in this same year he managed to submit his successful Ph.D. thesis *On a New Determination of Molecular Dimensions* to the University of Zurich. He also wrote a paper on Brownian motion, and followed this by two separate papers on Special Relativity. The second of these, submitted for publication in September, contained Einstein's thoughts on the nature of light and its transmission and led him to comment, almost casually, that "the mass of a body is a measure of its energy content." In addition, he stated "the law of conservation of mass is a special case of the law of conservation of energy." In the following year, he amplified this statement to be more specific as "In regard to inertia, the mass $m$is equivalent to the energy content—mc^2." Thus one of the most memorable formulas in the history of physics was born. Although Einstein did not strictly originate this idea by himself, he clearly ennunciated it in unequivocal language and subsequently over the years proceeded to give three different proofs of the famous $E = mc^2$ equation, as well as suggesting that the decay of radium salts could be a test of the proposition.

The result of all the above history is that, today, we are permitted to think of particle mass as a kind of frozen energy, figuratively speaking, like chilled moisture in the air congealing into snowflakes. The analogy is not entirely whimsical.

The constant "c" in Einstein's equation stands for the velocity of light in the vacuum of space, which calculates to about 299.79 million meters per second or 180,230 miles per second, in familiar American terms. If one transposes the terms algebraically, then:

$$\text{Mass} = \text{m} = \frac{\text{E}}{\text{C}^2}$$

The denominator of the speed of light squared obviously becomes a very big number, which simply means the atom can pack a huge amount of energy in an infinitesimal space. Notwithstanding, we are still talking relatively small units when dealing with most nuclear particles. In the early days of flat disc synchrotron accelerators, it became the convention to measure nuclear energies in electron volts. The latter unit is equivalent to 1.6×10^{-19} Joules or 1.6×10^{-12} ergs. If you ask how it is evaluated in the laboratory, we can only answer rather vaguely for now that it is the energy calculated from that required to change the direction of motion of negatively charged electrons moving with a known velocity from a straight to a curved path when passing between two plates having a known voltage difference or electromotive force and subjected to a perpendicular magnetic induction field of a known strength measured in Gauss or Webers per square meter. If you are a scientist, that answer will most probrably frustrate you as meaninglessly vague; but it will have to suffice for the purposes of our story here. However, some additional words of explanation for this energy unit convention used by nuclear physicists are in order.

The rest mass energy of a proton calculated by Einstein's famous formula is 938 million electron volts (938 MeV). Therefore, the rest mass of the proton could be considered simply 938 MeV / c^2 while keeping in mind the formula above. Just to confuse the novice, the scientists habitually omit the velocity of light squared divisor while knowing it should always be there anyway. It is supposed to be understood, which can be confusing to the layman. On the nuclear scale of things, it is usually convenient to deal with units of one mega-electron volt or MeV as an abbreviation for the normal energy measurements. It does happen, however, that when one is dealing with leptons or neutrinos, which have little or almost no rest masses, we may need to resort back to the tiny electron volt or eV. At the opposite extreme, the energies involved with the development of huge modern particle accelerator machines leads to the use of the Giga-electron volt or even higher. A table for the various increasing units used for describing the energy levels of high velocity particles encountered in magnetically confined beams is included here for the reader's convenience.

Abbreviation	**Full Name**	**Multiple**	**Value**
meV	Milli-electron volt	eV / 1000	10^{-3}
eV	electron volt	None	1
KeV	Kilo-electron volt	1000	10^{3}
MeV	Mega-electron volt	Million	10^{6}
GeV	Giga-electron volt	1000 Million	10^{9}
TeV	Tera-electron volt	Million million	10^{12}

Remember we are dealing with a sub-microscopic world. Just to keep our perspective, it is wise to keep in mind the analogy suggested by Professor Gerald Gabrielse of Harvard University. A large paper clip of perhaps one gram in weight falling from a height of one meter above the ground would deliver a kinetic energy of approximately 10^{17} electron volts or 10^{5} TeV.

Those of you who may remember your physics from high school days may practice, if you are so inclined, on the equation $E = 1/2\ m\ v^2$.

Armed with all the previous caveats and dull facts, the reader should now feel prepared to venture eagerly into the dynamic and intellectually stimulating world of particle physics. The writer earnestly hopes he or she will not be deterred, but that they will proceed with a modicum of skepticism about any excessively didactic assertions of the ultimate truth for all time. We do not introduce this peculiar caution merely because of the acknowledged frailty of the author's knowledge, but in agreement with the quotation of Isaac Newton given at the head of this chapter. All real scientists have developed, over the last two centuries, quite a rigorous definition of what may be considered "truth" in their occupation; and it has become a tradition in the conduct of their professional work. Generally speaking, an accepted theory or proposed law of physics becomes true only if a described physical experiment has been repeated by other people at a different time and at a different location on the surface of the planet and this other experiment, allowing for environmental conditions, has replicated exactly the previously observed results. The need for such a stricture sprang, for very good reasons, from experience with human nature and cannot be readily denied. There are, however, a good many people in the world who are sometimes impatient with such a rigidly restrictive credo and would sympathize with William Shakespeare and his play character "Hamlet." To quote the bard: "There are more things in heaven and earth, Horatio, than are dreamt of in your philosophy." (Act 1, Scene V) Scientists usually indulge in such speculation only in private and are more skeptical about ghosts.

The root of the philosophical argument is whether or not the operation of all phenomena and/or human experience can be satisfactorily verified in the experimental laboratory. If there is one field of scientific endeavor, apart from psychiatry, which is beginning to encounter serious difficulties with adhering to the traditional credo as it enters the twenty-first century, it is that of nuclear particle physics. The barriers of cost and technical feasibility for accelerators and detectors are becoming more formidable. The physicists well realize this, of course; but the optimists refuse to be discouraged. Time, they say, is on their side. Whether all their theories are quite valid or render any credible and convincing meaning to the ultimate purpose of human life in the cosmos remains a matter of personal opinion. Most scientists normally restrain themselves from expressing any such philosophical speculations in their professional journals, although they might venture some opinions to intimate friends in a private environment. It can certainly be stated that the current popular notion that all modern scientists are atheists is entirely false; though many may admit to being agnostics. The choice is inevitably left up to each human individual. Perhaps that, in itself, may be one of the immutable laws of the universe.

CHAPTER 1
THE ATOM AND ITS PARTS

Since the subject of this book is intended to be the parts and fragments of the atom and all the strange things can take place or originate in its interior, it seems logical to start with a highly simplified description of that entity. The reader will discover for himself or herself the various complications as he or she progresses; but it is sufficient to say that our picture of the atom given in this chapter and elsewhere was not revealed in a sudden burst of brilliant perception by any one individual. The vision of the atom which most often comes into our minds of a spherical billiard ball so tiny as to be invisible in any known microscope is ridiculously crude; but it suffices for the ordinary affairs of non-scientific people in daily life. The further subtlety that this billiard ball actually represents a miniature solar system of electrons rotating around a dense nucleus was probably a recognition of the earliest contribution of Neils Bohr of Denmark to the field of nuclear physics. We have acquired both these conveniently memorized descriptions as the result of an institutionalized cultural and informational lag in our educational system.

It is appropriate that the word *atom* or *atomon* derives from the ancient language of the Greeks and their philosophers (principally Leucippus and his student, Democritus) during the intellectual awakening of our western civilization. It is customary to credit Democritus with the invention of the idea of the atom. The word means "indivisible" or "uncuttable." Democritus was a wealthy and much traveled citizen of Abdera in an area once known as the Roman province of Thrace. He lived from approximately 460 to 370 Before Christ, which means, if these dates are nearly reliable, that he may have reached the age of ninety. That would have been an extraordinarily long time for any individual living that far back in history, but not impossible.

As Democritus expressed it himself: "These atoms are eternal and invisible; absolutely small, so small that their size cannot be diminished; absolutely full and incompressible, as they are without pores and entirely fill the space which they occupy."

He also asserted that the motion of atoms was in all directions as a sort of vibration; hence there resulted collisions and, in particular, "a whirling movement whereby similar atoms were brought together and united to form larger bodies."

Those are fairly specific words and might even be construed to mean that he anticipated "spin" or angular momentum in particles. School children in our own times, some twenty–three hundred years later, are taught that the atom is the smallest irreducible component of any given element among ninety–two or more varieties of mineral or chemical elements in the universe which still retain unchanged the identical chemical and physical properties of that element. Dividing the nucleus of any given elemental atom into smaller parts, if it were possible, either destroys that element entirely or else creates two new and different elements with different chemical properties. The known stable elements start with hydrogen, the lightest and smallest atom and end with plutonium, the heaviest and largest, ignoring several artificially created isotopes which are short–lived.

Democritus is said to have squandered his inherited fortune on travel and resided in Egypt for seven years, returning to his homeland a poor man. Inasmuch as Thrace was situated on the Aegean Sea just north of modern Greece and west of Turkey and the Bosphorus Straits, it seems quite probable he assimilated many ideas from the Persians and, indirectly, the ancient Babylonians. His interests were by no means limited to atoms, for he speculated upon his entire environment and the physical universe. One of his most cogent observations has been widely ignored for lack of full understanding. He asserted that: "Space or the Void had an equal right with reality, or Being, to be considered existent." He conceived of the Void as a vacuum - "an infinite space in which moved an infinite number of atoms that made up Being." In other words, the dark and empty space in which the stars are stationed like candelabra in the night gives us both the geometry of place or location and the size and shape of physical objects. This is an important insight and one which the cosmologists ponder.

The actual size of the atom frequently comes up for discussion. It was long considered impossible to "see" an atom with any optical device because of its extremely tiny size being lost in the wavelengths of the probing device. That barrier has even been broken in the laboratory now by using computer assisted imaging and a microscopically fine, pointed wire probe used to measure extremely weak electromagnetic field differences. The probe is made to scan the area of a crystalline metal surface by means of electro-mechanical devices. The picture of the surface of a deposited metal film is somewhat fuzzy and indistinct, it is true. Nevertheless, orderly rows of closely packed hemispherical domes become quite evident on the computer screen.

The size of an atom, of course, is an ambiguous idea in the first place. Not only is one attempting to measure something which varies with the mass of the element chosen; but the outermost boundaries of each atom are indeterminate by the very nature of electrons and their motion around the nucleus. We also assume a spherical shape, which is by no means certain or necessary. One can only offer the glib generality that their diameters range between 1.0 to 5.0 x 10^{-8} centimeters on average. The gold atom, for instance, is said to be approximately 2.6 x 10^{-8} centimeters in diameter.

The first sub-component of the atom to be recognized was the electron, and the laboratory apparatus which brought that particle to the attention of physicists world–wide was the Crookes tube. This device was so simple as to be almost laughable in our contemporary research laboratories; but it visibly demonstrated a distinctive phenomenom which demanded an explanation. The Crookes tube in its original basic form was a cylinder of glass with closed ends which had been evacuated of all air and which contained a metal anode or positively charged plate at one end and a negatively charged cathode plate at the opposite end. These plates were connected to the terminals of a high voltage electric storage battery by means of platinum wires passing through the glass ends. In principle, if one were to change the glass cylinder into a long, sinuous pipe and fill the pipe with a small amount of neon or argon gas at low pressure, it would become the ubiquitous and garish advertising sign so typical of our society.

The maker of this gadget, William Crookes, learned the art of glass blowing in a chemist's laboratory when he was young; and his invention made him eternally famous, even though he lived a long and productive life with many other accomplishments to his credit. Crookes lived in London from 1832 to 1919 and was one of sixteen children fathered by a prosperous tailor. Four years after his graduation from the local College of Chemistry he came to the attention of the contemporary sages of Stokes, Wheatstone, and Michael Faraday. The latter inspired the special admiration of Crookes. They all succeeded in partially diverting this keen young man into physical researches of an electrical nature, although he later became the editor and publisher of a chemistry trade journal.

A key contribution to the success of this cathode ray tube was Crookes' skill in developing a mercury pump which could achieve a very high vacuum in the glass bottle. With this ability and a curiosity about electrostatic attraction and repulsion, it is not surprising that the same man stumbled into the invention of the "light mill" or radiometer. Four horizontal arms having small metal discs attached to their extremities were soldered to a vertical shaft free to pivot or turn on nearly frictionless bearings. Each disc had a polished and shiny surface on one side and a carbon blackened surface on its back side. The entire device was enclosed in a spherical glass bowl emptied of all air. This gadget is now sold today as a toy and intrigues all people by its' whirling around when placed in sunlight.

In the years from 1876 to 1879 Crookes devoted much of his own time and that of his several assistants to the study of the mysterious green glow at the surface of the cathode electrode and the so-called dark space of gaseous ions nearer to the center of the glass tube. Particularly significant were the facts that high electric currents could cause the glass itself to become fluorescent and that unknown emanations susceptible to a magnetic field could be detected outside the glass walls of the tube by means of a zinc sulfide scintillation screen. The result of these observations was that Crookes adopted wholeheartedly the concept of Faraday's of a fourth state of matter and declared in his own publication that: "In studying this Fourth State of matter we seem, at length, to have within our grasp and obedient to our control the little indivisible particles which, with good warrant, are supposed to constitute the physical basis of the Universe.... We have actually touched the borderland where Matter and Force seem to merge into one another." Considering the age of electric power which was yet to come and that Albert Einstein had just been born in March of 1879, that statement was unusually prescient and not an exaggeration.

He continued on in an enthusiasm of mystical speculation about finding in the future "Ultimate realities - subtle, far reaching, and wonderful." As his enemies were wont to point out, Crookes believed implicitly in life after death and had an intense interest in the occult, medium séances, and what is called "parapsychology" in our times. It would seem he might make a good patron saint for the New Age cult of California.

The twenty years between 1895 and 1915 were a gloriously exciting period for physicists as they entered a new and unexplored field of research. Most textbooks give Sir Joseph John Thomson (1856 to 1940) credit for identifying the electron and demonstrating that it originated from, and was part of, the atom. Without intending to detract from his genius or accomplishments in any way, it needs to be said that the whimsical and haphazard ways of Fame in the world often unfairly minimize the contributions of many other dedicated scientists. There was a concerted effort made in those days in both Germany and England towards understanding these new phenomena and there was free and open communication between the scientists themselves—sometimes in spite of a state of international warfare.

J. J. Thomson, as he was more affectionately known by his associates, was born in Manchester, England, as the only son of a bookseller who died early. The young boy had been sent to Owens College with the intention of his becoming an engineer and he was able to complete his studies only with the help of scholarships. He excelled at mathematics and was coached by his professors to take entrance and qualification examinations for another scholarship at Trinity College in Cambridge in 1876. He never left there and became absorbed in what was then known as Natural History research. He eventually rose to become the first Director of the newly established Cavendish Laboratory. His obvious facility with electric and chemical apparatus, then scorned by older and more traditional English scholars as being uncomfortably close to "the trades," proved no handicap to his ultimately achieving renown and position.

Thomson returned his attention to the cathode ray tube and demonstrated that the stream of particles inside the glass tube moved very considerably slower than the speed of light. About this time Roentgen discovered X-rays, which attracted Thomson's attention also. The latter astutely perceived that measurements of the kinetic energy of particles afforded a much better method than potential energy for evaluating the mass. Thus bending the path of a corpuscle beam with a known and stable magnetic field was an indication of the momentum. If one could determine the particle velocity, then the mass became a simple calculation. By 1889 Thomson had made the distinction between what was soon to be named the electron and Michael Faraday's gaseous ions. He also maintained that these newly discovered particles were a unique atomic sub-particle of matter having a separate identity. His measurements closely determined the ratio of charge to mass (e/m) of this negative particle and found it to be on the order of one thousand times greater than the same ratio of the positively charged hydrogen atom, which we know today as the proton. This significant laboratory–verified result amazed and confounded all the other scientists at first announcement; but it eventually won him the Nobel prize in 1906.

At this time in history the atom was visualized as a diffuse sphere of positive charge through which the electron moved, possibly in circles and possibly not, being subject only to electrostatic forces. Thomson himself described the electrons as being embedded in a positively charged and amorphous matter of greater mass "like so many raisins in a plum pudding." The Cavendish Laboratories with its many brilliant graduate students had now become world famous; but it fell to Thomson's assistants and successor to prove this notion erroneous.

One of these was Ernest O. Rutherford, who was born on 30 August, 1871, in one of the remote "colonies" of England, the fourth child among twelve by hard–working parents on a New Zealand farm. Obviously alert and capable at an early age, he eventually obtained a scholarship to Canterbury College at Christ Church in New Zealand. He did well in all his subjects, but showed a special proficiency in mathematics. While an undergraduate student, he also acquired an interest in science and built a radio wave receiver. A second scholarship in 1895 permitted him to travel to Cambridge, England, for his graduate studies. There he survived a very rigorous and obligatory schedule in advanced mathematics imposed by the college administration. For a while he worked as a research student in the laboratory presided over by John J. Thomson, but ultimately received an invitation to come to McGill University in Toronto, Canada. A superbly equipped research laboratory had just been completed there through the generosity of a wealthy tobacco entrepreneur. It is said the same gentleman once declared that smoking was a disgusting habit. Nevertheless, it was a great opportunity for Rutherford and he threw himself with zeal into the phenomenon of natural radioactivity in the rare earths. This behavior was still not understood by the scientific world and Rutherford became fascinated with the alpha particle radiation emitted by radium, among other minerals. These alpha particles were soon to become recognized as helium ions which had split off spontaneously and for no apparent reason from the atoms of the parent radium.

Rutherford spent nine years at McGill University pursuing this research in collaboration with other eminent chemists. His book on radioactivity was published in 1904; but by that time he had become a busy and popular public speaker. This all helped to earn him a Nobel prize somewhat later in 1908. Meanwhile, he had become anxious to return to England; so he readily accepted an invitation from the University of Manchester in 1907 to direct their own excellent laboratory.

He continued to make very careful measurements of the speed and energy of the so-called "alpha" particles; but was quite aware of the "gamma" or X-ray emanations as well. With the help of a group of remarkably able and dedicated assistants such as Hans Geiger, Bertram Boltwood, and Ernest Marsden, Rutherford embarked on a thorough experimental program of using the relatively heavy alpha particles as "bullets" with which to bombard various targets. The famous Geiger counter was invented by the graduate student of the same name when they discovered that directing their particles into a glass cylinder evacuated of all air and containing a central and electrically charged wire could cause a discharge and thus record single events. From that point it was only logical to direct these natural projectiles emanating from radium at high velocities into various different metal targets inside a glass Crookes tube and see what happened. With relatively simple apparatus and the use of zinc sulfide scintillation detectors it became possible to study the scattered angles at which the alpha particles or their resultant secondary effects were deflected. This same technique has become a basic and valuable tool of particle physics ever since. By this time Rutherford was in possession of his delayed Nobel prize in chemistry (since the specialty of particle physics did not then exist) for his work at McGill University.

The tedious and long task of keeping all the measurements and records of the particle scattering process was assigned to Marsden. He was obliged to carefully check the ricochet fan pattern through a series of different angles measured from the axis between the alpha "gun" and the metal target on both sides of the glass tube. The great majority of the helium ions passed right through the selected ultra–thin gold foil target as if there was no obstruction present. Others glanced off in several random directions; but on some few occasions the deflected particles exceeded a 90 degree angle and bounced backwards towards the alpha particle source or before the front face of the target. He found that even at an angle of 150 degrees from the direction of travel to the metal target an occasional charged particle registered on his counter. This recoil result totally shocked everyone in the laboratory. Rutherford himself regarded it as almost incredible and has often been quoted as saying it was: " ...as if you had fired a fifteen inch artillery shell at a piece of tissue paper and it came back and hit you!" Rutherford's war service had been devising reliable range tables for military cannon and the like, as well as research in the projectiles, so his remark would have been quite characteristic. To follow his same analogy, the cannon projectile had been deflected by a very tiny steel ball bearing at the center of an atom. Thus it became evident the atom was not an amorphous piece of "pudding" after all.

The Cambridge experiments had most effectively demonstrated that the greater part of the atom was empty space; but that the positively charged nucleus in the center of the atom was very small, very dense, and very cohesive. The fact that Rutherford instantly recognized the full significance of the experimental results is a testimony to his intuitive genius. The proton, whose name came from the Greek word "protos" or "first," had been discovered and proven.

World War I emptied his laboratory of most of his younger assistants and diverted his science efforts to the Navy Admiralty offices, but near the end of this conflict he returned to his original work. This time, however, he used the alpha particles to bombard the molecules of hydrogen gas and demonstrated that the source of the counter scintillations were hydrogen nuclei or, in other words, protons. In 1919 J. J. Thomson retired from the directorship of the Cavendish Laboratories at the age of sixty–three and Rutherford, by unanimous approval, took his place. The older man continued his active interest in nuclear physics and lecturing with spectacular success, becoming a Peer or Lord in 1931.

One cannot avoid being startled by the singular coincidence that both these early giants in the field of physics in England, who really started the discipline of particle physics, depended for their first education and academic credentials on money scholarships extended to the gifted by examination. There is an important lesson thus exemplified and to be observed by all societies in the world.

Up until roughly the 1920 decade the line between chemistry and physics was quite blurred. It was most probably the study of radioactivity that gave physicists a legitimate reason to claim attention as an independent specialty. The chemical profession was by no means idle during this gestation period, however. It had already, in a real sense, pointed the way with Dimitri I. Mendeleev's invention of the Periodic Table of Elements in 1871 and the use of the "mole" weights for the quantitative analysis of molecular compounds. In both, atomic weights were assigned to the various different elements in terms of whole number multiples of the weight of the hydrogen atom with the Periodic Table organizing them all in columns ranging from chemically inert to highly reactive. Strangely, the calculated exact atomic weights came out not to whole numbers, but to uneven decimal fractions. These things all demanded some rational explanation. Valence, or the ability of atoms to combine or link with others, was quickly identified with the number of loose electrons in the outer shells of the atom; but it became clear that the atomic weights exceeded and did not correspond with the number of positively charged protons. Some other entity possessing mass was missing from their calculations. For a long time most scientists, including Rutherford, assumed that this extra mass could represent a tightly bound agglutination of negatively charged electrons with positively charged protons in the very core or nucleus of the atom, thus forming a neutral whole. This explanation seemed perfectly sensible and plausible for quite a while; but it was incorrect.

One of the first men to provide a clue to the possible existence of an independent neutral particle stable in its own right was Walther Wilhem Georg Bothe (1891 to 1957). He obtained his doctorate from the University of Berlin under the supervision of Max Planck. After graduation he was immediately caught up in World War I and was a prisoner of war in Russia for five years. Returning to his homeland, he went to work with Hans Geiger in a technical laboratory and taught physics for a living. While bombarding the light element beryllium as a target with alpha particles in 1930, they detected a mysterious radiation which was not in the least affected by any magnetic field. They had been expecting to get a yield of gamma rays, so such a behavior would have been normal under such circumstances. However, this strange "radiation" was stopped and readily absorbed by a number of different substances, including wax, to different degrees and evidenced different energies – a behavior not at all compatible with X-rays. The true nature of this unknown radiation remained a mystery to Bothe for a time. Notwithstanding, he continued his career to become the Director of the Max Planck Laboratories at Heidelberg in 1934 and receive a Nobel Prize in 1954.

The honor for the discovery of the neutron, which so puzzled Bothe and which became recognized as the third constituent of the atom, is generally given to James Chadwick (1891 to 1974) of Manchester, England. He was

Oxygen
Atom

8 electrons
8 protons
8 neutrons
Atomic Weight = 15.999
FIGURE 1.1

enabled to attend Manchester University by a scholarship and studied under Ernest Rutherford. Graduating in 1911, he continued at the same college for his graduate degrees. After World War I, he became a very able assistant to that famous man at the Cavendish Laboratories; and, while there, he heard about Bothe and his penetrating rays from beryllium. There was speculation even then that these secondary particles might be new entities. Chadwick switched to polonium as a source of alpha particles and conducted a great number of tests with a variety of materials as the target of their impact. He also carefully measured the recoil energies of the displaced particles; and the results clearly indicated to him that the mass of these neutral particles was extremely close to that of the proton. Emboldened by masses of data, Chadwick submitted a paper for publication in February of 1932; and the neutron subsequently became an established fact.

After spending seventeen years at the Cavendish Laboratory, Chadwick reached a disagreement of policy with his Director and superior which instigated a move. He had become dissatisfied with using alpha particles as the "bullets" in this smashing of atoms and had also become convinced that the new cyclotron accelerator machine developed by Ernest Lawrence offered higher possible energies and better future results. Rutherford, on the other hand, was reluctant to spend large funds on a relatively new device with some problems of operation and unproven worth. The result of this difference of opinion was that Chadwick somewhat reluctantly left in 1935 to accept a position with the less prestigious Liverpool University and start up his own research laboratory. Funds were made available to acquire a new cyclotron accelerator as the laboratory centerpiece feature.

This all happened just about the time that Chadwick received the Nobel Prize. He subsequently determined the mass of the neutron with great accuracy, received many honors, was knighted to become Sir James Chadwick in 1945 and lived to the venerable old age of eighty-eight years. Thus the three basic building blocks of the atom had become recognized, but that is only the very beginning of our story.

Chapter 2
Relativity and Light Rays

Few working scientists have much sympathy for those who try to interpret nature in metaphysical terms. For most wearers of white coats, philosophy is to science as pornography is to sex: it is cheaper, easier, and some people seem, bafflingly, to prefer it.

Steve Jones, "The Set Within the Skull" (a review of *How the Mind Works*, by Steven Pinker, Norton), New York Review of Books, November 6, 1997.

Albert Einstein's major contributions to new ways of thinking in physics gave an immense acceleration to the specialty of nuclear physics; and throughout his life he followed the developments in that field with keen interest and comprehension. It is sad and ironic that in his later years, because of his objections to certain assumptions on non-locality commonly made in quantum mechanics, many younger physicists unjustly dismissed him as a fuzzy old man who could not give up his reactionary classical attitudes and whole-heartedly join their new intellectual movement. The writer regards this convenient denigration as quite shallow and arbitrary; and it was a careless disregard for the man's real genius. Even today there are many eminent scientists and philosophers, such as John S. Bell, John Wheeler, and Abner Shimony and others, whose serious published papers keep alive an earnest discussion on the metaphysical implications of current quantum mechanics theory for our concept of reality. Many first–year graduate students sympathize. Whether the Quantum theory is entirely complete or not is still a matter of personal choice. The opponents on this debate have the undeniable argument that the mathematics work with uncanny precision and, moreover, what difference does philosophy or ontology mean to them? The paradox of older classical physics and quantum mechanics remains unsolved thus far; but the debate about the significance of the conflict rages renewed sixty years later.

In order to place Einstein's accomplishments in some historical perspective, it is important to realize that some standard theories in place during the so-called "classical" era were already recognized as inadequate to satisfactorily explain certain phenomena. The Brownian movement of tiny pollen particles in a liquid, the photo-electric effect of light rays on certain chemicals or metal surfaces, and the abruptly shifting wave lengths of the radiation emitted by a heated "black body" were all examples of inexplicable mysteries in the early nineteenth century. Perhaps the greatest puzzle of all was how light waves managed to travel from the sun through empty space to reach us. The famous Michelson-Morley experiments made to confirm the existence of the "luminiferous aether" serving as the vibrating medium demanded by the wave theories of J. C. Maxwell actually first took place when Einstein was less than eight years of age. Thus there was plenty of "food for thought" available for the penetrating minds of his coming generation.

It was perfectly natural for men to assume the existence of some invisible vibrating medium filling all of space to explain the propagation of light. After all, the assumption that light came in the form of electromagnetic waves beautifully explained light diffraction, the science of optics, and the refraction of light when passing through a transparent material such as water or a glass prism. As with sound waves in air or waves in the ocean, it seemed there simply had to be a carrier medium to be disturbed in order for the transmission process and interference phenomena to happen.

As early as 1881 Albert Abraham Michelson, then on leave from the United States Navy as a Master, was doing his post-graduate work in the laboratory of Hermann Helmholtz in Berlin. His particular research was on the speed of light, which was already known to be quite constant through empty space or even through a given transparent material. His own efforts were first directed towards establishing this constant of 2.9979×10^8 meters per second in a vacuum. Having demonstrated his credentials as an acknowledged expert on that subject, he turned to his attention towards verifying the existence of the supposed "ether wind" as the planet Earth moved through this assumed invisible medium. The basics of his actual experimental set-up are frequently described in standard physics textbooks. Essentially, he directed light rays in different directions (towards, sideways, or away from) the motion of the earth and, in each case, attempted to measure any detectable difference in velocity or delay in time by means of interference shadows as the rays were merged together on a detector screen by mirrors. Since the rays were coming from different directions, but had traversed identical distances, it was assumed that a "cross wind," so to speak, produced by the motion of the earth in its orbit around the sun at a speed of roughly thirty kilometers per second would result in a measurable time difference of arrival at the screen between the two. There was none.

After returning to the United States, Michelson found a job with the Case School of Applied Science in Cleveland, Ohio, and soon teamed up with a chemist named Edward Williams Morely from the nearby Western Reserve University. They repeated numerous experiments conducted with painstaking precision over years and could never find any evidence whatsoever to corroborate the widespread assumption of a surrounding "ether" medium in space. Using Michelson's own words in earlier correspondence while in Germany, "the result of the hypothesis of a stationary aether is thus shown to be incorrect." His first published paper making this assertion in November of 1887 was carefully confirmed. Predictably, all the principal scientists of the day were absolutely confounded. Even Michelson himself confessed to being somewhat disappointed. As Max Born has stated the problem, "vibrations without something which vibrates seemed unthinkable."

The riddle of these negative results preoccupied Michelson's thoughts for most of his remaining life. Seemingly haunted with fear of mistakes in his measurements, he repeated the same experiment over many years with minor variations until 1929. Moreover, a younger disciple of his at the Case Institute named Dayton Clarence Miller joined the quest, since the question was so crucial and challenged so many of the assumptions made in previous physics theories. Miller and Morely collaborated in 1904 to conduct similar tests on top of a mountain in order to counter arguments that the "ether" had been restrained from flowing freely in any experiments conducted in the basement of a building on the ground. The results were exactly the same as previous tests: no detectable difference in the speed of light rays, regardless of which direction from which they originated. It is recorded that, upon hearing this news while on a visit to the United States in 1921, Einstein uttered in German one of his many famous remarks which was to be memorialized in a stone engraving over a fireplace at Princeton University: "Subtle is the Lord; but malicious He is not!" It is this reversion to the religious training of his parents and childhood which incurs the disdain of some people of later generations.

The earliest and most ingenious explanation for the inability of the Michelson-Morely experiments to detect any "ether" drift past the moving earth was offered in 1889 by George Francis Fitzgerald, who thus unknowingly anticipated relativity. He contended "that almost the only hypothesis that can reconcile this opposition (to assumed behavior) is that the length of the material bodies changes, according as they are moving through the ether or across it, by an amount depending upon the square of the ratio of their velocities to that of light." In other words, the ruler with which you are measuring a given distance either shrinks or expands in length, depending upon its orientation with a given direction of motion. In spite of the fact that Fitzgerald still clung to the notion of a vibrating medium in space, as did most of his contemporaries, the simple mathematics drove him to an assertion which fully anticipated the consequences of both Einstein's Special Relativity and the necessary Lorentz transformations now associated with it.

This conversion may be expressed in simple classroom algebra:

$L' = L(1 - v^2/2c^2)$ when:

L' = The length of an object in another separate inertial system moving at a different velocity parallel to that of the observer making the measurement.

L = The length as measured in our own earth–bound inertial system.

v = The velocity of motion in the same direction as the measured length dimension.

c = The constant of speed of light rays in a vacuum or space.

Since the very concept of velocity involves the traverse over a distance in an interval of time lapse, it was logical for some scientists to begin having very long thoughts about the nature of time itself. Time was a subject which had long been taken for granted as absolute for all parts of the cosmos. As we have learned, the Michelson-Morely experiments directly challenged many of the classical assumptions of kinematics and the accepted electro-magnetic wave theories as well. Einstein himself stated the problem: "This invariance of the velocity of light (in empty space being assumed) was, however, in conflict with the rule of the addition of velocities which we knew so well in mechanics—I could not but realize that it was a puzzle not easy to solve at all."

The clue for the final solution, which he seized upon after a year of mental frustration, was *time*. He decided that "time is not absolutely defined; but there is an inseparable connection between time and the signal velocity."

The culmination of Einstein's pondering since the age of sixteen over all these abstract physical laws was a frenzy of four separate written articles accomplished consecutively in the space of only four months during the year of 1905 when he had reached the age of twenty–six. Such a feat of prodigious mental inspiration has seldom been matched and is the unmistakable mark of genius which occasionally appears in disciplines other than science. We may summarize his subjects in order:

(1) His first paper dealt with his new conviction that light was transmitted in small quanta of energy (now called photons) and was completed on March 17th. Much preliminary work had already been done on the emission of heat or light energy from a so-called "black body" by Ludwig Boltzman, Gustav Kirchoff, Max Planck, Heinrich Rubens, Ferdinand Karlbaum, and others. Part of their perplexity was that the light spectrum (or wavelength) from a heated cannon ball would, with relative abruptness, change from a very dull red to orange to white with a continued increase in temperature. Classical theory predicted a slow and gradual change with indiscernible incremental changes in hue or the wavelength of the radiant energy. The experiment showed a discrepancy with the observed facts. Einstein's contribution to this observed anomaly was a paper entitled *On a Heuristic Point of View concerning the Generation and Conversion of Light.* In this article he acknowledges the pragmatically obtained value of Planck's Constant and the erroneous, but then accepted, equation for the Rayleigh-Jeans Law relating to pressure and temperature for gases having different Avogadro numbers. He then went on to point out the various discrepancies of this formula with all laboratory observations and went through a series of his own calculations for the change in entropy of a specific number of gas molecules confined in a small volume. He finished with an equation that convinced him that light comes to us in small, discrete packets of radiant energy. This espousement of a radical concept, which appeared to go back to a notion once expressed and dismissed by Newton some two hundred years earlier, was described by its inventor with the starkly simple statement that: "monochromatic radiation of low density behaves in thermodynamic respect as if it consists of mutually independent energy quanta." Gone, at least temporarily, was the former dependence upon oscillating continuous waves in an "aether" medium. There were more convincing arguments for this decision to follow, for Einstein was thinking in terms of all physics being consistently applicable in all separately moving inertial states. In other words, he had "the big picture" in mind, including the known phenomenon of conversion of light into electrical energy commonly employed today in the hardware of the photo-electric detector tube. It should be realized that Einstein was deliberately rejecting the pure radiant wave theory of light.

(2) The second paper to appear, and the one which he had long been preparing as his thesis for a doctorate degree, was completed on April 30th. The title was *On a New Determination of Molecular Dimensions* and was a study of theoretical statistical methods for establishing Avogadro's Number (6.6×10^{23}). This constant is useful in chemistry for determining the number of molecules in a mole of any pure substance and ultimately the mass of any molecule and its estimated radius. Avogadro's constant is applied to the Ideal Gas Law for the relationship between pressure and temperature and is also valuable in chemistry in the study of molecules dissolved in a solvent, such as sugar in water. Thus Einstein's method is applied today to particle suspensions.

(3) His third paper on the Brownian movement of small particles in water when heated, which we had mentioned previously in Chapter 1, was completed on May 10th.

(4) The fourth paper to be presented in this single year of 1905, and the one which was epoch making in the history of science, was completed on June 30th, 4 months later. This was the famous Theory of Special Relativity. By the author's own description years after the actual publication of his ideas, there were two major clues which prompted him to seek a rational and mathematical justification of those phenomena which pointed to relativity. These were:

(a) The Michelson-Morley experiment, in addition to some others, demonstrated that light, when viewed from a single given inertial frame system in the universe, has exactly the same velocity in a vacuum—regardless of whether the light rays are being emitted from a body at rest or from a body moving in some direction at a uniform speed.

(b) If a nearby magnetic field is moved rapidly past an electrical conductor or metal wire, an electrical current is produced in that same conductor. Conversely, if the same conductor is moved relative to the magnetic field fixed at rest, then an electromotive force or current is also generated in that conductor in proportion to the velocity of relative motion. Maxwell's equations for the behavior of electromagnetic waves also evidenced this same constancy of the velocity of the emitted radiation as light, as, indeed, they should. Light and radio waves all behave similarly and are part of a very broad spectrum of wave lengths.

Einstein's solution for all these anomalies and several more required him to throw out a lot of historical mental baggage about local time and local space coordinates in a proscribed inertial system being true and absolute everywhere in space and to adopt whole–heartedly the now famous Lorentz transformation equations to kinematics and electro-dynamics.

For the non-scientist or ordinary college level mathematics graduate, these renowned and often-mentioned transformations can be made less awesome or intimidating by realizing that the original derivation of the principle evolved from the very prosaic and practical problem of calculating how much time it would take a ferryman to row a small boat with a paying passenger, across a wide river having a swift current, to the opposite shore and then return again with another passenger. Obviously, the ferryman's strength is limited to rowing through the water with the same maximum speed regardless of his direction and when first crossing he must head the boat diagonally against the current in order to fetch the pier on the far shore. In order to arrive at the proper destination on the opposite bank he must compensate for the set of the current. As seen from the jetty on the shore of his original departure, his speed going across the river will be measurably slower than his speed of the rowboat passing through still water, if one measures only the time lapse for his perpendicular crossing.

Keeping this analogy in mind, we can define the algebraic terms for the variable parameters and then discover what Lorentz had to say about the relationships between two separate and independent inertial systems in space moving parallel with each other but having significantly different velocities. This situation becomes especially relevant in astronomy, where the Andromeda galaxy may be moving extremely fast with respect to our own solar system. Realistically, the situation is seldom so simple, since our space has three dimensions and the systems involved may have a complex vector orientation with respect to their relative movement. However, the principle is still the same, only the mathematics merely becomes a trifle more involved. The transformations apply equally well to a jet–powered high speed airplane flying at a high altitude with respect to a telescope observer on the ground below. Those readers who are science fiction fans might better enjoy a picture of aliens from outer space riding in a spaceship powered by some form of gravity warp engine which is making a reconnaissance pass by our planet at a speed of several thousand miles per hour based upon our own measurement units. Meanwhile, earthling astronomers, having radio communication and synchronized clocks and being located separately in Hawaii and Germany, are observing the visitor's spaceship trajectory carefully.

Let x' = The dimension of a distance covered in a straight line along a chosen Cartesian coordinate axis by the separate (alien?) inertial frame moving at a very high rate of speed past and parallel to the direction of motion of the observer's on earth, which is a different material environment quite independent from the first and moving very much slower so as to become relatively at rest.

Let x = The same distance along a straight line in a parallel axis of an identical coordinate system, as seen and measured by the astronomers on the earth.

Let t = The actual measured time interval found by the observers on earth, using a radioactive Cesium clock, for the passage of a fixed point in the independent inertial system (the alien spaceship?) from the overhead meridian at Hawaii to the zenith meridian at the Germany telescope site.

Let t' = The time interval for this passage over the distance " x' " as measured with an equally accurate clock of their own by intelligent beings contained in the external inertial system (the spaceship?) from their own experience.

Let u = The uniform velocity maintained over distance " x " as calculated by the observers on earth.

Let c = The speed of light in a vacuum, which is constant.

The Lorentz Transformation equations from the earth-measured parameters to those experienced by the passengers within the external and independent inertial system then become, assuming motion in one dimensional direction only:

$$x' = \frac{x - ut}{[1 - u^2 / c^2]^{1/2}}$$

$$y' = y \qquad\qquad z' = z$$

$$t' = \frac{t - ux / c^2}{[1 - u^2 / c^2]^{1/2}}$$

Gone is an absolute fixed coordinate system and time frame for all matter in the cosmos. Other galaxies have their own. Everything is moving with respect to something else anyway, so relativity becomes universal. Lengths of physical objects, as observed from outside their own system, may shrink or expand. Time intervals and the clocks which measure them may seem to diminish or dilate—all depending upon their relative motion with respect to another outside system, whether the earth or solar system or another galaxy. Special Relativity applies to all things, great or small, when great velocities are involved. A little thought reveals that the significant divisor term in the transformation equations is the speed of light squared. That is an extremely large number, so it becomes evident that we should be concerned with very large velocities approaching that of light for the appropriate correction to become significant and discernible. The snail's pace speeds used in the everyday life of human beings make the relativity adjustments negligible for most of our ordinary activities. Thus for most of us the classic physics of Galileo and Newton remain satisfactory and unchanged for all practical purposes.

As we already stated, any clock device for measuring time intervals within one inertial system moving uniformly with respect to some other system regarded as static would operate more slowly and hence would yield different and shorter time intervals. This concept gave rise to the popular and oft-repeated fable of the space astronaut who completed a round trip voyage to the Andromeda galaxy at a speed near that of light, only to discover upon his return to earth that he was much younger than his twin brother who stayed behind. Moreover, as the velocity of a fixed group of objects, whether planets or molecules, increases towards a magnitude approaching that of light (commonly designated by the letter "c") all dimensions measured in the direction of motion shrink. The body itself would contract in size in the direction of motion and become almost a two dimensional "Flatland" plane. To borrow a picturesque image that the late Professor Richard Feynman loved to relate, as seen by a proton embedded in the metallic target inside the particle detector, the on-rushing colliding proton deflected from the circulating beam of a ring accelerator would look like a flattened pancake coming at it.

Special Relativity also had another bonus hidden within it for the experimental physicists. The collisions of particles coming from two different directions inside a cloud chamber, or other type of detector, create showers of "daughter" fragments which exist in our own macroscopic world for unbelievably brief instants of time (10^{-12} seconds). The contraction of time in the world of fast moving particles in the atom means these fragments are able to travel measurable distances in our laboratory, depending upon their residual energy, inside the cloud chamber before being photographed. Not only that, the original rest mass of any given particle which has been accelerated in motion to very high speeds increases with its velocity in accordance with the following formula:

Let m_f = The final particle mass as it hits another or a target after acceleration in a machine.

$$m_f = \frac{m_0}{(1 - u^2 / c^2)^{1/2}}$$

All these features of relativity conspire together to make our knowledge by means of "atom smashing" machines possible.

Einstein challenged all the conventional mechanical or, one might even say, materialistic notions for explaining electromagnetic phenomenon. The idea of a huge clockwork universe with all its stars running on the same absolute time scale with simultaneity between inertial systems moving with respect to one another had to be abandoned utterly. The velocity of light transmission through empty space was always constant and remained independent of the relative velocities in uniform motion by the light source versus the receptor system. Obviously these daring ideas upset many people and created a ferment of discussion among scientists. More importantly, these simple precepts led to a host of other significant consequences in physics which were equally revolutionary; although their final adoption frequently led to theoretical simplicity.

It should not be supposed that this extraordinary genius stopped thinking after his submission of the four papers just described. He continued later with a barrage of publications on fundamental physics, including the General Theory of Relativity and an explanation of gravitational fields. Those which bear mention as specifically applicable to our purposes, however, were those issued on September 27th of 1905 and July of 1916. The first was an extension to Special Relativity in which he pointed out that mass is an equivalent of the energy content of the atom or particle; and the second was a general discussion of the significance of light quanta. He there made an unequivocal statement that these discrete quanta have an energy content which involves Planck's Constant and that they also possess a momentum attribute evident in the phenomenon of photo-electric conversion of light to electric current in certain sensitive metals or compounds.

Thus the idea of the photon as a particle was born, although Einstein admitted that the older classical wave theories still worked well for optics. The name "photon" did not actually appear in print until 1926 and was invented by a chemist named Gilbert Lewis in a context not directly related to quantum mechanics. The habit of regarding the photon as a particle was acquired by physicists gradually over the years, although, as we will learn later, Nature herself is not nearly so amenable to such a concrete restriction as the word "particle" implies in our dictionary. Giving a name to something helps us humans to file the concept in some cross-referenced memory cells in our brain for later retrieval, but, giving the name "Gabriel" to an angel does not mean that we totally comprehend the religious mystery implied by the symbol. Photons are not really simple; but a name as a particle makes us think of it as a real "thing" in our environment and is thus reassuring. Photons most certainly exist and we certainly have no intention of denying that; but their combined attributes make them as mysterious as gravity and capable of behaving, under some circumstances, as if they were the waves of older classical concepts. We will learn why in a later chapter.

CHAPTER 3
HYDROGEN AND SPIN

Up to this point the reader has been beleaguered with history as an introduction to some very basic and general concepts common to particle physics. Figuratively speaking, it is time for us to venture a toe into the sea of facts which exists before swimming boldly into a very deep subject.

As we all most probably know, the hydrogen atom is the simplest and most basic atom in the entire Periodic Table of elements and consists in its most common form as a single proton surrounded by a single orbiting electron. It is not hard for us to perceive why it is the lightest element known in nature. Simple though this may be, its structure and behavior has commanded the intense attention of a long list of brilliant scientific minds. It also happens to be the most plentiful element in the entire cosmos and is the primary constituent of gigantic clouds of gas drifting in inter-stellar space. It is a primary component of the orb of stars themselves and of the water oceans which dominate the surface of our own planet.

Hydrogen is rarely found in its pristine elemental form on the earth. At what we conventionally call "room temperatures" it combines quite slowly with oxygen to produce water vapor; but reacts most explosively if assisted with an open flame, spark, or catalyst. It also combines very quickly with the halogens, such as chlorine or fluorine, to form hydrochloric or hydrofluoric acids. One of its greatest commercial uses is in the manufacture of ammonia (NH_3) for a variety of molecular compounds with such commonplace uses as fertilizers and household cleaners. Perhaps most important to us, hydrogen in close company with carbon, oxygen, and nitrogen becomes one of the basic building blocks for a huge family of organic molecules. These range from wax and paraffins to petroleum and kerosene to alcohol to the protoplasm of most life forms on this planet.

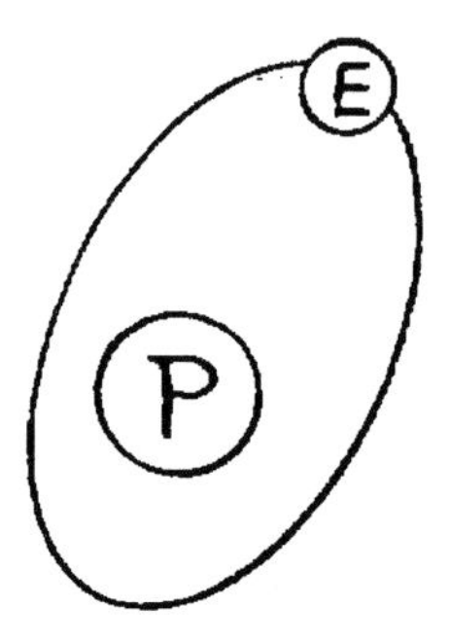

THE HYDROGEN ATOM

Hydrogen can be liquefied and stored cryogenically at the extremely low temperatures below minus 252.8 degrees Celsius or minus 423 degrees Fahrenheit. This led to the idea of using it as a liquid fuel for rockets to carry satellite probes into space, although the ever present danger of leaks or boiling and explosion discouraged this application for manned capsules. The atomic weight of hydrogen is 1.00797 and the fact that this number is not exactly unity is an indication that a small quantity of heavier isotopes do exist naturally. Being the lightest substance known, with a gas density of only 0.00561 pounds per cubic foot at freezing temperatures, it has been tried in past history as a buoyancy medium for balloons and dirigibles often with unhappy results. A cloud of diffuse hydrogen atoms will soon condense to a gas consisting of closely bound pairs of atoms (H_2). Observation of this simple fact led to the discovery of a general peculiarity of all atoms and particles, as we will learn. With a chemical valence of one and an atomic weight of substantially one, it would seem hard to make things more complicated.

One might easily assume that gravitational attraction between the individual atoms could account for this pairing conglomeration phenomenon satisfactorily. Indeed, in the vast clouds of hydrogen of many light years in dimension that have been photographed in inter-stellar space it is gravity which may gradually start the process. However, gravity is a very weak force on a scale compared to electric charge attraction or electro-magnetism; and the aggregation of atoms into gas molecules takes place far too rapidly for the scientists to accept gravity as the sole explanation. Not only do hydrogen atoms behave in this manner; but so also do other gases such as oxygen, nitrogen and molecular carbon dioxide. Pondering upon this remarkable affinity of one atom for another despite the lack of ionization or any electric charge, scientists eventually came to an unexpected conclusion. Virtually all particles, from atoms to electrons to photons, possess a singular characteristic known as angular momentum—or in more simple language, spin.

Probably the first scientist to observe the results of spin, well before it was fully understood, was Pieter Zeeman (1865 to 1943). The son of a Lutheran minister in the Netherlands, Zeeman became a professor at the University of Amsterdam from 1897 until his retirement. As an ardent admirer of H. A. Lorentz and Michael Faraday, Zeeman was well aware atoms could be deflected in their trajectory by a strong magnetic field. He therefore decided to study the spectroscopic spectrum yielded by burning sodium metal in a flame which was subjected to a magnetic field. He came to the conclusion in 1896 that the "D" lines of a negatively charged sodium atom were indeed broadened and that, moreover, these light beams were polarized. That is, all the light waves were restricted to the same dimensional plane. These experiments required a prodigious amount of patience and tedious repetition in order to produce a statistically significant result and to obtain good photographs or pictures of the effect. Any slightest vibration caused by people walking in the corridor outside his laboratory or traffic in the street outside of his building blurred and ruined about twenty–nine out of every thirty attempts.

At this point in history the dominance of chemists in the experimental laboratories of the world was still strong; and thus there was a great deal of interest in the determination of actual weights of atoms in either the vaporous or gaseous state. It had already been discovered that drawing ionized atoms down an air evacuated glass tube was a tool to discover their velocities and that they behaved very much like light beams. A young man in his mid-twenties named Otto Stern (1988 to 1969) became interested in this technique. He was the son of wealthy and indulgent parents who were grain merchants in what is now southern Poland. Stern could afford to choose his own career avenues and consequently he had attended many lectures by Arnold Sommerfield and had studied the molecular theories and statistical mechanics of Boltzman. Stern decided to concentrate his work on silver atoms and to discover what effect, if any, a magnetic field placed perpendicular to the trajectory of such heavy atoms might have. As the leading theoretician of the time, Sommerfield had predicted that certain atoms, such as hydrogen, silver, and the alkali metals, should possess what he called a magnetic moment (symbol p_M) having a magnitude in accordance with the following formula:

$$P_M = \frac{eh}{4\pi mc}$$

when: e = the electric charge
h = Planck's Constant
m = the rest mass of the electron
c = the velocity of light

Obviously, the rotational moment would calculate out to be extremely tiny; but a moving particle's behavior in a directionally oriented magnetic field might show some evidence of deflection which would confirm his theory. Stern felt rather clumsy with the actual manual work of assembling the laboratory apparatus, so he recruited the assistance of an associate at the Institute of Experimental Physics at Frankfurt, Germany, named Walter Gerlach. Their efforts culminated in 1921 with the successful measurement of a magnetic moment for the silver atom and the publication of five papers on the subject. They had passed a stream of hot silver ions through a strong magnetic field before being deposited on a plate of glass and were astonished to see the build-up of two black bands of silver with a narrow space separating them. It was clear the atoms were being deflected into two different directions according to some individual plus or minus polarity. In a search for a simplistic picture of this angular momentum feature which may actually be a wave mechanics phenomenon, the idea arose that all, or almost all, atomic particles spin around an intrinsic axis like a child's toy top. The idea seemed childishly bizarre to the early scientists and in fact it does not really do justice to the quantum-mechanical mathematics of angular rotational moments.

Other scientists jumped onto the bandwagon. A team at the University of Leiden composed of Samuel Abraham Goudsmit (1902 to 1978) and George Uhlenbeck (1900 to 1988) took up the puzzle of the Zeeman effect with light spectra from atoms. Historically, this was the period when the entire theory of Quantum Mechanics and quantum state "numbers" was being formulated. Thus it happened in 1924 that Wolfgang Pauli, who had already become eminent in that specialty, proposed his "exclusion principle" and that the condition of the electron inside an atom could be stated mathematically by four quantum numbers associated with its degrees of freedom. He left the fourth number unstated, however. Uhlenbeck quickly jumped to the conclusion that the angular momentum of the electron had to be either plus or minus Planck's constant over two pi ($h / 2 \pi$), which implied a spin of one half. His team published their novel supposition in November of 1925. Some said their paper deserved a Nobel prize; but original merit does not necessarily result in world fame.

By coincidence, an American graduate student from Columbia University named Ralph Kronig also proposed the idea that the spin number could be the fourth quantum number to Wolfgang Pauli, who was conceded to be an eccentric genius in physics. The senior scientist brusquely and impolitely dismissed the notion as ridiculous nonsense, as he was famous for doing, thus destroying any further discussion and the student's self-confidence. Nevertheless, the trend of events moved towards substantial confirmation of the idea; and Pauli eventually recanted and apologized to Kronig. Scientific consensus gradually led to the conclusion that spin in particles (whether photons, electrons, protons, or entire atoms) was the most reasonable explanation for a whole series of perplexities which had originated in optics and spectroscopy. The idea ultimately led to the "shell" atom structure proposed by Neils Bohr. Pauli then formalized his famous "Exclusion Principle" for quantum mechanics, which stated that for the case of electrons orbiting around the nucleus no two particles having the same quantum number state could occupy the same space. Different spins, however, could make this allowable. Thus the maximum number of electrons permitted in the lowest and bottom energy state orbit for an atom in its stable rest state had to be only two with positive and negative spins respectively.

This implies that they must be anti-symmetric in their wave mechanics form with a magnetic moment sum of zero. Both electrons could not possess an identical direction of axial spin, which is not to be confused with their orbital momentum.

The initial idea of spin is deceptively simple. If one adopts, accurately or not, the convenient visualization of a particle of matter as a hard ball of spherical or oblate volumetric shape containing mysteriously "frozen" energy, then spin or angular rotation about some axis becomes very similar to the rotation of the earth about its north to south axis. Although it is quite possible for a particle to have a spin of zero, most fall into a category of either multiples of one half $h / 4\pi$ (such as 1/2, 3/2 or 5/2) or multiples of one (l, 2, 3, etc.), depending upon whether they are fermions or bosons. Spin, as we have thus far described it, is customarily denoted by the letter "J."

Determining and expressing the direction of rotation of any given particle can be confusing. How does one make sense of such terms as "up" or "down," or "plus" and "minus" or "clockwise" and "counter-clockwise?" Strict determination in the laboratory depends on knowing the direction, line of travel, or translation of the particles through space when they have been all polarized by a known magnetic field. In that case it may be assumed that the great majority of them are all spinning in the same direction as seen from the direction towards which they are moving. If they are moving vertically with respect to our local coordinates and we image ourselves as peering through a very powerful microscope from above looking down upon the top of the electron or other atomic particle, then the rotation may be described as going either clockwise or counter-clockwise. Physicists go one step further in this picture, which was adapted from a very old-fashioned textbook rule for determining the direction of a magnetic field around a wire carrying an electric current. If the thumb is pointed in the direction of motion of the electrons or particles, then wrapping ones fingers, figuratively speaking, around the particle would indicate the direction of spin. Thus the thumb would indicate either "up" or "down." Most physicists are content to specify negative or positive and let it go at that. Another complication in the real

POSSIBLE STATES OF A QUANTUM PARTICLE IN A SCALAR FIELD

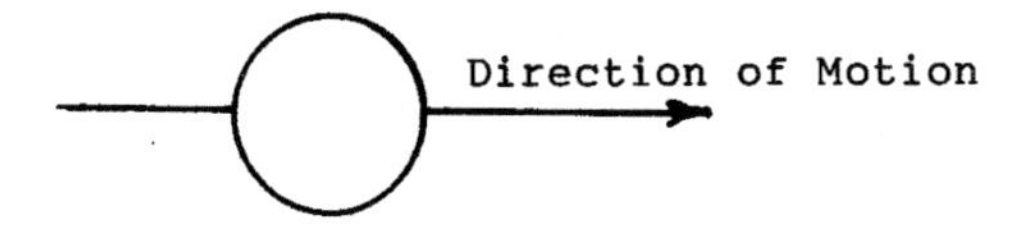

At Rest or Zero Spin

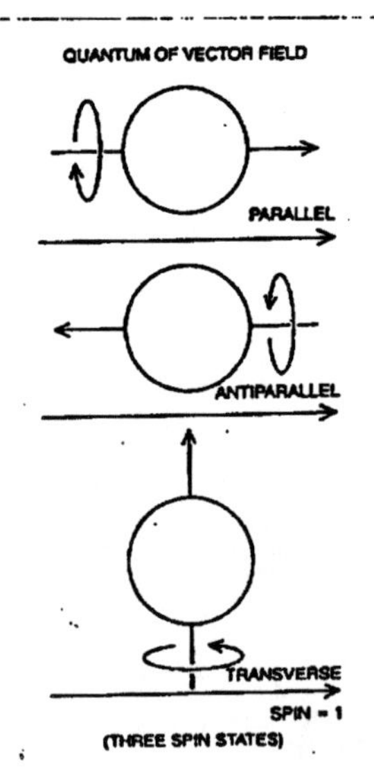

laboratory world is to know the orientation of the spin axis with respect to that of similar adjacent particles. It may be parallel or perpendicular or any angle in between. On a more whimsical note, one correspondent who had a sense of humor, wrote in to the *American Journal of Physics* that he liked to think of all the hydrogen atoms in the water molecules of a snow bank near the front door of his home as slowly precessing in the weak magnetic field of the earth. If nothing else, this admission demonstrates that physicists keep their mind on their work a lot.

We can now understand another reason why hydrogen gas molecules form so readily. In the singlet state there is a very strong interaction between two adjacent atoms of hydrogen when their electron spin states or magnetic moments are of opposite polarity. Once in very close conjugation, a third mutually attractive force comes into play since protons and neutrons mysteriously bind together as an atom nucleus. The atoms mate to form a bound molecule having a total spin number of zero or complete neutrality. All this says nothing of about the spin direction of the protons, however. In a mixed state it might be possible for a given proton to have a spin axis parallel to the direction of motion and its companion electron have a spin anti-parallel. This line of thought leads to mathematical complications; but it also encourages dreams by the experimenters of altering the physical properties of some elements at will.

As we mentioned very briefly, those particles which have a spin number which is some multiple of one half are classified as "fermions" after Enrico Fermi of Italy and later the United States. These include electrons, protons, positrons, muons and other long–life particles which constitute matter. They all must obey the Pauli Exclusion Principal, so no two fermions can occupy the same space in the same state at any instant of time and, if it were not for the strong force, would remain unsociably hostile to one another.

On the other hand, those particles which possess a total whole number or integer for their spin number are called "bosons" after the Indian physicist Satyendra Bose. This implies that the particle's wave function is symmetrical (either plus or minus). The boson is exempt from Pauli's Exclusion Principle. That is, two bosons can compatibly occupy the same limited space and thus become tightly packed. Strangely, it is possible to classify entire atoms as either bosons or fermions and not merely the nucleons or electrons. One of the most abundant atoms found in space and the stars is helium gas composed of the normal He4 with two protons, two neutrons and two electrons. Since each of these particles individually has a spin of one half, the total of all the spins together adds up evenly to three, which makes He4 a boson. An isotope of helium (He 3) does exist, however, in which there is only one neutron. In that case the sum of all the spins comes out to two and one half, making the He3 gas atom a fermion. This difference in the total spin number has been found to create real differences in the physical properties and behavior in the two gaseous isotopes just mentioned, especially in the liquefied or super-cooled state. He 3 has a different heat conductivity and when placed in a centrifuge in its cryogenic state it exhibits a most quixotic behavior, including creeping up over the edge of its container and having the liquid break up into numerous small independent eddies. In other words, its viscosity is altered.

It is important for us to realize that our visualization of a spinning body, large or small, may be a convenient, but somewhat erroneous, picture of magnetic moment. The quantum number property of spin becomes significantly more complicated in the mathematics of particles with the introduction of a quantity called "isotropic spin"—usually designated as "I." Scientists are usually careful to say that a given particle *behaves as if* it were rotating in space. This is a fine point and apt to be disturbing to the lay reader who is hoping for something familiar in his own experience. Once again, our language and imaginations are demonstrated to be inadequate to adequately describe the fundementals of the reality of matter and how it comes into existence. In fact, just why infinitesimally small particles should possess a magnetic moment is not easily explained. Nevertheless, we shall stubbornly continue to cling to our simple mechanical analogy from which we can derive a modicum of mental comfort.

There is another aspect of spin. Identical particles having identical angular momenta in both directional orientation and magnitude cannot be distinguished from one another. No measurement can be made to determine separately any different physical properties from another. If two electrons have identical spin states, then all we know is that they both cannot be found at the same identical point in space. Since all normal electrons have the same negative charge magnitude, one would expect the electrostatic repulsion to keep them apart in a free state anyway; but the spin characteristic seems to have been added just to make certain. This fact is most fortunate for our own existence, since it guarantees the separate identities of all the various atomic elements and their distinctive behaviors in forming molecular compounds which otherwise might not happen at all.

When spinning particles collide, the direction of their initial spin can affect how they scatter after the impact. This makes their behavior more complicated, as anyone who plays billiards or tries to hit a spinning baseball with a bat learns. Moreover, the spin of an individual particle, as we have previously defined it, should not be confused with the total angular momentum of an atom or its entire nucleus. Electrons have a rotational orbital velocity around a proton-neutron nucleus, as well as a very tiny mass. This results in an additional and verifiable momentum of an entirely different sort. One can easily understand why specialists in quantum mechanics talk about adding up all the combined effects of the quantum numbers and why they all happily greeted the modern computer as an invaluable tool.

Another unexpected aspect of spin came to light in 1990 during the research experiments of Frank S. Stevens and others at the Lawrence Berkeley Laboratories in California. They were studying "super-deformed" nuclei with distorted shapes, such as a football or a misshapen potato. This occasionally happens in an accelerator machine when a much smaller nucleus used as a projectile at slow speeds hits another larger one slightly off center. Using the customary calculations of macroscopic classical physics for rotating elastic bodies, the scientists fully expected that the resulting spin of the conglomeration could be almost any value of spin, depending upon the masses involved.

They were wrong. What actually happens is that the deformed "glob" gradually reverts back to the quantum rule of multiple integer spins by regularly emitting photons of gamma ray energy. No cheating allowed! Quantized angular momentum appears to be an immutable law for atoms. Just why this is actually so is not well understood. The metaphysical implications of creation have fascinated human minds for centuries, of course, but some of the secrets of our material existence still remain a mystery.

We ordinary humans derive another benefit from spin of a rather indirect and secondary nature. This is the phenomenon of Nuclear Magnetic Resonance or "NMR." The basic theory is that the spinning mass of a rotating particle, in addition to generating an angular momentum which we have already learned about, also produces an electric charge which is distributed uniformly over the assumed spherical surface. This rotating electric charge creates a small magnetic field symmetrical about the axis of rotation and proportional in magnitude to the cross-sectional area of the particle. It thus becomes a magnetic dipole. Whether it is a nucleus of protons and neutrons or an entire molecule of atoms, the sum total of all the unbalanced spins results in a certain feeble, but measurable and detectable, magnetic field. Early experimenters in 1945 discovered that if atoms of hydrocarbons, such as wax, were placed in a strong and fixed electromagnetic field and then subjected to radio wave fluctuations in the applied electromagnetic field, all the atoms jumped back and forth between various spin states. Much more important, an externally applied radio frequency of a specific number would result in an energy absorption or emission by the atoms which may be measured.

Today this phenomenon is used in hospitals as a substitute for X-rays in the examination of human tissue with far less damaging effects and less risk to the flesh or reproductive organs. The patient is placed inside a long cylindrical tube or elliptically contoured pipe in cross section which runs back and forth on a trolley carriage. This tube is then moved in between the poles of a very large and powerful magnet. Measurements of the resonances created by a radio wave source which can be rotated around the person in the cylinder create a computer-assembled picture of the biological tissue being examined with far better resolution than formerly possible with X-rays. Moreover, there exists the added feature of computer generated three dimensional views on a television monitor screen for the diagnostic physician. Because of the more benign aspects of Nuclear Magnetic Resonance, it has become a universal tool for brain scanning. The physicians in the medical industry prefer to call it Magnetic Resonance Imaging or "MRI." The only difficulty with the procedure, aside from the large expense of the equipment and the need for highly trained diagnosticians, is that many people suffer a severe attack of claustrophobia once inside the confining narrow cylinder and must be sedated with tranquilizing drugs.

As we have stated before, the concept of spin as a rotational moment like that of a toy top is a comforting idea and gives one the reassuring feeling that we comprehend the reality of the workings of atomic particles in a no-nonsense and practical way. The use of this mental metaphor in the mathematical analysis of particle behaviors actually results in fairly reliable predictions for experimental results in the laboratory. Thus scientists are convinced they are justified in their assumptions. However, when one later encounters arguments about whether the electron is a point source of electrical charge or a tiny packet of energy radiating in some insubstantial field or else, even more abstract, merely the high statistical probability of finding the phenomenon known as the electron in a given location in space at a given time, then the reality of our mental picture begins to leave us. That difficulty with our pictures derived from daily living and the mathematics of quantum mechanics is fundamental. Our "common-sense" and experiential names for particle attributes gradually become more and more inadequate to describe a world beyond our own macroscopic and material existence. As one progresses further into this specialized study, terms used such as "strangeness" or "color" or "flavor" become increasingly inadequate for a meaningful description of particle attributes. The scientists know this full well; but limitations of our language continue to hamper them.

Chapter 4
Two Simple Atoms and Two Riddles

4.1 The Deuteron

The deuteron is simply a heavy isotope of the hydrogen atom consisting of the combination of one proton and one neutron. Two deuterons in chemical combination with a single oxygen atom form "heavy water." The Greek prefix "deuter" refers to the two heavy nucleons in the relatively rare massive variation of the hydrogen atom. Heavy water was very seriously considered as the surrounding moderator in atomic energy piles during the very early development days, since it was a natural element which would absorb or slow down the neutrons being emitted by special uranium isotopes and thus allow fission to take place.

The deuteron has a positive spin of one and an electrical charge of one. It is believed that this two–nucleon atom has a prolate shape similar to a football standing vertically on one of its conical ends with the two central nucleons vibrating ceaselessly to and fro across the shorter axis. The hydrogen gas made with two deuterons in partnership which form the molecule is quite logically called "deuterium." The existence of the positive electric charge inside each deuteron means that they all may be focused and accelerated by electromagnets into a beam. Such fast moving deuterons may be aimed at a target of frozen heavy water ice with the strange result that "slow" neutrons are produced in large numbers and the greater part of the kinetic energy passes into scattered protons. This circumstance drew the attention of scientists in 1940 when the idea of a nuclear fission bomb was just beginning to be taken seriously. Heavy water suddenly became the object of intense interest, as well as desperate military excitement; and it was no longer a mere laboratory curiosity. "Slow" neutrons could stimulate the fission of the uranium isotope atom into more neutrons and thus promote swarms of more neutrons in adjacent uranium fuel in a chain reaction.

The World War II story of the efforts expended in manufacturing heavy water and simultaneously preventing the German Nazis from succeeding in accumulating a supply for their own development of a bomb is a spy thriller in itself. The story includes dangerous reconnaissance air flights over Scandinavia and Europe and even more risky bombing and commando raids in an effort to destroy the German heavy water production facility at Rjukan in occupied Norway. Because the production of hydrogen gas molecules from water by their separation from the oxygen atoms by electrolysis requires huge amounts of electricity, the Germans decided to locate their heavy water plant inside the Vemork hydroelectric generating station in a very mountainous area roughly 110 kilometers west of Oslo, Norway. During the invasion of Norway by German troops in 1940, the importance of this key facility was well realized by the loyal Norwegian army and their enemy invaders both; but the resistance was soon overcome and the Germans took control of the electric power plant facility for their own purposes.

The British secret service was already quite cognizant of the importance of heavy water in atomic bomb research then; and plans were made in London to destroy this manufacturing facility at all costs. Bombing from airplanes did not, at first, seem at all feasible. Knocking out the generating station by wholesale destruction would be costly in planes and lives and would also wreak hardship on the captive and friendly Norwegian population. More important, the Allies lacked really specific information on the exact location of the electrolysis laboratory in the enclosed complex. There was no clear and precise target. Commando raids by expatriate Norwegian and English soldiers were chosen as the first means of sabotage.

A salient early attempt was a small task force carried in two gliders towed by a single Halifax bomber at night. In order to avoid the German detection radar, the bomber came in low over the mountains. The raid was badly botched. The gliders landed in rough and boulder–strewn terrain in the wrong place, causing injuries and some deaths of the glider occupants immediately. The tow line had been dropped too early at low altitude, whether by confusion in geography by the bomber pilot or because he feared hitting the mountain ridge and wished to quickly gain altitude is not certain. The few commandos who escaped injury in the crash set off on their mission to the Vermork dam; but were soon discovered. All were executed by the Germans.

Meanwhile, a Norwegian patriot named Jomar Brun, who was the Chief Engineer for the hydroelectric plant, had managed a daring escape from his country and sought sanctuary in England. He there willingly described to British Intelligence the plans of all the buildings in the complex and their functions in meticulous detail. Armed with this new information, a second demolition party was organized and dropped by parachute near the site. The men were equipped with skis in order to traverse the winter snow, weapons, and small packages of plastic high explosive charges. Four men managed to break into the plant area, locate the heavy water separator tanks and place their destructive charges successfully. Their bravery was incredible. For more details the reader is invited to refer to *The German Atomic Bomb* by David Irving and *The Making of the Atomic Bomb* by Richard Rhodes (both published by Simon and Schuster).

Heavy water was eventually supplanted by graphite as a moderator and returned to its peacetime neglect temporarily, but not until millions of dollars had been spent on its quest. The reason for its abandonment was not merely the high cost and difficulty of producing it by electrolysis or gaseous diffusion. Strangely, it proved entirely too successful in absorbing neutrons. Consequently, smaller amounts of it were required in atomic energy piles than anticipated; and the neutron chain reaction thus encouraged had a tendency to proceed at such a pace that the reaction was extremely difficult to control and presented safety hazards. The danger of an involuntary heat explosion and of the workers in the vicinity being "hoist with their own petard," as Shakespeare would have it, was a sobering deterrent to its use. Fortunately for the United States and the entire world, Hitler himself never became convinced of the practicality of the atomic bomb or its immediate value to his Panzer tank divisions, which were then overrunning Europe. Perhaps another factor which may have contributed to this blunder was that some of the most knowledgeable German scientists, such as Erwin Schrodinger, may have refrained from expressing any enthusiasm for the possibility from an ideological reluctance to give such a terrible weapon into the hands of a paranoid dictator. There is much speculation about this subject.

Deuterium is crucially important to us in another way which seldom comes to our minds. The astronomers are convinced it is an important agent by which the stars generate their prodigious radiant energy in the early stages of their formation. As gravity pulls a vast cloud of hydrogen gas drifting in space together and compresses the protons close together in an increasing density, some protons may combine together through the strong force to form a deuteron nucleus. By the so-called "weak interaction" one of the protons may shed a positive electron and a neutrino to become the usual neutron. Thus:

$$p + p = (p+n) + e^{+} + \nu$$

The positron escapes to combine with any nearby free negative electron into instant annihilation with the release of a photon of gamma ray energy. The real importance of the remaining deuteron nucleus is that it goes on to combine with other deuterons in the heart of the new star to form helium, which ultimately fuses into lithium. The entire process, which is not quite as simple as described, is called the "burning" of hydrogen with the by-product being the release of atomic energy into heat and light. Thus the deuteron is assumed to be one of the key elements in the atomic furnace which powers our own sun.

4.2 The Alpha Particle

As we have already learned, the alpha particle was the focus of a great deal of attention and curiosity during the very early investigations of atomic behavior, starting with Becquerel's discovery of natural radioactivity in 1896. The alpha particle is an incomplete atom, since it is actually one atom of helium which is deprived of its two normal planetary electrons, making it a stripped nucleus of that element. One reason it is included in our "particle zoo" list at this early stage is its extreme importance in the evolution of atomic theory. It is one of the principal by-products of natural radioactive decomposition of the unstable and heavier elements, radium and polonium being the best known. Ernest Rutherford used polonium as the source of his alpha particle research in 1911.

Not surprisingly, the symbol for the alpha particle is the beginning letter of the Greek alphabet, "alpha" or "α." The normal helium nucleus contains two protons and two neutrons in close cohesion, giving it an atomic mass number of four and making it one of the very heaviest particles used in research laboratories for bombarding targets. Moreover, the absence of electrons when it is first emitted from the radioactive parent makes it easily detectable by the positive charge and susceptible to magnetic deflection or acceleration. He^4 is one of the most stable atoms found in the entire universe (helium is called a chemically "inert" gas) and consequently is very plentiful in stars. Isotopes of this element exist with the common one having only one neutron instead of two, but these are not important for our purposes. The alpha particle is doubly ionized, having an electric charge of + 2, and a spin of zero. This last statement requires an explanation. In its rest state of minimum energy for He^4 it is assumed that one neutron has a positive, or clockwise, direction of spin and its companion neutron possesses a negative, or counter-clockwise, direction of spin. The two protons will be in a similar situation and have opposing directions of spin. Thus the plus and minus signs all cancel one another out, leaving a net magnetic moment of zero. Alpha "rays," when they are ejected by a radio-active element, mostly have a velocity in the order of one twentieth of the speed of light, with slight variations depending upon the material of origin. This is considered slow. Alpha particles possess a relatively large potential energy because of their large mass; but, in spite of this, they are not very penetrating. An ordinary sheet of paper can frequently stop them. Notwithstanding, the importance of this nucleon for early discoveries of atomic particles was incalculable. The bombardment of gases or metal foils enclosed in glass tubes in the beginning of the twentieth century opened the proverbial "Pandora's Box" of puzzles and insights about which famous people of science debated.

When alpha particles pass through a sealed glass tube containing a gas the large positive charge carried by them, as one might logically expect, robs electrons from the gas molecules and thus creates an intense ionization. This, in turn, increases the conductivity of the gas to any electric current which might pass between two opposed electrodes having an electromotive voltage difference. In effect, this arrangement makes a very simple detector for any radioactivity nearby.

Alpha particles may be produced artificially in the laboratory quite easily. Back in 1932 J. D. Cockcroft and E. T. S. Walton bombarded lithium metal with accelerated protons (essentially hydrogen ions) and discovered that the struck lithium atom disintegrated into two alpha particles. This was evidence that atoms could be "split."

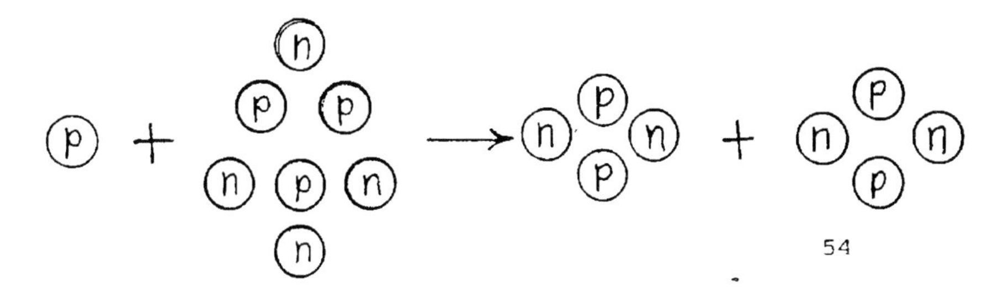

No great energies or accelerated velocities for the protons were required for this reaction either, since the initial proton was absorbed by the lithium atom quite readily. As early as 1940 it was recognized that there existed some kind of resonance values of energy at which an alpha particle could enter and destabilize another atomic nucleus quite readily. The calculations for determining this energy level are today a fairly commonplace exercise in quantum mechanics; although the theory needed to evolve over a long period of trial and experimental verification before its reliability had been demonstrated to the satisfaction of all. All these experiments and precise measurements served to prove again and again that mass is merely another form of locked–up energy.

There was, however, an anomaly about the radioactive emission of alpha particles which bothered the scientists for quite a while. As early as 1909 Ernest Rutherford, with the help of very able assistants named Ernest Marsden and Hans Geiger, had demonstrated that alpha particles derived from sources such as radium and other radio-active elements had energies as high as 8.78 MeV and they were repelled by the positive electrical charges in the nucleus of metal atoms. This proved, incidentally, that the protons were all located in the center of the atom in a bunch. On the other hand, it was later observed that these incomplete helium atoms often erupted spontaneously from the interior of the uranium atom with energies as low as 4.19 MeV. Their relative weakness could also be confirmed by the longer time that it took uranium to fog up or register on photographic film. The question was: "How did they get out?" In 1928 George Gamow, along with Edward Condon and Ronald Guerney, offered the startling conjecture that this was an example of "quantum tunneling" permitted by the then only recently announced notion of "The Principle of Indeterminacy" demanded by the new theory of Quantum Mechanics. They were ahead of their time, actually; but this subject is getting ahead of our story and the causes of radioactivity really demand a much more complicated explanation which we will save for later. Nevertheless, the idea today applies to commercial Josephson connections for fast electronic computers.

Imbued with an enthusiasm for Gamow's suggestion that the tunneling escape of alpha particles might mean that these same positively charged protons would not require the extremely high kinetic energies anticipated to overcome the repulsion of protons in the atomic nuclei of a target material, an electrical engineer named John Cockcroft and an Irishman from Dublin named Ernest Walton set about building a proton accelerating machine of modest 300 kilovolt power. This was later raised to 800 KV. The apparatus was constructed in a high ceiling room inside the Cavendish Laboratories in London; and its appearance was weird enough to satisfy the most imaginative science fiction writer. They employed accumulated static electric charge to produce "lightning" discharges of direct current between two very large hollow copper balls in order to achieve bursts of the high voltage electrons necessary to ionize hydrogen atoms. The latter were then directed down a vertical glass tube towards a negatively charged cathode and target of lithium metal. In 1932 they accomplished the fission of lithium and proved Gamow correct. They also helped to usher in the new age of atomic particle accelerator machines, although the apparatus today has the status of a quaint museum piece.

4.3 Nucleons

Both the deuteron and the alpha particle possess a dense, compact, and very small center mass containing nucleons; and these represent by far the greatest amount of mass in any given atom. The word "nucleon" is a rather vague and general term which includes both protons and neutrons indiscriminately. The word is intended to suggest the concept of a massive core of tightly packed protons and neutrons around which there exists a cloud of tiny and

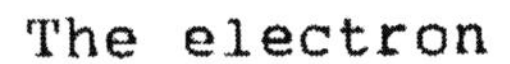

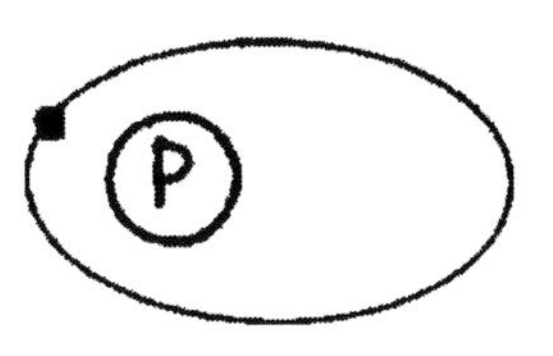

The Hydrogen Atom

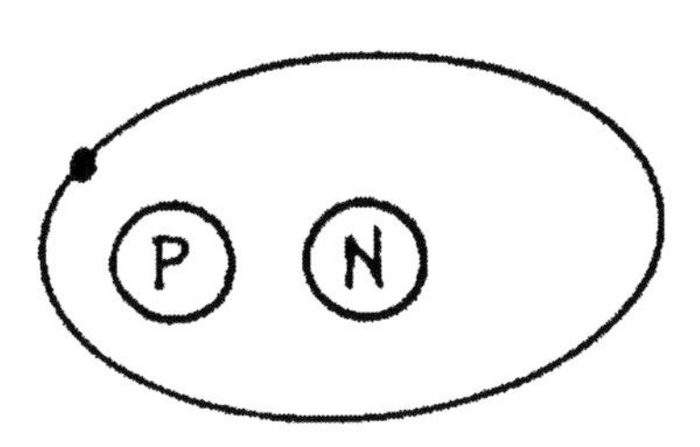

Deuterium

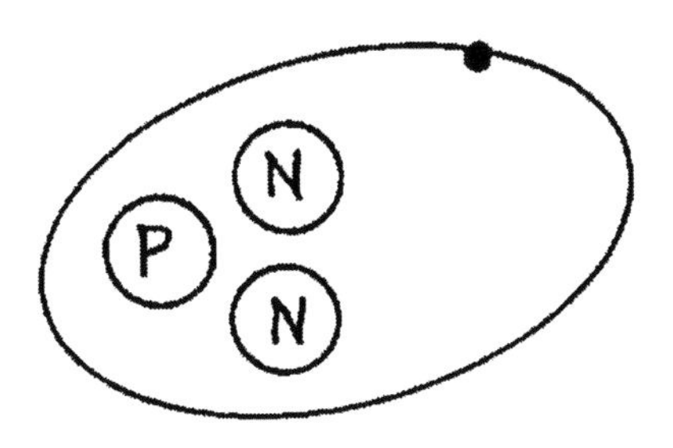

Tritium

FIGURE 4.1

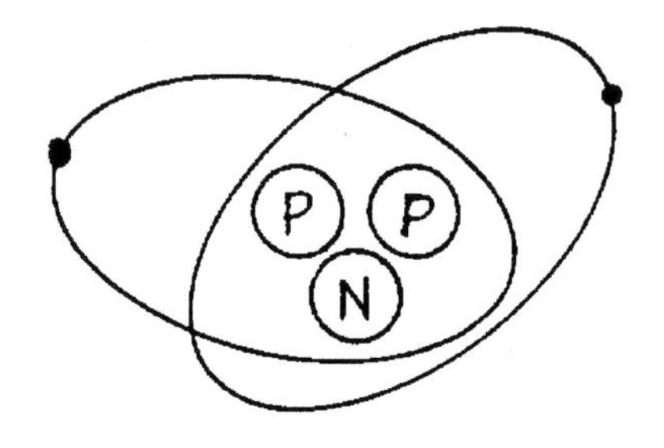

Helium 3

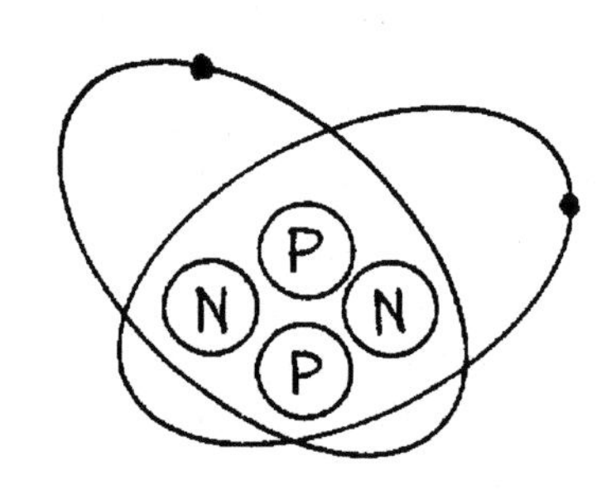

Helium 4

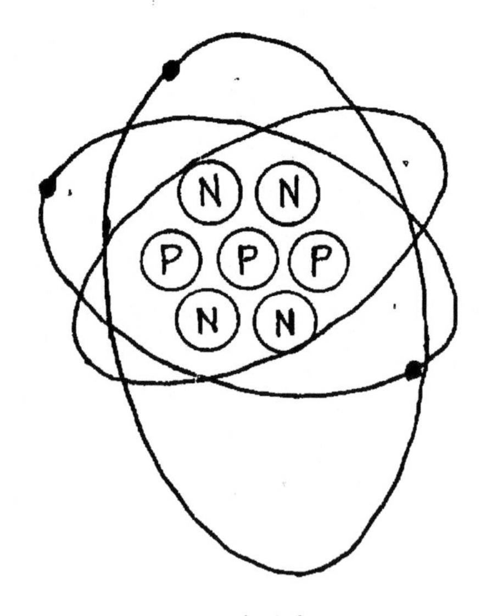

Lithium 7

FIGURE 4-2

extremely light electrons. A nucleon can thus be either a proton or neutron; and such ambiguity is not particularly helpful in describing the internal forces which may take place in the center of an atom. The term, in fact, is a vestigial remnant of the beginnings of atomic physics when no one really knew what lay at the heart of the atom.

In most discussions or artistic renderings of the shape of a nucleus (or even the atom) it is tacitly assumed that it is spherical. While this may be true for a preponderance of atoms which are lighter in mass than lead, a few may possess an oblate shape like a football; and several more, in all probability, have a prolate shape like a spherical balloon which has been squashed partially flat between two parallel plane surfaces. With respect to the concept of the size of a nucleus, it is important to realize that the boundries on an atomic distance scale are not sharply defined. The nucleons may spin individually and are also free to move or jiggle somewhat within tiny limits imposed by nature.

It might be said that the diameter of a nucleus varies between the smallest size at 7.87×10^{-14} to a maximum of 2.75×10^{-13} inches. These dimensions are sub-microscopic and beyond the capacity of most people to visualize at all. Nevertheless, atomic scientists are accustomed to using them routinely in their calculations.

4.4 The Riddle of the Strong Force

There are several mysteries concealed inside the atom. Many physicists like to assert they are well on the way to solving them (at least mathematically); but others are less convinced, as yet. Not the least of these mysteries is the answer to why, of the hundreds of particles known and discovered to date, the proton and the electron should uniquely remain by all odds the most absolutely stable combination and, so far as anyone can determine, eternal in our own material world. The proton has a slight edge over the neutron in durability. All others are susceptible to spontaneous decay after varying half-lives—either into protons and electrons or back into energy. Even the relatively long lived neutron is not immune to this assertion.

Regardless, a moment's reflection will reveal a much more immediate question. What is it that makes the protons and neutrons in a nucleus bind together so tightly in so small a spherical space? Angular momentum, or spins in opposite directions, is certainly inadequate to explain this intense affinity between proton and neutron; and any force which overcomes the natural electromagnetic repulsion between the positive charges of two or more protons confounds the imagination. With slight understanding, the early scientists simply invented the term "strong force" or "strong nuclear interactions" for this unknown cause. They also realized that it sprang into being only when the various nucleons came within sub-microscopically tiny distances from one another. The strength or intensity of this mutual attraction had to far exceed the known strength of gravity. This is despite the fact that the strong force increases exponentially in inverse proportion to the distance apart of the particles in complete similarity with the behavior of gravity.

Both the deuteron and the alpha particle are relatively simple examples of this problem, which accounts in part for the author's choice to introduce them in this chapter. For real complexity one should consider the task of calculating the necessary nuclear forces to bind the nucleons in the heavy atoms beyond Bismuth where there may be eighty four protons and one hundred and twenty eight neutrons all in one aggregation. Protons individually have an positive electromagnetic charge (or potential energy) exactly equal to that of the negative electron. This electromagnetic force is of respectable magnitude and can exert a force, whether attractive or repulsive, upon another charged particle over a considerable distance, atomically speaking. Thus two protons in close proximity to one another inside a nucleus can exert a very significant mutual repulsion upon each other which, in the absence of the strong force, would cause the nucleus to fly apart immediately.

For purposes of illustration, it has been said that two protons separated from each other by a distance of 2×10^{15} or 7.87×10^{14} inches experience a mutual electromagnetic repulsive force on the order of ten (10) pounds in Newtonian terrestial terms. When one reflects that the mass of the proton is a mere 1.672×10^{-30} grams, then a force of that magnitude becomes astronomically huge and beyond our familiar experience. Theoretical physicists have spent countless hours and published hundreds of papers trying to establish a firm mathematical origin for the strong force. Such a force is obviously *not* electrical in nature, since it acts upon all nucleons regardless of any lack of charge. The attractive force on protons used in our example above would need to be roughly one hundred times the repulsive electromagnetic force between two positive charges which it overcomes, or an estimated 1000 pounds. That fact in itself is astonishing; but there is another strange peculiarity of the strong force which denies almost all classical or Newtonian laws with which most of us are familiar. Its range of action or distance over which it may act is not only very tiny; but terminates at this outer boundary quite suddenly and sharply. It has been likened to the walls of a well. Once a proton or neutron manages to cross this action limit it becomes free to move off and find another existence of its own elsewhere. This is utterly unlike the behavior of gravity or a magnetic field, both of which become weaker and weaker to fade away gradually with increasing distance.

Thus the strong force may be thought of as only acting when the protons and neutrons are virtually touching each other. Its effective range of 10^{-15} meters (meaning fourteen zeros before the one meter) is extremely short.

For those of us who once learned in school to visualize the atom as a bunch of infinitesimally small ping-pong balls scrunched together, it might be natural to picture an escaping nucleon as similar to a rocket reaching the escape velocity from the earth and dropping its gravitational chains to continue its acceleration into outer space. However, such an imaginative metaphor is false and misleading. We know both the gravitational force and radio waves attenuate inversely as the square of the distance from the source of emanation; but these forces nevertheless continue to act over very vast distances measuring thousands of miles or more. Mathematically, one may say that their intensity declines asymptotically. Not so with the nuclear strong force, so far as we may judge. There is a sharp boundary limit. The nucleons are trapped in an energy well, as we indicated previously, and physicists use this concept in their applied mathematical analogies. However, a decline in the strength of this binding force does occur as nucleons move away some larger distance away from the atomic core. In the case of polonium or uranium atoms this does help to understand the phenomenon of natural radioactivity. Since the strong force is essentially limited in magnitude by the number of nucleons in tight propinquity, it follows that the very much larger and heavier atoms with more electron shells would be more susceptible to radioactive decay. It can be argued that this weakening of the strong force in the heavier atoms found in Nature may be overcome by simply adding more neutrons in the atomic core with the result of diminishing the total repulsive force of the charged protons. However, both protons and neutrons are fermions, as we have learned, and have spins of one half. Consequently Pauli's Exclusion Principle forbids that they occupy the same space or have the same energy state. Thus superheavy atoms going beyond the roughly 108 known atomic elements become difficult or impossible. The Grand Design seems contrived to prevent such a possibility, although collapsed and dying neutron stars may be an exception to this rule. Moreover, the fact there is a limit to the range of the strong force is most fortuitous for all us humans, for otherwise all the particle matter of the planets would agglutinate into a super-dense ball of condensed matter the size of a golf ball or smaller.

Hideki Yukawa proposed in 1935 that the strong force may be mediated by a meson having a certain definite mass. Although he picked the wrong name (muon), his logic was correct and led to the pion. Thus the strong force is presently thought to arise from the interaction or active exchange of these pions between the much more massive nucleons. Other scientists today believe the mesons themselves to be a secondary effect from "gluons" inside each proton or neutron; but that story must wait until later chapters.

4.5 The Riddle of the Weak Force

The alpha particle presents a second riddle to scientists by its very existence. Why should heavier atoms such as radium, polonium, cesium, or uranium (and quite a few others) spontaneously throw out fragments of themselves at a statistically steady rate and a measurable velocity, leaving behind a chemically different element? What is the mechanism that instigates this self-destruction of the larger element atoms. The phenomenon is known as radioactivity, of course. The word "radio" is archaic and somewhat misleading, for this process of disintegration generates both particles and X-rays or higher energy gamma rays; but not the much lower frequency radio waves with which our century is familiar for daily communication. Again, for lack of a suitable explanation, the early scientists invented a cause and named it the "weak force." Once again, the origin of the weak force at first defied a ready explanation.

Today it is thought this force is carried by a certain group of bosons, which, as you may remember, carry only integer number spins. These multiples of Planck's Constant can be zero, one, two, or even three. This feature allows them to comfortably inhabit the same small space and ignore the Pauli Exclusion Principle which applies to the fermions. These weak force carriers are usually called in the trade "intermediate vector bosons" and instigate reactions between fermions inside the atoms. Since the weak force is presumed to act over a smaller radius distance than the strong force, it follows that during their fleeting independent existence of nano-seconds they must be larger and/or heavier than the carriers of the strong force. This statement may seem perversely obscure to the reader; but it is the consequence of the Uncertainty Principle of Quantum Mechanics. The allowable range of a force carrier must be inversely proportional to the mass of this exchanged boson particle. Thus the weightless photon can carry energy over inter-stellar distances undiminished.

However, we are getting too far ahead of ourselves. It suffices to state now that the theoretical physicists have been busily preoccupied for decades now with the mathematics necessary to prove the strong force and the weak force are merely different aspects of the same thing. We should be content to leave them at their work for now; but merely reflect that if the weak force only comes into play at distances less than 10^{-18} meters this indicates that all the particles which we have been discussing must approach the dimension of a single point in space to make realistic sense.

In all the nuclear transformations which the reader will encounter in this book it is important to recognize that the Law of Conservation of Energy is expected to apply. Masses may come and go, change, or even disappear into energy. In any spontaneous decay process starting from repose, the sum of the masses of the daughter products cannot exceed that of the original parent particle. They may be less, however, since energy may be radiated in the form of photons or imparted to the fragments in the form of increased momentum. The only way that this rule may be contradicted is by the infusion of Kinetic energy into the system from an outside source, whether natural or man-made.

Thus, in the calculations made by the scientists to make sense of their laboratory observations of fragments resulting from collisions of particles with a target of significant mass, the rule of Conservation of Energy is an important tool in research.

Chapter 5
Photons and Duality

Using language carelessly, one might be inclined to categorize the photon as just another particle with all the rest; but that would be an error. The photon is unique and basic to the structure of the universe with an intimate relationship with time and energy. Both those concepts involve the very mystery of all existence. Since the photon has no measurable mass, it cannot logically be called "matter." They travel by the billions through the empty regions of interstellar space at a constant and incredible speed, although slowing down somewhat when passing through transparent mediums such as glass, air, or water. They come into existence by the agitation of electrons inside atoms or by the destruction of matter and are emitted in separate, discrete bundles of energy. A better word for them might be corpuscles, just so long as we do not think of them as tangible tiny balls. The visible evidence of their existence to us humans is light when they impinge upon atoms in the retina of our eye. Despite all that has been previously said, photons can sometimes behave as if they had certain characteristics of matter, which makes their nature clearly ambiguous to our limited understanding. They are the messengers of creation.

The symbol used for the photon among scientists is the Greek letter "gamma" (γ). Photons have no electrical charge whatsoever and are generally considered to have a spin of one, which automatically makes them bosons. That the property of spin should exist in the photon is a demonstration that an intrinsic angular momentum must originate in entirely non-material fields in space, which is a difficult concept to adopt. Strangely, physicists believe photons can possess the attribute of momentum, which, as we mentioned in our Introduction chapter, is a characteristic of mass and usually is associated with particles of physically tangible matter.

Since we started out by saying the photon had no rest mass, this seems to be a flat contradiction. Nevertheless, the possession of momentum is perfectly consistent with Einstein's Relativity prediction that large and massive stellar bodies can bend or deflect light rays. This fact has been observationally confirmed by astronomers many times.

The energy "E" traveling in one direction with the speed of light "c" has a momentum magnitude "p" of E/c associated with its motion. This curious dichotomy is evidenced by the commercial sale of small scientific toys consisting of a hollow glass sphere evacuated of all air and containing four thin metallic vanes mounted on a vertical shaft having tiny end bearings. When exposed to strong sunlight the vanes rotate silently within the transparent sphere with no other mechanical attachment or energy source than the light. Notwithstanding this assignation of the attribute of momentum to what seems a radiant energy process, we should not assume the photon becomes also endowed with the other properties of material corpuscles.

Thus far we have been discussing the photon in the context of a corpuscle or "packet" of energy, which is an idea quite harmonious with the late 1600's. Some scientists today have dismissed this idea quite contemptuously as the "fuzzy ball" theory. Yet, as both Newton and Einstein pointed out, there are perfectly credible reasons for accepting this viewpoint to explain at least some of the tricks for which the photon is capable. The phenomena of a photo-electric current engendered within certain metals, such as selenium or germanium, and the fluorescent emission of certain chemical compounds, such as zinc sulfide, serve as a standard demonstration. To the surprise of the experimenters, the maximum velocity of escaping electrons excited by a beam of light was independent of the intensity or brilliance of the light, although it does vary with the light frequency or color. A stronger light ray will, it is true, increase a photoelectric current somewhat; but this is merely the result of a greater number of photons arriving at the surface in a given instant. The probability of a photon impinging upon a random atom at the correct frequency or energy level necessary to release it from its outer orbit around the nucleus has been increased by a greater number of photons arriving over time. All this implies a corpuscle action.

The corpuscular aspect of light was confirmed in 1923 by the American Arthur H. Compton. He directed a narrow beam of X-rays, which are another form of light transmission, into a cloud chamber of water vapor. He noticed that the radiation on the exit side of the cloud chamber had lost energy; but that the electrons inside the chamber had gained energy in a manner suggestive of individual impact action.

The energy exchange from photon to an electron in an atom is an "all or nothing" process in which the photon disappears and the electron jumps up to an orbit around the nucleus having a greater radius from the center. In some cases the energy acquired by the electron may be enough to enable it to escape from the surface of the target material entirely and join others, equally free, to produce a measurable current, providing it avoids capture by another surrounding atom. The direction of the free electron's movement may be controlled by an externally impressed electromotive voltage. This is the principle of a simple photo-electric detector cell. The phenomenon of electromagnetic energy exchange between electron and photon is reversible, inasmuch as the shift of an electron to a lower orbit level may release a quantum of energy in the form of a photon. The significant peculiarity of this change of energy level within the atom is that it always takes place in a discrete, tiny jump called a "quantum." It is not a smooth or gradual process; and this simple fact of energy changes in jumps confounded all the classical physicists of an older era.

The amount of energy which may be carried by a single photon varies over an extraordinarily wide range. For visible light this may be only a few electron volts; but for penetrating X-rays the energy may be on the order of one thousand electron volts. By all counts, however, gamma rays produced by streams of protons hitting our upper atmosphere from extra-terrestrial origins at velocities approaching that of light are the most powerful ever measured. These bursts of cosmic rays originate from different areas in our sky from unknown point sources far out in space, including quasars. Their existence confused the readings obtained from the space "spy" satellites launched by the United States around 1968 for the purpose of monitoring nuclear bomb tests by Russia as part of the international Nuclear Test Ban Treaty. Point sources of gamma rays are also suspected to be produced by large massive objects falling into "black holes;" but their precise mechanism for this production is still unresolved. All we can do is plot these point sources on the map of the celestial sphere and speculate.

Notice that in the discussions given above we abruptly and unfairly introduced without explanation the idea of frequency in light, which is a wave concept. Most people now living at the beginning of the twenty-first century have heard this word so often in their daily experience with radio or television as to immediately and subliminally accept it as familiar with little further thought. Yet this concept is entirely foreign to our picture first presented of the photon as a corpuscle or "bullet" of energy. Notwithstanding, the "wave" nature of light became apparent long before particle physics came into existence. It is derived from the mathematical sine curve representing an oscillatory or vibratory degree of amplitude for energy, such as that displayed by the pendulum in an antique clock. The justification for the wave theory can be demonstrated by the separation of white sunlight into a spectrum of rainbow colors by means of a simple glass prism of triangular cross section. The same effect may be accomplished by another means if a diffraction grating or glass etched with dozens of isolated parallel lines separated from one another by even and microscopically small distances.

The science of optics developed the wave theory of light from studies of the speed of propagation of light in different transparent mediums and the phenomenon of refraction. Color spectrums and diffraction, as well as interference patterns obtained with monochromatic light beams intersecting at angles, could all be explained by considering light as formed in waves. Perhaps the earliest and most convincing argument for the wave theory was the double slit experiment developed by Thomas Young (1773 to 1829) in England in 1800. Descriptions of it have been repeated in thousands of physics textbooks ever since. A bright, monochromatic light source is set up for projection on an opaque screen located not more than one hundred centimeters (39 inches) away which has in its center a narrow slit no more than 2/10ths of a millimeter wide. A second opaque screen is then set up at an adjusted distance of plus or minus three meters (that is to say, a yard). In the case of this second screen, however, it bears two slit penetrations separated from one another by a distance of less than one millimeter. In other words, these two narrow slits are extremely close together and represent some multiple of angstrom wave units. Beyond all this apparatus is positioned a third partially reflective target surface, such as a white wall or cardboard sheet.

If a screen or shutter is lowered to cover one of the two slits on the second barrier away from the light source, then one sees on the last screen a pattern of light which one might well expect on a target at which a soldier had sprayed bullets from a machine gun. The light intensity may be at its greatest along the centerline of the open slit and gradually fade towards the outside edges of the target illumination pattern; but the notion of individual photon "bullets" is still applicable. However, when both narrow slits on the second screen are allowed to remain open a spectacular change appears. A well defined pattern of alternately black and white horizontal lines appears and extends to an angle of at least three degrees on either side of the light beam source. This phenomenon is called "interference." It is easily explained by assuming that light behaves as waves that are bent or slightly deflected around sharp edged barriers, just as the waves of the sea turn slightly as they pass a stone breakwater. Since light always travels at the same velocity in dry air (or any other consistent medium) and since each wave front as it passes through the two slits simultaneously must travel slightly different distances, then it follows that they must meet or intersect each other at very slightly different times or out of phase. The black lines represent that exact point where the top or peak amplitude of one wave is entirely negated by the deepest trough amplitude of the wave coming from the other slit. They cancel each other out perfectly at intervals which depend upon the wave length of the monochromatic light source. The black and white bands would appear to have very much more fuzzy edges for white light. (See figure 5 - 1).

The conventional symbol for wave length is the Greek letter lambda (λ), whereas its frequency of maximum amplitude in a given time interval is often designated by "f."

Shortly after Young's experiments a French engineer named Augustin Jean Fresnel (1788 to 1827) took up the study of optics and eventually, in 1814, prepared a paper for publication on the aberration of light. He demonstrated that the diffraction phenomenon could take place along the edges of the shadow made by any opaque obstacle, whether a plane surface or circular hole, providing that the boundary edges are thin and sharp. He also called attention to the circular polarization of a light beam and correlated all of this with the mathematics for wave action. Later, during his quiet career under the reign of Napoleon (he had been a supporter of the Bourbon kings when young and thus adverse to public exposure), he went on to invent the famous Fresnel lens used in maritime lighthouse beacons all over the world in order to focus the light from gas or kerosene lamps into an intense beam. These lenses were a large cylindrical array of curved glass prisms of varying cross-sections artfully cemented together in a careful pattern which would refract the light rays at the upper and lower boundaries into a horizontal center. This device was used for a century or more until an electric arc lamp and polished parabolic reflector on a rotating table came into competition.

It remained, however, for the genius of James Clerk Maxwell (1831 to 1870), who was the scion of a wealthy and land-owning Scottish family, to assemble all the various observed facts and recognize the intimate and harmonious relationship between light and the propagation of electromagnetic forces or energy. He was the founder of the Cavendish Laboratory for Experimental Physics at Cambridge University and had long been impressed by the "lines of force" ideas of Michael Faraday. In 1873 he published a treatise on "Electricity and Magnetism" and there introduced the concept of energy transmission having a vibratory frequency and bearing a close mathematical similarity to a harmonic oscillator. Later laboratory experiments by Heinrich Hertz proved him correct. These developments ultimately brought the realization that light was only one small portion of a vast spectrum of radiant energy frequencies which ranged all the way from heat and infra-red through the various colors of light and ultra-violet and beyond to radio waves, X-rays and gamma rays. The last represent the highest energies now known. Today we have learned to speak of electromagnetic energy transfer by its frequency, or number of full cycles of any steady wave per second. Radio "ham" amateurs are apt to talk in terms of frequencies most often; but they sometimes lapse back to the older custom of waves and chatter about 20, 40 or 60 meter wave bands.

In as much as the speed of light and/or radio waves through a vacuum is an identical constant, then it follows that:

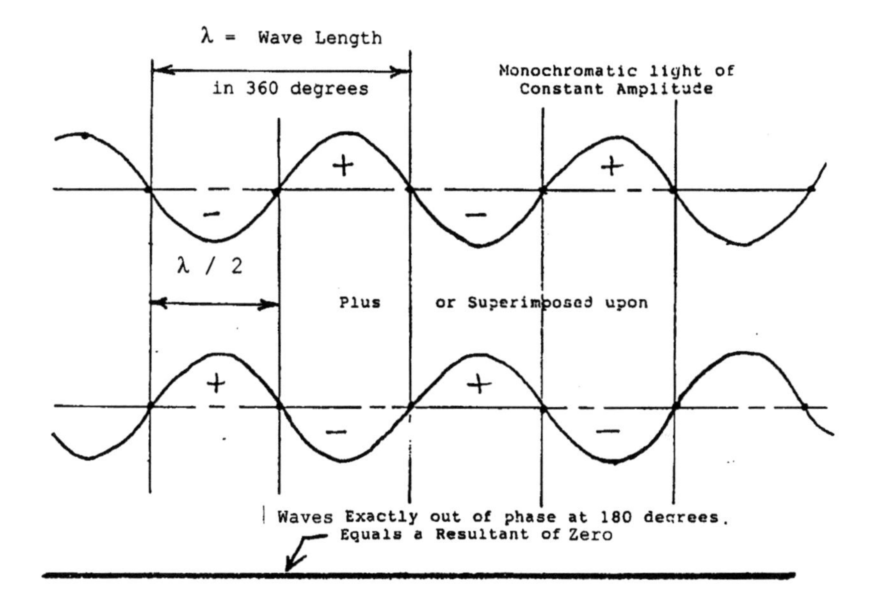

A Black Line or Darkness

Figure 5 - 1

$$f = \text{Frequency} = \frac{\text{Speed of Light}}{\text{Wave Length}} = \frac{c}{l}$$

The energy contained in a certain quantum of light is proportional to the wave frequency in accordance with the equation given below:

$$E = h\ f \qquad \text{when:}$$

h = Planck's Constant = 6.63×10^{-34} joule–seconds
f = Frequency of monochromatic light beam in cycles per second or Hertz.

For the benefit of the more fastidious reader familiar with physics, the energy content of a red light is 3×10^{-12}; whereas the energy content of a violet light quantum is 4.5×10^{-12} ergs. We are obviously dealing with very small numbers; but they are significant nevertheless. Ultra–violet frequencies pack more energy.

Hidden in the equation given above is a very strange and perplexing fact. In spite of the fact that the rest mass of a photon is assumed to be zero, photons can carry momentum. The implication for electrical engineers is that a powerful radio wave source consisting of numerous photons is capable of exerting a slight push or force on a large antenna array.

The actual moment theoretical momentum of a photon, however infinitesimal, is given by the equation below:

$$p = h / \lambda \qquad \text{when:}$$

p = the momentum value
h = Planck's constant
λ = the monochromatic beam wavelength.

It should be remembered that Maxwell also pointed out that the field strength or magnetic flux of an electromagnetic signal diminishes or attenuates in inverse proportion to the square distance from the source of energy to the receptor, as with the case of gravity .

On the other hand , the luminosity or brilliance per square foot of receptor of a beam of light or radiation may be considerably reduced due to a factor called dispersion by dust, gas, or moisture. These facts are quite important and do not help to explain why electromagnetic waves from stars can be detected over vast cosmic distances of millions of light years, whether by the retina of the eye or by a tuned oscillator circuit in a radio. As Sir Arthur Eddington pointed out in his lectures back in 1927, this reality is exceedingly strange. The only reasonable explanation for the peculiar facts just described appears to be that the electro-magnetic stimulus radiated by a light source remains for all time concentrated in tiny bundles or packets which we call photons . Only an energy quantum slightly greater than the miniscule quantity described by Planck's Constant is required to stimulate an electron to leave its orbit in a particular atom at the receiving end and thus become available to initiate a response in the human eye or a photo-electric cell or radio amplifier .

By this time the reader should have become fully aware that our explanations to account for the commonplace phenomenon of light, whether corpuscular or wave motion, exhibit an ambivalence or, worse yet, represent an inadmissible dichotomy. The scientists have been fully aware of this perplexing situation for more than two hundred years. It is small wonder that the 1911 edition of the Encyclopedia Britainnica evaded the issue in its introduction to the subject by stating that "Light may be defined *subjectively* as the sense impression formed by the eye." That was not much help, of course. Other people have said that photons are best thought of as discrete and tiny bundles of radiant energy coming in different vibratory frequencies and greater or less numbers. Werner Heisenberg, who was one of the founders of quantum mechanics, described it as a "wave packet" or "a wave-like disturbance whose amplitude is appreciably different from zero only in a bounded region. This region, in general, is in motion and also changes its size and shape."

A more modern definition liable to be found in a scientific dictionary of today typically comes down hard on the wave theory, stating that "the photon is a quantum of a single mode of wave length, direction, and polarization of an electromagnetic field." The reason for their decision to adopt this viewpoint is understandable, since the only rational and coherent way to treat the phenomenon of light mathematically is by means of wave mechanics. Since they were quite unable to fully resolve this riddle about light behaving as corpuscles under one set of conditions and undulatory waves under a different circumstance, the scientists gave a name to it. They called it "duality," thus giving many people the opportunity to believe they could answer the questions of the curious. The corpuscle and wave behavior have never been observed simultaneously, however. Not only is the dualism of light (or any other form of electromagnetic energy transport) no longer contested, but it remains an embarrassing gap in our comprehension of the universe. Many efforts have been made to explain this in terms of probabilities and interacting corpuscles.

In 1923 the French genius of particle physics, the Duc Louis de Broglie, had the audacity to propose the seemingly outrageous notion that, if photon corpuscles could behave as waves, then why could not electrons behave as waves also? We know electrons possess the property of mass, however slight;

and, because of that fact, they have long been regarded as material particles. More specifically, de Broglie proposed that a free electron moving at a speed less than that of light should have a wave length computed by the following formula:

$$\text{Electron Wave Length} = \frac{\text{Planck's Constant}}{\text{Electron's Mass} \times \text{Velocity of Electron}}$$

He turned out to be correct. Not only photons, but also electrons, protons, and even neutrons sometimes behave as waves. To those of us accustomed to dealing in their daily life with such intractable objects as trees, rocks, bricks, and asphalt pavements this proclivity of physical entities to behave as waves requires a considerable mental and psychological adjustment. Nevertheless, there are a great many cases which are demonstrable in the laboratory and support this assumption. X-rays, when passed through certain crystals, egress from the other side and give a diffraction pattern, as one might expect but it was the behavior of X-rays when reflected from the surface of a crystal which attracted the fascinated attention of the scientists. The phenomenon, now known as "Compton Scattering," was established in 1924 by Arthur Holly Compton (1892 to 1962), who was one of four children of a professor of philosophy at the College of Wooster in Ohio. Compton had secured both his Masters and Doctorate degrees from the University of Princeton in 1916. During World War I he worked for two years at the Westinghouse Electric Company in East Pittsburgh, Pennsylvania, and there became absorbed with the behavior of a beam of X-rays directed upon metallic crystals of magnetite and silicon steel, both magnetized and unmagnetized. Of particular interest was the reflection from the crystal surface of the X-ray photons by what he presumed were electrons in the atoms. A very rough analogy to these experiments would be like hurling large handfulls of small peas at a pile of regularly spaced large rocks having uniformly arranged plane surfaces at consistent angles. Some of the X-ray photons were deflected or scattered off at various angles; but their frequency and wave lengths were not the same as the original incident beam and varied with the angle of reflection, as expected with a wave behavior. Moreover, this was inconsistent with the picture of quanta of energy having no linear momentum. On the strength of receiving a National Research Council fellowship award, it became possible for Compton to spend a year from 1919 to 1920 at the Cavendish Laboratory at Cambridge University in London with the famous J. J. Thomson and Ernest Rutherford as one of their disciples. This was a tremendously stimulating event and introduced him to the current intellectual ferment on atomic physics. Returning to the United States, Compton became a professor and head of the Physics Department at Washington University in Saint Louis, Missouri. There he resumed his research experiments with X-ray scattering; but with the difference that he now regarded the reactions as one of a simple collision between a free electron and a photon having an energy of "h f," as postulated by Planck, and a linear momentum "h f / c," as postulated by Einstein. In other words, Compton had embraced the quantum theory and abandoned strict older and classical explanations for the effect. He even went further and assigned a quantum wave length for the electron of "h / m v" in complete harmony with Dirac. The entire process could be explained by assuming that an individual photon impacts with an electron at rest and gives up some of its momentum and energy to the electron, which then takes off. The photon, on the other hand, recoils off in another direction; but, having less energy now, possesses a different frequency and wave length altogether. This is all simple quantum mechanics; but new to many physicists in 1923 when Compton reported his findings to the American Physical Society. There was great furor and considerable opposition to his ideas; but in 1924 a public debate on the validity of his published results was called in the summer of 1924 in Toronto, Canada, by the British Association for the Advancement of Science. Compton's experimental results and ideas were completely vindicated and he shared a Nobel prize in physics three years later for the discovery of what is now universally known as the "Compton Scattering" effect. However, the explanations of the particle physics adherents did not quite settle the question of duality entirely.

By a strange coincidence a thirty six year old physicist named Clinton Joseph Davisson (1881 to 1958) had just started work with the engineering department of the Western Electric Company (which was then a subsidiary of the old Bell Telephone Company) at about the same time that Compton had gone to work for Westinghouse Electric. Davisson was the son of a painting contractor and school-teacher living in Bloomington, Illinois, and had with many difficulties and interruptions achieved a Ph. D. from Princeton University in 1911 in physics and mathematics. His employer put him to work studying the emission of electrons from certain metals when bombarded with a stream of electrons. In particular, he was measuring the angles at which the secondary electrons were scattered, which was a conventional laboratory procedure. The accidental explosion of a flask of liquid air had oxidized the surface of one of his nickel targets; so in order to clean the surface Davisson placed it in an electric furnace for a time. This apparently changed the crystalline structure of the metal and also the results of his experiments. Order seemed to be appearing from chaos; but Davisson did not quite understand why. With great serendipity and cleverness he then took a vacation in the summer of 1926 to go to Oxford, England, in order to attend a conference of the British Association for the Advancement of Science. There he discussed his perplexity with electrons with Max Born and also heard about de Broglie's ideas. The association with other physicists of stature who were involved with the new quantum mechanics opened up whole new intellectual possibilities. Returning to the United States, Davisson set to work, enlisting the aid of an associate named Lester H. Germer, in a systematic study of the diffraction of electrons by a single crystal of nickel. The emitted secondary electrons were definitely behaving as waves and fully confirmed de Broglie's prediction. Duality was still alive and well. The fact that electrons could be focused and used in place of light rays, which had a longer wave length made the later invention of the electron microscope possible and revolutionized the pharmacological, medical and micro-biological laboratories all over the world. It also made the computer "chip" possible.

Max Born (1882 to 1970), who was a renowned mathematician and theoretical physicist, worried about the meaning of the duality dilemma a great deal. He took the problem so seriously that he wrote the following statement, which should surprise, if not shock, almost anyone:

"Hitherto, we have always spoken of waves and corpuscles as given facts without giving any consideration at all to the question of whether we are justified in assuming that such things actually exist."

As one of the participating founders of quantum mechanics, for which he was rewarded with a Nobel prize in 1954, Born knew and accepted the basic premise that it was impossible to determine by any laboratory experiment or calculation with any exactitude both the position and velocity (or momentum) of any particle at any given instant of time. Hence he went on to his assertion that it was also beyond our capability to produce a proof that we are actually dealing with corpuscles or waves. Thus the "Principle of Complementarity" espoused by Neils Bohr was born. The ultimate solution to this problem, in so far as Born was concerned, lay in statistical probabilities as the unifying concept. Many later scientists agreed. For those readers who still wish for a philosophical or ontological meaning we can only refer them to Max Born's own final words:

"If even in inanimate nature the physicist comes up against absolute limits, at which causal connection ceases and must be replaced by statistics, then we should be prepared, in the realm of living things (and emphatically so in the processes connected with consciousness and will) to meet insuperable barriers where mechanistic explanation, the goal of older naturalistic philosophy, becomes entirely meaningless." Some people derive comfort from the fact that proof exists that the universe is not entirely mechanistic. Others become disappointed atheists. All must admit that unknown probabilities are there none-the-less.

Figure 5.2
48 Iron Atoms On A Copper Surface
Surrounding Electron Waves

Courtesy of: IBM Almaden Research Center in San Jose, California, as example of Low Temperature Scanning Tunneling Microscope.

Chapter 6
The Atom Revisited

"The law of things is a law of Reason Universal; but most men live as though they had a wisdom of their own." — Heraclitus of Ephasus (540 to 475 B.C.)

This seems an excellent time for us to pause for a review of what we do and do not know about the atom and its three principal components. In a previous chapter we have discussed those parts only in very general terms; but have hinted at certain aspects of the atom's structure which present many profound puzzles to the nuclear scientists. Moreover, our simplistic picture of the atom as a tight conglomeration of tiny billiard balls, some larger than others, has been proven to be totally inadequate for any serious understanding.

The author has also been guilty of glibly mentioning the subject of quantum mechanics in a few places with no serious explanation of what is actually entailed in such allusions. The name "quantum mechanics" makes it sound to the uninitiated as if the scientists had reduced the behavior of the atom to that of a mechanical erector set toy with utterly predictable and exact rules which involve Euclidean geometry, vectors and forces. All one needs is the handbook. Nothing could be further from the truth; but our restricted subject now should serve as an introduction to a subsequent chapter and a hint of difficulties yet to come.

There is also the question of the actual size of the atom. We have all been taught that the atom is so tiny as to be sub-microscopic and almost unimaginably minute; but the amazing fact is that the atom is mostly empty space as compared to the nucleus of protons. There have been many textbook analogies invented in order to suggest to the lay person some reasonable mental picture of its structure. If we may be allowed some poetic license on the scale of actual dimensions, it might be said that if the ordinary atom were to be expanded or enlarged so that the diameter of the outer electron ring was equal with the mean diameter of the earth (7918 miles), then the nucleus at the center might be the size of a large orange or grapefruit. The actual size of a large atom in our macroscopic international units may be from one to two times 10^{-8} centimeters (10^{-10} meters). This last number in the parentheses is a significant number. For one thing, it happens to coincide with the Angstrom unit of 1×10^{-10} meters commonly used in optical spectroscopy for wave lengths of light.

Anders Jonas Angstrom (1814 to 1874) was the son of a church minister and received his doctorate in mathematics and physics from the University of Uppsala in 1839. He became fascinated with astronomy and was thus led into his life's work studying the light spectra of the sun, the aurora borealis, and, eventually, heated metals. By 1872 he was a world recognized expert in the new technique of spectroscopy and the study of the light color signatures for various elements. Among his many discoveries was that hydrogen formed a huge part of the sun's mass.

Just how or why Angstrom chose 1 x 10-10 meters for his basic unit of measurement for light ray wavelengths may not be certain; but he clearly recognized this point in wavelengths as a significant short wave barrier. It is intriguing to speculate how much he knew or suspected about atoms at that time; but it is remarkable that his arbitrary unit calculates out to a corresponding frequency of 3×10^{17} cycles per second. This happens to coincide roughly with the range where ultra-violet light ends and the long wave X-rays begin. Moreover, this point also identifies the wavelength where the perturbed or excited atom starts to radiate these X-rays. Angstrom was astute enough to recognize a significant change.

Actually, a chemist from Manchester, England, had already endorsed the idea of atoms well before Angstrom had been born. John Dalton (1766 to 1844), the son of a weaver and early Quaker, became a teacher at the age of twenty–eight and devoted much of his career to the study of gases and their expansion under heat. He was also prescient enough to predict the opposite of this behavior and the ultimate liquefaction of gases under extremely cold temperatures, which were then unattainable. He became convinced atoms were the explanation for chemical compounds and, furthermore, proposed in writing in 1805 the idea that a structure of atoms could account for the absorption of some gases in water. Thus the unseen atom began as a rational deduction and not a direct observation.

There is another aspect of the atom's size which should be mentioned. When we start thinking about something that small, we should abandon the associated idea that all the electrons circle around the nucleus in peaceful celestial harmony much like the planets around the sun or the moons around Jupiter. Inside the atom the normal laws which we attribute to Isaac Newton do not always apply. This does not mean that Newtonian physics are not still useful; but almost everything inside the atom is jiggling and vibrating around like mad. In relation to three other forces involved, gravity is quite insignificant. In fact, the ratio of the attractive gravitational force between two electrons to their mutually repulsive electromagnetic force is calculated to be roughly one to 4.17×10^{-42}. That is too small a number for us to imagine. When we get inside the atom, strange things happen with which we have no similar experience in our life. In fact, when things get to be as small as to be only 10^{-33} centimeter, which is known as the "Planck Scale," it has been proposed by the theorists that all the internal forces in the atom combine into one field. Nothing that we now know can reliably predict what happens then. In order for us to avoid further confusion, it might be wise for us to begin now with what we do know about the principal basic components of the atom.

6.1 The Proton

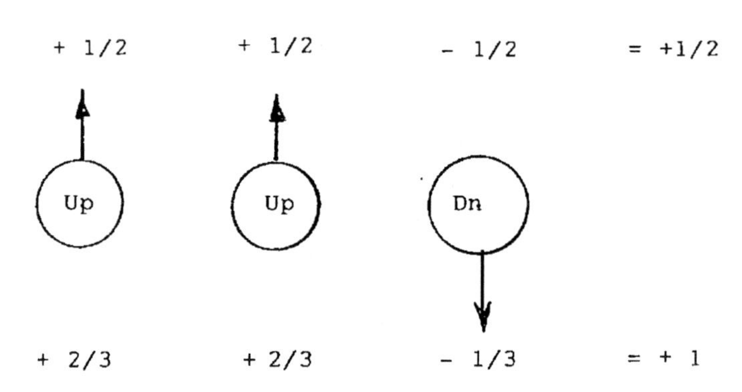

The figure shows the composition of proton in quarks.

The accepted symbol for the proton is the small typed letter "p."

The mass of the proton is 982.27 mega-electron volts (MeV), which approaches one billion electron volts. This makes it 1,831 times the mass of the electron.

The size of the proton is roughly estimated to be over 1.6×10^{-13} centimeters in diameter. The size of the nucleus of any given atom is larger, of course, and may range to a few digits times 10^{-12} centimeters, depending upon which element one is considering. One of the best measurements of the size of the proton, insofar as it has any really fixed radius, was completed in the 1950 decade by Robert Hofstader at Stanford University in California. Risking an explosion in a potentially dangerous experiment, he aimed a concentrated beam of electrons at 800 MeV energies into liquid hydrogen and analyzed the scattering effect. Since electrons are electromagnetically attracted to the positive proton, but are entirely unaffected by the strong force, they become extremely undeviated projectiles with a potentially greater aiming precision. His results indicated the positive electric charge was located essentially near the center of the proton sphere.

The proton carries a positive electric charge of plus one (+1), meaning that it is electromagnetically equal to and exactly balances the negative charge of the electron.

As the reader already knows, the proton has a spin ("J") of one half (1/2), which makes it a fermion.

The fundamental nucleus of the most prevalent isotope of hydrogen is identical with the proton. This fact induces some physics writers who create textbooks to occasionally use the small letter "p" for hydrogen gas atoms being bombarded with other particles, which can be mistaken for momentum by the uninitiated.

As we mentioned in Chapter 1, Sir Ernest Rutherford, while still experimenting with alpha particles at age forty–eight, made the first observation of a nuclear reaction produced by the collision of two nuclei at a relatively low velocity. He incidentally reconfirmed the existence of the proton as a separate and unique particle. While bombarding nitrogen gas with alpha particles emitted by a radioactive element, he detected the positively charged proton as one of the by-products. A possible scientific explanation for this reaction might be:

One alpha particle + Nitrogen atom yields an Oxygen isotope + Deuterium

At least, that is the way in which one physicist described it, however abstrusely.

It is interesting to learn that one free and independent proton removed from an alpha particle has an incremental difference in mass of 6.70 MeV from the same proton bound up in the nucleus of an atom. The difference is a measure of the binding energy or "strong force" necessary to hold it in tight aggregation within the nucleus. $E = mc^2$ again!

The proton is the most stable of all known particles in our earth environment, excluding any fusion reactions which may take place deep within the sun. Notwithstanding all apparent evidence, however, it is widely believed by many physicists today that all atomic particles are inherently unstable to a greater or less degree and eventually must spontaneously disintegrate into fragments of lower mass. This may or may not be true; but it is a tacit assumption in particle physics today. The belief in such an inevitable decay has obviously been encouraged by the proliferation of newly discovered and exotic particles which have resulted from the invention of very powerful accelerator colliders in the past fifty years or more. The great preponderance of these particles have extremely short lifetimes. Another encouragement to applying the same ultimate rule to protons is the theories developed by the astronomers over the years to explain the "burning" of the hydrogen atoms deep inside our sun. The trick to this last statement is that they must invoke the "Uncertainty Principle" of quantum mechanics in order to explain the disintegration of the proton into electrons and neutrinos. One of several experiments in the world to confirm the decay of the proton is known as the Irvine-Michigan-Brookhaven (IMB) collaboration. This is a joint agency venture which went into operation between 1981 and 1982. It is located in the bottom of a salt mine near Cleveland, Ohio. Nucleon decay, whether proton or neutron, is expected to release some charged sub-particles having high energies and velocities approaching the speed of light. These fragments produce a visible "Cerenkov" radiation in water. This is the same eerie blue light which one may see at nuclear power stations where the spent uranium fuel rods are temporarily stored in a protective bath of boron treated water contained in what are ironically called "swimming pools" with grim humor. Thus it becomes logical that the researchers elected to use a huge storage tank containing some seven thousand tons of very pure water. The surrounding walls are covered with photo-sensitive detectors connected to electric multipliers for the purpose of photographing and recording any light tracks. Locating this gigantic apparatus deep underground is for the purpose of shielding the tank from cosmic radiation. Even so, a collision with a stray neutrino from space may occasionally simulate a deceptive incident.

The Japanese were not to be out-done and constructed their own detector for proton decay in an active lead and zinc mine near Kamioka. The entire apparatus is located in a tunnel which extends deep underneath Mount Ikenayama; and this fact coupled with the presence of lead ore provides a better shielding from background radiation. The money thus saved was spent on better photon sensor devices covering a larger surface area of the tank walls. No incontrovertible evidence of proton decay has yet been detected.

One well may wonder why there is all this very expensive preoccupation, funded by governments, with nucleon decay. The simple reason is that all the symmetry and Non-Abelian gauge theories which physicists use today with such confidence and enthusiastic credulity demand that both protons and neutrons self-destruct sometimes. It is a matter of no small aggravation to the scientists that the proton seems unexpectedly reluctant to do so. The lack of demonstrable proof of this prediction after more than ten years of patient operation of the experiments casts a great pall of doubt over the complete validity of both Super-Symmetry Theory (S.S.T.) and the Grand Unification Theory (G.U.T.). Much hinges upon this debate, for the authenticity of the Standard Model and quarks also require the ultimate decay of the proton as another proof. Consequently a great deal of money and time has been spent on these underground detectors of pure water surrounded by scintillation counters without success thus far. No one can yet state unequivocably that they have witnessed the decay of the proton. Even if their detector does fire, its signal is already suspect because of the possibility of the impact of a cosmic ray particle or neutrino with a water molecule. Great faith and patience are the support of those experimenters. When challenged by people outside of their field of endeavour, the physicists respond that their calculations indicate a mean lifetime for the proton of from 10^{30} to 10^{34} years. That seems like an eternity to us humans and, in all probability, may exceed the expected lifetime of our entire galaxy. Since the average life of a proton may exceed that of our sun and planetary system, the entire argument may seem tautological to the ordinary person. It is just as well for us mere humans that the basic atoms in our organic molecules last a great deal more than the few days or seconds characteristic of nuclear fragments which we will learn about soon.

6.2 The Neutron

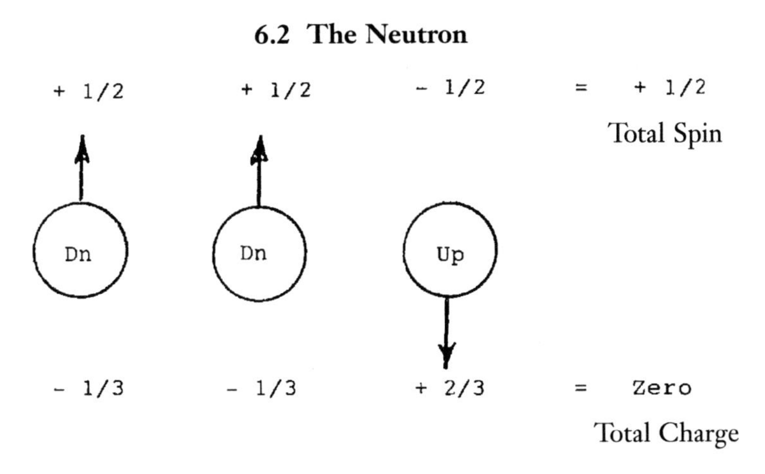

The conventional symbol for the neutron is the small typed letter "n."

As the name implies, it is electrically neutral and carries no charge whatsoever. It is strange; but, in spite of being neutral, it actually has a magnetic moment of minus 1.93 μ N.

The neutron has a mass of 939.57 mega electron volts or MeV, which is just a tiny bit more than the proton. This circumstance caused Erwin Heisenberg and others during the early history of quantum mechanics to form the quite reasonable speculation that the neutron might be an amalgamation of an electron with a proton. With this thinking one might have expected the proton to be actually heavier than the neutron in order to account for the energy needed to create its positive electrical charge. As things stand in reality, it is impossible for the proton to decay into a neutron and some other particles, such as a positron and a neutrino, for example, because this reaction would be asked to create a new mass greater than its own original mass. Such simplistic conjectures based upon conservation of energy considerations only proved fruitless in the end.

In short, this disparity in masses presented the scientists with another riddle; but an important one. Both particles must be regarded as fundamental in their own right; but interacting with one another when bound inside a nucleus by the strong force.

The neutron has an internal angular moment or spin "J" of one half, thus making it a fermion similar to the proton.

The size of the neutron is presumed to be very close to that of the proton, since its observed cloud chamber tracks and scattering angles in experiments with crystals or particle impact all closely approximate that of the proton. A magnetic field, of course, deflects them very little. One should realize that physicists today are extremely leery of specifying sizes of particles. It is not merely because these entities are so very small as to not be observable or easily measurable; but the very mental concept of a particle as being a hard, firm ball of "matter" contradicts the more modern idea of their being more like a fuzzy ball of vibrating energy surrounding a point source.

Likewise, any discussion of the mean lifetime of the neutron leads us directly into a puzzling anomaly. Just so long as the neutron is cosily nested within the core of an atom and restrained to remain there among its powerful neighboring protons by the strong force, its lifetime is prolonged and indeterminate. If, however, it is knocked loose or separated in some way into a free and uncombined state, it will spontaneously decay and disappear into fragments within fifteen minutes! Careful experiments have led to the accepted value for the average half-life of the neutron under these special conditions of 10.69 minutes. Harking back to the days of experiments with alpha particles, this behavior along with the emission of electrons radioactively has become known as "beta decay."

On the other hand, the question of the life expectancy of the neutron inside the atomic nucleus was of great interest to the theoreticians. One of the very earliest attempts to measure this value was made by Reines, Cowan, and Goldhaber at Los Alamos, New Mexico, in 1954. The site of the experiment was a cave known as "the icehouse" which had been excavated into the north wall of a canyon for the purpose of providing some shielding from extraneous radioactivity or cosmic rays. These experiments continued through 1957 with Kruse replacing Goldhaber; and the final consensus was that the captured neutron was good for 4×10^{23} years. Not to worry!

Insofar as the effects of gravity upon the free neutron are concerned, all experiments which have been made could find no difference in its behavior than that expected in our own macroscopic world.

6.3 Beta Decay

Hundreds, perhaps even thousands, of learned papers have been published on this subject; and its study has led to a better understanding of the weak force. In spite of the convention of its name, however, Beta decay has very little to do with the normal electrons in atomic orbit.

This phenomenon was a great puzzle to physicists as early as 1907 when Lise Meitner and Otto Hahn at the University of Berlin took up the study of beta rays emanating from radioactive minerals. They were followed by James Chadwick at the University of Manchester. The subject captured a great deal of attention for twenty years before it began to be understood. A large part of the problem was that the electrons emitted from such elements as radium, polonium, or thorium covered a large and continuous spectrum of energies. These electrons do not originate from the cloud surrounding the atomic nucleus. In addition, all the balance calculations for these energies to meet the Conservation of Energy law seemed to defy solution.

Due to the action of the weak force, a single neutron does occasionally spontaneously decay inside the nucleus of a radioactive element. However, it can only do so when it is possible for the remaining atom after the event to have less mass. Beta decay originates with the disintegration of a neutron into a proton plus an ejected negatively charged electron and an invisible neutrino. The electron is a newly created particle and one must invoke special relativity to account for its energy. The new proton thus created by a single change may alter the chemical properties of the original atom and change it to the next higher element on the periodic table. The Standard Model for all elementary particles successfully explains this behavior as an interaction between the electromagnetic and weak forces; but this is somewhat incomplete in the sense that it ignores the strong force and gravity as participating in the event. This small detail bothers some physicists quite a bit.

Beta decay sometimes occurs as an internal adjustment inside a heavy and complex atom in order to balance the binding energy with the electron orbit shells to achieve greater stability. The atomic number may increase by one; but the mass remains relatively (but not entirely) unchanged. Beta decay or nuclear fission result in an ejected or freed neutron, which then spontaneously disintegrates into a proton and newly formed electron. These two last mentioned particles should theoretically fly out in exactly opposite directions. This would be a requisite for the Law of Conservation of Energy, as we indicated previously. In fact, they do not exactly oppose each other in line back-to-back in experiments. This circumstance was the second clue to the necessary existence of the neutrino, the first being the unbalance in the conservation of energy balance equations.

The freed neutron rapidly undergoes decay with the excess energy distributed among the decay products in the form of motion or velocity. Thus:

Neutron yields proton + electron + anti-neutrino + 0.782 MeV.

$$n \rightarrow p + e^- + \bar{\nu}e$$

This is called a "weak" interaction. The strong force is not involved; and, furthermore, the electron is immune to the strong force anyhow.

Double Beta decay is extremely rare; but it can occur if, and only if, two neutrons decay simultaneously in some heavy isotope of an atomic element and result in a lighter and more stable element while still observing the Law of Conservation of Energy. Examples are Selenium with 48 neutrons into Krypton with 46 neutrons.

All these special conditions leave the proton and electron as the only really permanent and durable constituents of matter with the neutron as an "also ran" contender.

The entire question of Beta decay was vigorously discussed by a galaxy of famous pioneers in nuclear physics at the Seventh Solvay Conference in Brussels, Belgium, in 1933. Enrico Fermi (1901 to 1954), of nuclear bomb fame, later wrapped up the entire subject in a brilliant amalgamation of ideas which he published in January of 1934.

Fermi was the son of an administrative clerk in the Italian railroad system and his mother was a school teacher in Rome. He was an excellent student in the local public schools; but was also an inveterate autodidact and did a prodigious amount of reading on his own initiative and under the encouragement of a friend of his father. It has been said that by the age of seventeen he already had a comprehension of physics equal to that of a graduate student from a university. At any rate, he went to college in Pisa and obtained a doctorate degree from the University of Pisa in 1922. His conspicuous talent for mathematics obtained for him Fellowships with Max Born in Gottingen, Germany, and Paul Ehrenfest at Leiden until 1924, when he returned to Italy as a lecturer. Through a competition and with recommendations from prestigious teachers he was finally awarded a newly established chair in theoretical physics at the University of Rome in 1927. During the interim between then and the Solvay Conference of 1933 Fermi had become well known for his articles on tensor analysis in General Relativity and something vaguely known as quantum number statistics back in those days of early nuclear physics.

Nonetheless, he had switched his interest to quantum mechanics and had learned about spin and the neutrino from Wolfgang Pauli. Thus his magnum opus in 1934 on Beta decay was both a summary and exposition of the weak interaction and the acknowledgement of the existence of a "weak force" as well. He received the Nobel prize for all this in Stockholm in December of 1938. By this time, however, history had made his remaining in Italy impossible. The dictator, Mussolini, had already launched his war on Ethiopia and the Nazi German disease of persecution of the Jews had spread to the Italian government. Fermi's wife was Jewish; so, consequently, he had accepted an offer from Columbia University for refuge and left from Sweden for the Unites States directly.

It is strange that Fermi's intense interest in the behavior of neutrons led him to the radioactivity of uranium and, only later, to the possibility of a self-sustaining chain reaction. Already back in 1935 he and his staff at the University of Rome had discovered that neutrons, when passed through a substance containing hydrogen such as paraffin, acquired an increased efficiency for producing artificial radioactivity in certain elements. Thus neutrons, when they are freed from the nucleus of the atom and happen to be slow moving, or else alternatively are retarded in velocity by a deliberately introduced moderator, can cause the atom of a special isotope of uranium to break apart into two pieces of less total mass. This reaction releases other free neutrons as well and also a burst of raw energy from the original binding energy inside the atom. This fact was first discovered by Otto Hahn and Fritz Strassman in Germany in 1938 and is known as nuclear fission.

This writer still vividly remembers sitting as a young teenager in the crowded auditorium of the Franklin Institute in Philadelphia in 1939 listening to Professor Fermi as a newly arrived celebrity from Europe. It was one of the Institute's wonderful Christmas season lectures by guest scientists, which were annual events before World War II. The announced subject of the lecture was the emission of multiple neutrons from the isotope of uranium 235. The excitement of the speaker over the possibility that a chain reaction could be started with the release of great energy was evident; nor did he attempt to hide it. How all this got past the military censors I do not know; but I do wonder how many people in that audience, including myself, really grasped the full significance of what this volatile Italian was telling them. His language was measured and clearly directed towards professional physicists; and the idea of an atomic bomb was then inconceivable to ordinary people. Moreover, the population of the United States was at that time rather complacent and the fear of a war with Nazi Germany, which all Europeans dreaded, had really not penetrated their collective psyche to any great extent. Fermi, notwithstanding, went on to the University of Chicago to build the first nuclear reactor, moderated with blocks of graphite, in a small space underneath their sports stadium. The reactor went critical successfully in 1942; but only he and a handful of nuclear physicists alone knew the secret of an awful power waiting to be unlocked into an unstable political environment. The rest of us remained ignorant until 1945.

6.4 The Electron

The conventional symbol for the electron is the lower case letter "e," although sometimes the symbol is made more specific by using "e-" in order to make a distinction from its anti-particle or positron. Had the positron been discovered first, which is extremely improbable, the electron might have become the "negatron."

The electron has unique characteristics all its own and often defies a clear and absolute description, much like its second cousin the photon. It frequently demonstrates a behavior more like a wave packet or electromagnetic probability wave source than the hard, material particle which we are apt to picture.

It has a mass of only 0.511 MeV (divided by c^2 as being understood, as usual). This is only 1/1837th times the weight of a hydrogen nucleus and translates to a miniscule 9×10^{-28} grams in our larger worldly units. Since it can approach the speed of light in cosmic space and even in powerful particle accelerator machines and since its primary function seems to be carrying an electric charge, some people wonder why it has any mass at all. Because it has the smallest mass of the three stable and elementary components of the atom it bears the category name of "lepton," meaning the "lightest." Being conservative academicians at heart, physicists tend to stay with the traditional Greek - perhaps out of deference to Democritus.

The electron carries a constant negative electrical charge of minus one (-1), which is exactly equal to and opposite in polarity to that of the proton. Considering the concentration of such a strong electric charge in such an insubstantial entity and minute space, there is cause for wonder at the beautiful simplicity of nature. This fact that the negative charge of the electron exactly balances in magnitude that of the proton should not be casually accepted merely because we may have been taught this fact in High School. Why could it not have been one third or two thirds of the charge of the proton? This kind of symmetry seems to be a basic part of the laws and formation of the universe; but we shall all have to wait for the final version of the Grand Unified Theory or Theory of Everything to learn why.

Robert Andrews Millikan (1868 to 1953) succeeded by a series of painstaking measurements between 1909 and 1917 in determining quite exactly the magnitude of the electric charge on the electron. His experiments are often described in detail as the perfect example to any scientific experimenter for the need for meticulous precision, ingenuity, and dedicated patience. The general principle which he used was to ionize or add a negative static electric charge to a droplet of oil or glycerin and then balance its weight against the force of gravity by means of a variable direct current voltage difference between top and bottom contained in an air–tight cylindrical container. During this process he needed to determine the actual weight of the approximately spherical droplet by timing carefully its free fall at terminal velocity in a viscous fluid (namely air in this case). All of these steps took hundereds of hours of diligent observations in order to accumulate pages of meticulous statistical data. The final results were figures which were all multiples of the same number, which was 1.6×10^{-19} coulomb for a single electron. Of course, the actual number registered depended upon the number of negative free electrons being carried by the particular drop of oil.

A coulomb represents the total electric charge in a current of one ampere flowing in a conductor past a given point for one second. This computes into the accumulated total charge of about 6×10^{18} individual electrons. It becomes clear we need to adjust mentally to the very small numbers that scientists use as a matter of routine in their calculations for atomic behavior.

The magnetic moment of the electron is 1.001159 μ N.

The electron has a spin "J" of one half (1/2), which actually means one half of the quantity known as $h / 2 \pi$ or the abbreviation "h" when "h" is Planck's constant.

Now that you have learned this we will try not bother you with it again. It is a quirk of nuclear particles which took the scientists a long time to discover and accept. All electrons are supposed to spin about an axis (or so it would seem) at exactly the same rate; but the direction of rotation may be clockwise or counter-clockwise as seen from the direction in which they are translating. It must be one or the other. As you also learned, a spin of 1/2 means the electron must be classified as a fermion.

The actual size of the electron is an indeterminate thing. It is a curious fact that in most texts concerning particle physics the actual physical size of an electron or a specific number for its radius is seldom definitely quantified. The subject is often avoided for very good reasons. In the very early days of particle physics (meaning prior to 1922) a notion of the size of the electron came into discussion which, although quite wrong and misleading, persisted in science literature for some time. The formula suggested was based entirely upon electromagnetic considerations:

Radius = r = Approximately e^2/mc^2 = 2.8 x 10-13 centimeters when the symbol "e" stands for the charge of the electron here.

It was quickly recognized that this resulted in an electron radius conflicting in magnitude than the entire atomic nucleus, which was an absurdity. The average radius of the electron's orbit around the nucleus, however, was conceded to be 5.3×10^{-9} cm.

As late as 1939 the accepted equations in quantum mechanics for the radius resulted in infinities unless "normalized" by inserting known observational numbers. Thus a debate raged among physicists whether the electron had any really firm dimension of size or whether it was a "point charge." No one knew, and their ignorance was an acute embarrassment to the profession. Werner Heisenberg suggested as a reason that the type of observations made on the electron and the photons necessarily used in these observations all affected it. The great facility of electrons to constantly absorb and then emit photons of various energy levels implies an easy transformation of its energy content. Also basic to the confusion was the assumption that an electron in motion in an orbit created a surrounding magnetic field which had to act upon itself in some manner. Any assertions as to the actual physical size of the electron may be an exercise in futility. If the reader were to search the literature diligently he might be able to contradict the above statement with written estimates.

Perhaps second only to the proton, the electron is the most stable particle in the universe. Its estimated mean life is 2×10^{22} years or somewhat shorter than that of the proton. Although it is considered possible for the electron to decay into a photon of gamma energy, no one has sat around long enough waiting to see an evasive electron spontaneously self-destruct; so this life figure is theoretical only.

In as much as the masses of the proton and electron both remain the same for each and, for all practical purposes, constant and their electric charges are identical, although opposite in polarity, it occurred to some scientists (including Wolfgang Pauli) that the electromagnetic force which binds them together also ought to be a constant. They went to work with the following equation for what they called the Fine Structure Constant or alpha:

$$\alpha = \frac{2\pi e^2}{hc} \quad \text{or} \quad \frac{e^2}{\hbar c} = \frac{1}{137.03599}$$

When: e = The electric charge of the electron.
c = The speed of light in a vacuum
h = Planck's Constant

Many people speculate that this uniquely small number proves the universe is not merely an accident but was deliberately built in accordance with a mathematical plan. Others simply accept it as evidence that we are closer to understanding the reality of Nature. Today arguments abound that this "magic number" is not the same constant for the "Z" boson and, in any event, may have been slightly different in the very early time of the "Big Bang" formation of our universe.

Electrons can, and usually do, exist in a free state moving with varying velocities through vacuum, air, gases, or a metallic conductor whenever subjected to an electromotive or electrostatic voltage difference. We were all taught in school that an electric current in a copper wire is the headlong rush of masses of electrons through the interstices of the copper atoms from one end to the other under the pull of an electromotive force. It probably would be more precise to picture them as hopping from one atom to another very rapidly and thus displacing an electron in the outer shell of an individual copper atom so that the second electron is compelled to find another home down the line. All these examples tacitly involve the convenient assumption that electrons are infinitesimally tiny globules or corpuscles of matter rushing about under electromagnetic attraction or repulsion. Abraham Pais, a physicist himself and historian, rather contemptuously calls this the "marbles" stage of thinking about the atom. The actual behavior of electrons in the laboratories often belies this simple picture.

Chapter 7
A History of Quantum Mechanics

> *"Today there is a wide measure of agreement, which on the physical side of science approaches almost to unanimity, that the stream of knowledge is heading toward a non-mechanical reality; the universe begins to look more like a great thought than a great machine. Mind no longer appears to be an accidental intruder into the realm of matter."* - Sir James Jeans in The Mysterious Universe in 1935.

The subject of quantum mechanics is sufficiently complicated and bewildering as to preclude anything in this type of book except a brief sketch of the history of its development and very general allusions to the concepts involved. It is an intensely mathematical study of a high order. There exist a large number of popular books in print on the subject, many of which have been written by knowledgeable people with graduate training in physics. The reader is urged to attempt a few selections from among them which appeal to his taste and suit his depth of interest as a supplement to this simple introduction. One of these which this writer found helpful is *Understanding Quantum Mechanics* by Roland Omnès.[1] It is a meticulous analysis of the difficulties, philosophical, and otherwise, one may encounter in Quantum Mechanics. He or she will be wondrously rewarded with an incredible glimpse of the architecture of the universe.

Nevertheless, for the writer to omit any explanation whatsoever of the significance of quantum mechanics here would be analogous to writing a manual for the repair of current automobiles with no mention of the Otto cycle and the basic principle for the operation of the four–cycle gasoline engine. To put it in perspective, the evolution of quantum mechanics calculations which came from the close study of the atom brought about a total revolution in scientific thinking quite equal in importance to Isaac Newton's

1. Roland Omnès, *Understanding Quantum Mechancis*, (Princeton University Press, 1999)

Philosophae Naturalis Principia Mathematica and Albert Einstein's *Special and General Relativity*. John Archibold Wheeler, who spent a lifetime thinking about the quantum theory and pondering its mysteries, wrote with total sincerity: "I continue to say that the quantum is the crack in the armor that covers the secret of existence!" A subject of that significance can only be approached by us ordinary citizens carefully and gradually with very modest expectation of complete comprehension. With that warning the reader who finds this chapter tendentious is quite at liberty to skip ahead if he pleases.

The narrow specialty of what is known today as "quantum mechanics" had its earliest beginnings in chemistry and, more specifically, the technique of the spectroscopic identification of elements in molecular compounds by means of their light emissions under activation by heat. This specialized study of different light spectrum produced by various substances, either directly or else superimposed upon the emissions of another element when heated or burned, had become an extremely useful tool in the qualitative analysis of compounds. Although the practitioners of the day may not have understood the reasons for the phenomenon at all, they were quite aware they were looking at different frequencies or wavelengths of light. They also recognized that each individual element had its own unique signature in a spectrum with the highest intensity radiation separated into bands of differing wave lengths. By his late middle age Johann Jakob Balmer (1825 to 1898) had become a leading expert in spectroscopy, although it had started for him as an avocation late in life. Balmer was a kind of lone and unsung genius who was content with a comfortable family life with wife and six children in Basel, Switzerland, where he taught at a girl's secondary school. He was the eldest son of a Chief Justice and had actually obtained his doctorate in geometry in 1849. Despite his relative obscurity in scientific affairs of the day, he obviously was keenly intelligent and soon recognized an orderly numerical relationship between the spectrum lines for both hydrogen and helium. Although Angstrom had measured and established the differences between wave lengths previously, the regular numerical order had escaped the perception of that pioneer and others of that time.

Simply by playing with the numbers Balmer hit upon the key to the puzzle and developed a simple approximate formula for the four then known frequencies of the spectral lines for hydrogen. Each wavelength could be written as a single fundamental shortest wavelength multiplied by the difference between the squares of two whole numbers. In 1885, when he was approaching the age of sixty, Balmer wrote his first scientific publication announcing his discovery. His relationship can be expressed by the equation given below:

$$\lambda = \frac{hm^2}{m^2 - n^2}$$

The symbols may be defined as:

λ = The wave length of monochromatic spectrum line observed.

h = The fundamental shortest wavelength nearest to the ultra-violet color range, taken here to be 3645.6×10^{-8} centimeters.

n = The numeral 2.

m = Any of the whole numbers 3, 4, 5, 6, etc.

As a playful example for the reader's understanding, let us arbitrarily let m = 4:

Substituting:

$$\lambda = \frac{3645.6(4)^2}{(4)^2 - (2)^2} = \frac{3645.6(16)}{(4)^2 - (2)^2} = 4861.3 \times 10^{-8} \text{ centimeters}$$

The reader should refer to Figure 6.1 in order to compare the results with different "n" numbers, remembering that the frequency is equal to the speed of light divided by the wavelength. "c" or the speed of light is taken as 2.9979×10^{10} centimeters per second.

Balmer actually went on with substituting the "m" numbers up to m = 7 and thus predicted a new hydrogen wave length which he had not yet seen. A friend at Basel University told him it had already been discovered by others. Nevertheless, the reasons for such an astonishing regularity remained a mystery. His formula, however, becomes less reliable as one approaches the high frequency or ultra–violet color range of the spectrum where the lines become closer together and less distinguishable.

Later Johannes Robert Rydberg (1854 to 1919) of the University of Lund refined Balmer's formula to a more general form shown below:

$$\frac{1}{\lambda} = R[\frac{1}{n^2} - \frac{1}{m^2}]$$

n = The numeral 2 for the hydrogen atom spectrum.

λ = The wave length in centimeters or angstrom units.

m = Any number of the series 3, 4, 5 or 6.

WAVE DESIGNATION	WAVE LENGTH IN CENTIMETERS	FREQUENCY IN CYCLES PER SECOND
Balmer Fundamental	3645.6×10^{-8}	8.22×10^{-14}
m = 6 δ	4101.3	7.31
m = 5 γ	4340.5	6.908
m = 4 β	4861.3	6.168
m = 3 α	6562.8×10^{-8}	4.568×10^{-14}

FIGURE 7.1

Balmer's Table Of Wavelengths For The Hydrogen Atom Spectra.

R = Rydberg Constant = $\frac{\pi^2 q_e^4 m_e}{2k^2h^2}$

qe = The electron charge.

me = Rest mass of the electron.

h = Planck's Constant.

k = Was assigned a value of 1/2 by Neils Bohr.

Today the value of the entire Rydberg Constant is assumed to calculate out to be 109678 per centimeter or 1.097×10^7 per meter, depending upon which units are selected. We may satisfy ourselves that the equation works satisfactorily easily enough by substituting a value for "m" of 4.

$1/\lambda = 109678 \{1/4 + 1/(1/4)^2\} = 109678 (3/16)$

$1/\lambda = 20{,}564$ or $\lambda = 4861 \times 10^{-8}$ or 4861 Angstroms.

It was precisely this numbers puzzle which fascinated Neils Hendrik David Bohr (1885 to 1962) and which appeared to him as a challenge. He deliberately chose hydrogen as the simplest of all atoms in the Periodic Table and felt that if the classical physicists could find no way to explain its behavior then the situation was indeed pathetic.

Bohr was one of three children of a father who was a professor of physiology at the University of Copenhagen in Denmark and of a mother who was heiress to a mercantile banking fortune. The progressive education of the children was assured and his younger brother, Harold, became an eminent mathematician in his own right. Neils entered the University of Copenhagen in 1903. His doctoral thesis in 1911 was on the classical electron theory of metals; but this study did little to resolve any of his perplexities. The failure of Maxwell's electromagnetic theories to explain the magnetic properties of metals and photo-electric emission convinced him that the traditional approaches to the atom were inadequate.

Bohr went to Cambridge, England, in October of 1911 in order to meet and work with J. J. Thomson, whom he greatly admired. It is probably fair to say that the old gentleman did not respond immediately with equal enthusiasm. Bohr was a large bear of a man having great physical energy and athletic abilities who was somewhat given to philosophical speculations on some occasions and whose English diction was still halting and awkward at that time. Worse yet, on his first meeting with Thomson, Bohr impetuously pointed out an error in the text of his host's book on the *Conduction of Electricity in Gases*. Nevertheless, he attended many of the lectures, including some by Ernest Rutherford, and made friends with the other younger physicists there at Manchester. He left England a year later in order to get married.

It was Bohr's hope to find some reconciliation between the new quantum ideas of Planck and Einstein and the more traditional notions of the relationship between the electron's energy in a stable orbital track and its time of rotation or angular velocity around the nucleus (wn). In this effort he was by no means alone. Thus he conceived the mechanical state of the electron in its motion around the nucleus as having a binding energy (En) which would follow the general form given below:

$E_n = knh_n$ when:

k = Some appropriate modifying constant.
h = Planck's Constant.
n = An integer number representing different orbital levels.

Although Balmer's numbers formula for the spectrum had already been known for twenty-five years, Bohr stumbled upon it only fortuitously, probably at the suggestion of a friend. He instantly recognized its compatibility with his own ideas.

Bohr was well aware of the adherence of the traditionalists to their Newtonian and Maxwellian approach to electron orbits; but he was looking for new clues that would incorporate the proposals of both Albert Einstein and Max Planck with regard to quanta. He discovered it in Balmer's spectra lines and Rydberg's Constant. True, there were still problems to be solved. Bohr's system did not work very well at all for the helium and the next heaviest atom; but we must remember that particle spin and Pauli's Exclusion

Principle were still not known. Meanwhile, things were going well for him in his private life in Norway.

Essentially, Bohr started with the flat assumption that electrons were confined in the atom to very definite tracks or circumferential orbits around the nucleus and that they did not radiate energy continuously while there, as the classicists assumed. More than that, they could only move from one orbital ring to another by either gaining or losing a quantum amount of energy which depended upon some integer number times Planck's constant. This change in energy levels was accomplished by the absorption or emission of a photon quanta of energy. Each of these various energy levels represented the momentum of one or more electrons moving around the circumference of an assumed circle with a variable radius "r." Obviously, the circumference of this theoretical circular path must be 2πr. If one wishes to adhere firmly to the picture of an electron as a little hard ball traveling at great speed around the nucleus, then the trick is to correlate these separate and different "ring" tracks with Balmer's wave length calculations for different energy levels. This approach leads one directly into harmonic oscillations of "string" frequencies similar to those of a pendulum where the quantity "2π" occurs naturally. The similarity between wave equations and those for pendulums was already familiar to the mathematicians.

By this time, in 1912, Bohr had achieved the position of Assistant Professor at the University of Copenhagen. He used the following year's time to expand his concepts furiously and to show their relationship to the scattering of other charged particles, including the electron and alpha particles, while passing through a gas or material medium while subjected to an electromotive voltage and perpendicular magnetic field. His earlier work with Thomson and Rutherford had already convinced Bohr the normal hydrogen atom had only one electron; the calculations which he made on the spectrum scattering measurements substantiated his opinion. To summarize, he attempted to demonstrate the relevance of his description of individual electron energy levels to known laboratory results and ultimately published his conclusions in 1913. Bohr's formula for the hypothetical radius of the hydrogen atom (rH) worked out to the approximation shown below:

$$r_H = \frac{h^2}{4\pi^2 q_e^2 m}$$

Although by no means accurate, the relationship of the radius formula to the Rydberg Constant becomes evident without even doing the necessary algebra. It was also Bohr's unequivocal assertion that the hydrogen atom at its lowest possible energy level or ground rest state does not radiate any energy.

His proposition of concentric electron shells did not meet with instant acceptance. In all justice to the skeptics of the day, this was the beginning of what has been called the quantum "numbers game" in which various numbers were assigned to categories of properties and "rules" offered for their combination in order to predict how atoms were expected to behave without any precise concomitant explanation of why they should do so. Schrodinger's work was an example of this. Many did not regard this approach as reliable science having a demonstrable mathematical proof. It was too empirical for the classicists. Moreover, there exists today a genuine perplexity about the electron's behavior while trapped in its bound state within an elliptical orbit. Michael Faraday and others had earlier established to everyone's satisfaction that any electrons in motion, whether passing through a conductor or streaming in a cascade through empty space or a gas, carried a negative electrostatic charge with them which produced a radiating field of electromagnetic energy in a direction perpendicular to their direction of motion. That is the fundamental principle of transformers and electric motors in our daily life today. It is exactly this effect which makes the ordinary solenoid in an electric starter for the automobile possible. In the established classical physics, this radiated field in interaction with other atoms or particles in its immediate environment represents a loss of energy or drag in velocity. Hence the question arises of why one should expect a solitary electron inside an atom not to have the same behavior and not to exhibit a loss of kinetic energy which would gradually, but inevitably, cause it to spiral downward into the nucleus? How does one explain the fact that this "death spiral" does not happen and what exactly is the mechanism for the orbital electron's balancing act?

Various conjectural explanations were invented. Since the photon is a boson and thus can hop about between other particles with little restriction, it has been suggested that a photon of energy becomes a carrier of electromagnetic force between the electron and a proton in the nucleus of the atom, thus continually sustaining the electron in its orbit. In more modern times it is now considered that Bohr's idea of definite orbits associated with the angular moment of rotation is feasible only in very large outer orbits. These are limiting cases where the classical rules may still be applied. The deviation from classical theory occurs in very small or tight orbits below the limiting cases; and one is no longer justified in picturing any accurately defined orbit. That still may not answer the question to the reader's satisfaction; and he or she is quite correct. The real answer is the Uncertainty Principle, which we will encounter later.

The representation of the atom championed by Neils Bohr was an early example of the contradictions between the new quanta concepts of electromagnetic radiation by photons and the older field theories of Maxwell and

Faraday. The doctrine of electrons remaining stable in fixed orbital shells of discrete and different energy levels until the gain or loss of photon packets of a precise energy inspired outraged cries of nonsense from many scientists at first, including Sir J. J. Thomson. Fortunately, Bohr's proposition received prompt and enthusiastic endorsement from Arnold Sommerfeld, who was seventeen years his senior in age and a respected expert in spectroscopy and luminous wave lengths. Not only that, the world renown French mathematical genius Henri Poincare´ (1854 to 1912) had written his last treatise and published memoir just six months before his death. In this paper Poincare´ announced his personal conclusion that Max Planck's quanta argument for electromagnetic radiation phenomena had to be correct. He thus effectively turned his back on a lifetime of experience with the established classical notions of causality and continuity of actions; but not without some severe qualms. As he expressed it himself: "It is hardly necessary to remark how this new concept of quanta differs from what we imagined up to this point; physical phenomena would cease to obey the laws expressed as differential equations!"

Bohr eventually presented all his ideas in a published paper in 1913. Ernest Rutherford, the successor of Thomson, invited him to return to Manchester, England. Bohr accepted and stayed there for two years. By this time his fame had spread world-wide. The University of Copenhagen promoted him to the rank of full Professor of Physics in 1916. The citizens of Denmark wanted to make certain that their new notable would remain in the homeland and consequently a new building designated the Institute of Theoretical Physics was created for him with the financial aid of the very successful manufacturer of Carlsberg beer. Bohr moved into the new facility in 1921, and it rapidly became a Mecca for the young students of the new "quantum" revolution.

Meanwhile, Arnold Johannes Wilhelm Sommerfeld (1868 to 1951) contributed his very considerable influence to the new science. He was the son of a physician in Konigsberg in what was then known as Prussia. His college education there followed the traditional pattern for a German Protestant family with the concomitant activities of beer drinking and dueling with both fencing foils and sabers. In spite of all that, the youth was extremely intelligent and showed an exceptional talent for mathematics. After graduation from the city university, he enlisted in the Army in 1892 for the compulsory year of military service at an age of twenty three. Surprisingly, he enjoyed the military life enough to volunteer for subsequent eight week long periods of army exercises in the reserves for each of five years in a row. He eventually became a lieutenant, though quite short in stature, compensated for it by sporting a waxed moustache and swagger (not quite the traditional picture of an academic pedant). Nonetheless, he had decided on mathematics as a career and went to Gottingen, where he obtained an initially minor job in the

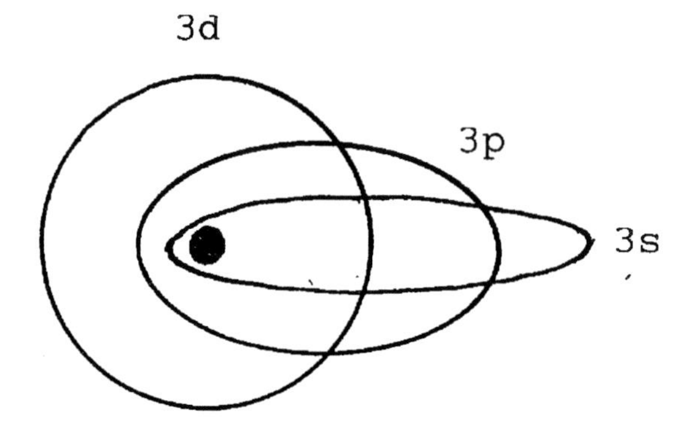

FIGURE 7 – 2
Three Permitted Orbits For Electron
About The Hydrogen Nucleus

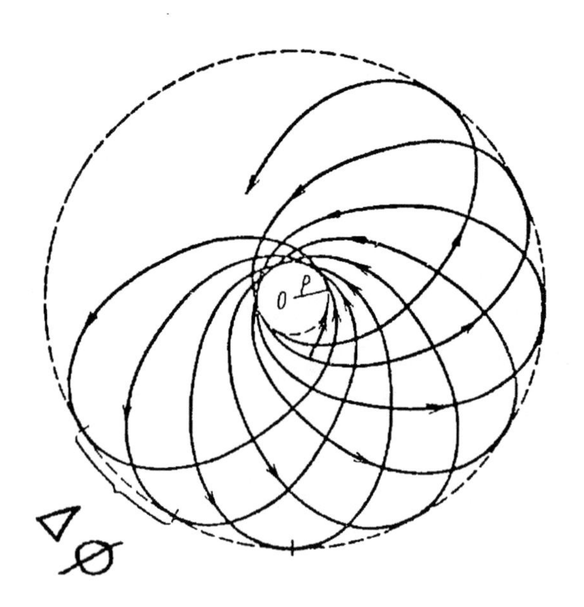

FIGURE 7 - 3
The Precession Of An Elliptical
Electron Orbit About the Nucleus

Mineralogical Institute at that university in 1893. The next thirteen years were extremely productive and earned him an enviable reputation in the engineering world; although his own research work interests gradually shifted towards the dynamics of electrons. His own publications on physics in the years of 1904 and 1905 resulted in a "chair" at the University of Munich, which was no small achievement at the age of thirty-eight. There a research institute was made available for his direction. Sommerfeld met Einstein for the first time in 1909, and they became mutually congenial correspondents on the new quantum theory implied by light radiation and Planck's Constant.

In particular, Sommerfeld became preoccupied with the fact that the Balmer lines, when magnified, were actually two separate lines divided by a microscopic space or gap. He called this the "fine structure" and introduced in 1914 the idea that the orbits of the electrons might actually be ellipses, the perihelion of which precessed in a circle around the nucleus of the atom. This meant that Einstein's relativity theories could affect the mass of the fast moving electron at different points in its orbit, not to mention the possible disturbing effects on electrons in other shells that could exist in other elemental atoms. He perceived that these possible different elliptical orbits could have different eccentricities which could form a family which corresponded with the integer numbers 1, 2 or 3 in the Balmer formula as multiples of the angular momentum. (Refer to Figures 7 - 2 and 7 - 3.) The greater the eccentricity of the orbit or the distance of the perihelion from the nucleus, then the greater became the relativistic effect on the mass of the electron and also the resulting precession motion. All this was a new and ingenious picture of the atom; and in 1915 Sommerfeld publicly announced the dimensionless Fine Structure Constant or Interval Factor, to become known as "a" to physicists.

$$\alpha = \frac{2pqe}{hc}$$

p = The ratio of circumference of a circle to its diameter - a constant.
qe = The electric charge of an electron - a constant.
h = Planck's Constant.
c = The speed of light in a vacuum - a constant.

The Fine Structure Constant was intended to be part of the mathematics to correct the calculations for the Balmer Lines for hydrogen. In fact, it gave quite reasonable results for the helium atom also and became a very fundamental assumption for a starting point in the development of quantum mechanics. Given a charge of one for the electron, then the reciprocal of "a" or 1/a calculates out to 137.036. Much ado was made in both the public press and scientific journals during Sommerfeld's life about this strange "magic" number. It fascinated both the scientists and numerologist mystics alike, since it contains four well known physical or geometrical constants in combination. Many people regard it as evidence that there is some mysterious mind-planned order to our reality and a restriction to the apparent infinite chaos of the cosmos.

Sommerfeld's work correlated with the already known "Zeeman effect" for the separation or splitting apart of spectral lines by the introduction of a magnetic field near the observed atoms. Thus Neils Bohr, Max Born, and Arnold Sommerfeld together supplemented each other and enormously impressed the general scientific fraternity. They also built a cadre of brilliant and enthusiastic disciples for the new physics. Many of them were Germans who knew and corresponded with each other regularly and together invented the new quantum mechanics discipline at the very beginning of the twentieth century. It was sad and ironic that Sommerfeld, in spite of his Prussian training, later was arbitrarily deprived of his University position and intellectual honors by the emergence of Adolph Hitler and the prejudice against Jews in the Nazi political party and thus became estranged and alienated from his own native country.

There was one colorful personality who fully and capably participated in the excitement of finding a new mathematical basis for the science of the atom and who was not in the least German. Louis-Victor Pierre-Raymond de Broglie (1892 to 1987) was born in Dieppe and lived in France all his life. As his full name implies, his family took their ancestry and their long history of military and political leadership very seriously indeed. In the later years of his life, after his older brother had died, Louis Victor inherited the ancient French title of Duke and the deferential courtesy title in France of Prince. Louis was emboldened by the example of his older brother, Maurice, during his undergraduate studies at the Sorbonne and decided after considerable hesitation to break away from the ancient family tradition of statesmanship and choose a career in science. His brother had preceded him with distinguished work in spectroscopy, radio and X-ray radiation; but the radical ideas of Max Planck and Neils Bohr resolved the decision for him. His dissertation in 1923 at the University of Paris for a post-graduate degree in physics not only created a great academic furor; but also made history. He had long been pondering upon Einstein's assertion that light quanta or photons behaved as an electromagnetic phenomenon. De Broglie had a trained mathematician's mind, and his own particular penchant was fitting known mathematical equations or relationships as a theoretical tool or method of thought to laboratory observed phenomena.

He had been puzzling over Bohr's new proposition that the frequencies for spectroscopic lines of colored light were due to one, two or four electrons stationed at particular energy levels in specific fixed orbits around the nucleus;

but at different radii. The idea occurred to de Broglie that this situation was mathematically similar to different musical notes having different vibratory frequencies. In other words, he thought of the electron as similar to the plucked string under tension and fixed at both ends, such as on a violin or guitar, and having one or more neutral or stationary node points in between. Moreover, such standing waves had unit changes in their momentum. He also remembered that the momentum of the circling electron in different diameter orbits had to be whole integer multiples of some fundamental constant, namely Planck's Constant. It thus seemed perfectly logical to de Broglie to picture the electron as vibrating like a circular resonant string in its given orbit with either zero or an integer number of node points around the circumference. (See Figure 7 - 4). This simple idea with its remote connection with musical harmonics and violin playing, became a bedrock principle of quantum mechanics.

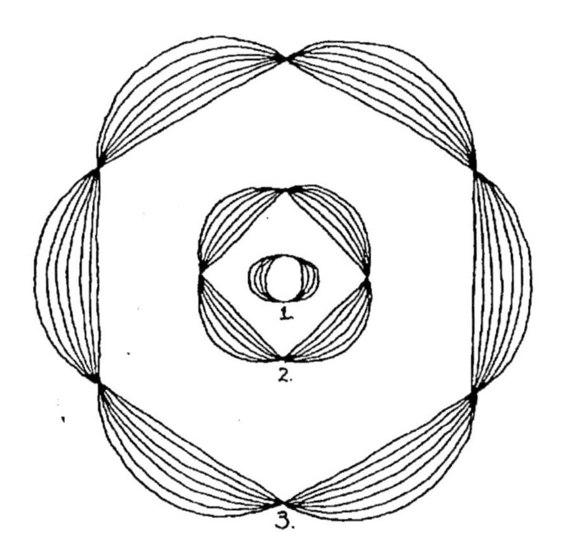

FIGURE 7 – 4
Louis de Broglie's Dance of Electrons

If one decides to thus regard the electron as purely a wave phenomenon (and not a particle at all) and assigns some appropriate wave length to this electron, then only those orbits around the nucleus which can accomodate a single or integer number of complete wave lengths for this electron in a certain ring or elliptic would be possible. The consequence of this idea is that only certain definite radii are feasible in order for the electron to conserve its rotational energy. Hence it becomes necessary for the electron to either absorb or emit energy in the form of a photon in order to produce the skipping between rings. For the absolutely lowest or rest energy of an electron in the hydrogen atom the equivalent wave length must be exactly equal to the circumference of its orbit. For the next highest or second and outer orbit two identical wave lengths may fit exactly into the circumferential distance traversed in one complete revolution.

In effect, all de Broglie said was that the relationship which we stated in Chapter 5 for photons that their energy content had to equal Planck's Constant times their frequency (f) or $E = h f$ could also be applied to the apparent behavior of electrons. The dualism of sub-atomic matter was once again confirmed to the consternation of the traditional classicists of the past and the puzzlement of ourselves.[2] One justification for the assertion that Maxwell's equations for waves in a field can be applied to electrons is the assumption that electrons move at or very close to the speed of light and that their moments of motion (mo) are susceptible to the same mathematical treatment as the photon.

De Broglie quite daringly presented his proposition for his thesis to the faculty of the University of Paris, although he cautiously added the qualification that his computations would hold for "a fictitious associated wave" assigned to electrons. The examining professors were rather shocked and confused by such a very unusual presentation, since all conventional wisdom then was that the electron was indubitably a material corpuscle. Moreover, they were extremely dubious about all this business about resonating "strings." One advisor friendly to de Broglie, however, sent a copy of his thesis to Einstein in the mail with a request for his comments. The latter, in substance, replied that the idea was not all that crazy and, furthermore, stated "I believe it is the first feeble ray of light on this worst of our physical enigmas." The young student not only was awarded his Doctorate in 1924; but subsequently received a Nobel Prize in 1929.

The justifications for these ideas were elegantly expressed in his thesis; and it became a tremendous boost to general credence in the truth of Bohr's basic picture of the atom. There was only one serious catch. It inevitably brings one back to duality and to the necessity of regarding the electron, under some circumstances, as merely a probability pilot wave rather than a particle. Sub-microscopic pictures possible today show concentric waves around the nucleus of an atom. Nevertheless, Balmer's spectroscopic lines for radiant energy, Bohr's integer numbers for the electron orbit shells, and

---

2. Scientists customarily designate a wave frequency by the Greek letter " nu " or sometimes lower case " upsilon. " The reader will notice that, since we are not talking about forces or spring constants here, the author has taken the liberty to substitute the English lower case letter " f " for simple recognition. This is really not so terrible a desecration of convention as might be sup-

the X-ray diffraction phenomenon by crystals were all handily explained as compatible. Opposition to quantum wave mechanics was slowly beginning to crumble among the younger physicists. De Broglie's wave equivalent to a material particle ultimately became a starting point for the development of an entirely new science.

Expressed mathematically, this meant that:

Wave Length $= \lambda = h / m v$ when:

v = velocity and m = the mass of the particle

The quantity "mv" is obviously momentum, which thus may include the photon.

Meanwhile, another actor, who was a student of Sommerfeld's, took the stage in this drama. Wolfgang Pauli (1900 to 1958) was the son of a professor of colloid chemistry at the University of Vienna. Even during High School, it became evident that he was extraordinarily brilliant. He was given to reading Albert Einstein's General Relativity works during classroom hours with no adverse effect on his grades. Having decided on a career in theoretical physics very early, Pauli decided to take his undergraduate college studies under Sommerfeld at the University of Munich in Gottingen. At age nineteen his competency had already been discovered, so he was asked to write a long article on the subject of relativity for the *German Encyclopedia of Mathematical Sciences*. He completed his baccalaureate requirements by 1921, when Sommerfeld immediately made him an assistant. Here Pauli met and studied under Neils Bohr for the first time and was introduced to the shocking flaws in the old classical physics concepts when applied to the atom. He was awarded his PhD "summa cum laude" by 1922, which was an astonishingly short time. During the following year he traveled to Copenhagen in order to help edit Bohr's own publications; but eventually returned to the University of Hamburg in 1923 to take a physics teaching position there as a *Privatdozen*. He succeeded in publishing five papers in the next two years in that position. It was then that he recognized that the electron had a fourth quantum number, namely spin or the angular momentum which we introduced to the reader in Chapter 3.

As the reader certainly must realize, at this point in history the scientists were preoccupied with mathematically describing the structure of the atom with particular attention to the electron and its orbital shells around the nucleus. They were also attempting to correlate these calculations with spectroscopically observed wave lengths and, additionally, to explain the Zeeman phenomenon, observed as far back as 1896, of the splitting of the lines into fine subdivisions with different optical properties. Not only that; but when the light spectrum patterns of heated atoms are observed under the influence of a magnetic field exerted in a direction perpendicular to the path of the light rays, strange things happen. The spectrum lines for a monochromatic light beam have three different waves, the middle one oscillating parallel to the direction of the magnetic field and the other two oscillating perpendicular to the field. This effect was one of Pauli's earliest interests, second only to Einstein's Relativity.

By this time all the young and ascendant physics majors after World War I, at least in Europe, knew where the new frontier of intellectual challenge lay, nor were they indifferent to the probability of building a reputation or even possibly earning a Nobel Prize. Many of these young men knew each other, some were close friends of the same age and quite a few took weekend or vacation hiking trips into the Alps mountains together. All attended seminar conferences whenever possible and corresponded together. The evolution of the quantum theory became a collaboration between a group of dedicated personalities with a gradual development over a span of sixty years. The full story has not yet led to a fully satisfactory completion.

Thus it happened that a twenty-four year old professor at the University of Munich, who was named Werner Heisenberg, together with two Netherlands friends at the University of Leiden, who were named George Eugene Uhlenbeck and Samuel Abraham Goudsmit, all worked together. An expatriate student from Columbia University in the United States named Ralph deLaer Kronig was also a member of the group. They all conferred and more or less mutually agreed that not only must the electron have a property which is now called "spin," or rotational momentum, about some axis, however oriented; but also that this angular momentum must have a value of one half. That is:

$J = \text{Spin} = 1/2$ (in units of $h / 2\pi$)

It was presumed that this rule most probably also applied to protons.

Wolfgang Pauli wrapped all of this up in 1925 when he published his first really major and original article in the German *Journal of Physics*. Here he stated the now famous dictum that the electron rotates about its own axis and that no two electrons having the same quantum state (and this means also the same direction of spin) can exist together in the same space in an energy orbit of an atom. This statement became known in all subsequently printed textbooks as Pauli's "Exclusion Principle." It was further modified and extended to apply to any two identical particles, whether electrons, protons or neutrons, which have the same four "quantum numbers." These so-called quantum numbers, which are complex quantities (often designated by the letters n, m, l and j), are not necessarily simple numerical attributes, such as rest mass, or electrical charge, or even momentum. Spin may be the simplest one of all four. The number "n," for instance, might actually represent the complex function shown below:

$$n = \frac{B}{k} = \frac{m_0 Z e^2 k}{\hbar^2 4\pi\varepsilon_o} \qquad \text{when:}$$

m_o = Rest Mass of Particle
Z = Nuclear Charge or Atomic Number or Number of Protons
e = Electric Charge (usually that of the electron)
k = Boltzman's Force Constant
$\hbar$ = (h / 2p)
ε_o = Permitivity Constant for a Vacuum

Most of us can derive little significance from this piece of arcane information except perhaps wry amusement; but it should suffice to demonstrate why particle physics is not a national pastime. It may also be a cause for no small wonder that there are about seventy five universities in the United States which are still turning out physics majors with doctorate degrees who are expected to understand it. It was the atom bomb and the siren promise of cheap energy which later fueled a huge interest and fascination with the subject. For the participants in our present story, however, it was the pursuit of knowledge for its own sake.

If one chooses to regard the electron as an extremely tiny charged magnet spinning about an axis of rotation and having north and south magnetic poles at each extremity of this axis and thus creating its own magnetic moment as it circles the positively charged nucleus, it becomes evident that it not only may generate its own magnetic field, but must simultaneously be also constantly interacting with other electrons in close proximity in other orbital shells. Recall also that due to the Special Relativity effects its mass must constantly be changing slightly as it whizzes at high velocities around its elliptical orbit. These same elliptical orbits, as we have learned, are themselves slowly precessing around the nucleus. Knowing all this, it must become apparent to the reader just how complicated and intricate the calculations for quantum mechanics may become. Today some calculations only become feasible by the use of an electronic computer. Thus Pauli's "Exclusion Principle" helped greatly to reduce the infinite possibilities and explain why electrons are locked into specific positions in the various energy level orbits. Not at all incidentally, it makes the entire subject of chemistry possible.

At this point a young and talented student of Sommerfeld's and very close friend of Pauli's emerged as a key figure in the development of quantum mechanics. Werner Karl Heisenberg (1901 to 1976) was the son of a professor of ancient languages, who was also an expert on Greek philology, who had moved to Munich in 1910. The boy was an accomplished musician from an early age; but his aptitude for mathematics led him to theoretical physics at the University of Munich. With the strong endorsement of Professor Arnold Sommerfeld, and over the objections of the official Department Head for Experimental Physics, Heisenberg received his doctorate in 1923 at the unprecedented age of only twenty–two years. He then moved to Gottingen in order to study under Max Born, eventually becoming the latter's assistant and staying there for three years with only a seven month interruption in order to visit Neils Bohr in Denmark in 1924.

Although Bohr and Heisenberg mutually admired each other's intellect and became friends, often debating their ideas late into the late hours of the night, the fact is that the younger man became disenchanted with any attempt to explain the spectroscopic emission of fine lines or the Zeeman splitting by simple astronomical or "Celestial" analogies. That is, he lost all confidence in the notion of treating electrons as tiny bodies traveling around the nucleus at different speeds in various different radii of increasing dimension. This classical approach, to which Bohr was attached, appeared to work moderately well for the hydrogen atom and even the alpha particle having a single electron; but Heisenberg became more preoccupied with electrons as waves. The scientists were at a complete loss to explain mathematically any of the more complicated atoms.

It happened that in late May of 1925 Heisenberg became afflicted with a very severe case of "hay fever," or asthma, and requested a fortnight's leave of absence from his university duties in order to take a prolonged vacation in a rented cottage by the sea on the island of Helgoland, which is north of Germany and west of Denmark in the North Sea. During that period of completely solitary life he devoted himself to intense thought about the problem of electron "jumps" and quantum radiation. As he later confided, there were some starting points which he assumed for his approach. One of his firm convictions was that the Law of Conservation of Energy had to apply. Another was that he would deal only in numbers or attributes of the electron which he considered as experimentally observable and thus not engage in speculative assumptions on things which were invisible and not susceptible to actual measurement in a laboratory. Some of these values which he used might include the electron mass, its electric charge, angular momentum or spin, and the integer numbers series for the spectroscopic observation of light frequencies. A third and most important assumption was his firm belief in Louis de Broglie's thesis that the electron behaved as a wave inside the atom or in conformity with our now familiar $E = h f$ equation.

The actual physical positions of the electrons inside the atom were left to assignment to various concentric shells of increasing circumference and radius from the nucleus center. Little or no attempt was made to giving any specific measurements of the space between orbits. Electrons could jump from one shell to another by the absorption or emission of a photon of energy; but could not and did not exist in the space between the shells. These orbital rings were customarily designated by the letters "K, L, M or N." One, two or four electrons could exist in a particular ring as one approached the outside diameter of the atom, whatever that may mean.

It should be remembered that Heisenberg was a highly educated young man who was quite proficient in advanced mathematics, which had been his original career choice. Thus he was completely at home with the Fourier series for wave periods and amplitudes, as well as "perturbed" or non-harmonic oscillators. Moreover, he was accustomed to using calculus derivatives and Hamiltonian functions.

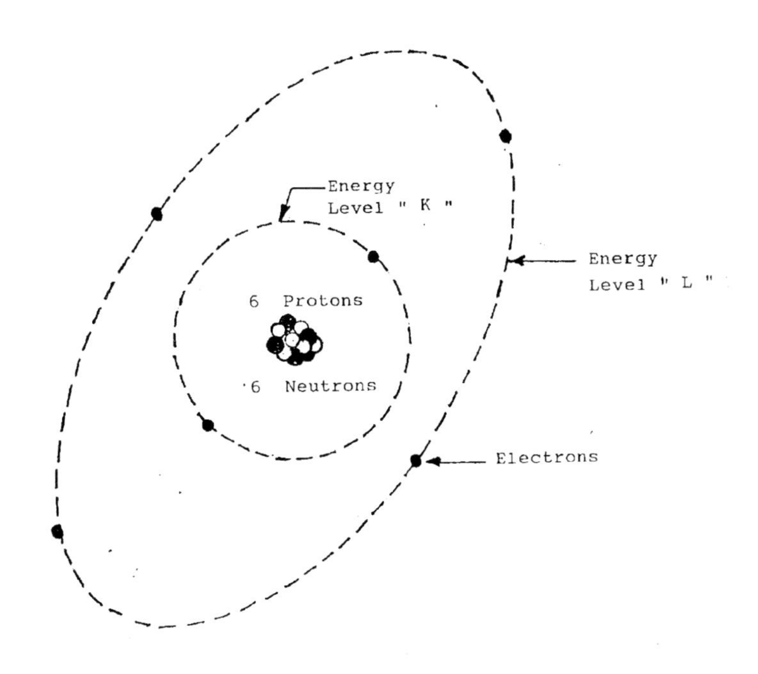

FIGURE 7 – 5
Schematic Diagram of Carbon Atom with Valence of 4

This last mathematical short-cut was invented by an eccentric Irish-Englishman named Sir William Rowan Hamilton (1805 to 1865) around 1835. Hamiltonian functions have a correlation with physical reality in a concept known as "action." Light photons, for instance, in passing from one coordinate point in a space system to another select that path which takes the shortest possible time and energy consumption. In common vernacular "the shortest distance between two points is a straight line." The true situation becomes rather more complicated when a beam of light is reflected from a mirror or refracted by a glass lens, however. A Hamiltonian function, designated by the symbol "H," expresses a rate of change with respect to time of a moving body or particle as an ordinary first order differential equation involving given position coordinates and momentum of the particle. Velocity is skipped altogether; but the big attraction for its use is being a first order partial differential equation. None of the above information really helps the lay person to fully comprehend the thought processes of Heisenberg; and, in fact, his ideas when first submitted were regarded by some of the older physicists trained in classical formulae as arcane gibberish.

Nevertheless, Heisenberg is usually conceded to have been the first theorist to open the door to a successful quantum mechanics theory which became the standard basis for all atomic research in the last half of the twentieth century. In order to describe what he actually did during that epic two weeks alone in June this author can only offer a very crude and simplistic description. Heisenberg, himself, did not elucidate until later.

As a start, he compiled a series of tables of significant numbers in a logical arrangement of rows and columns in a two dimensional array. He kept in mind that there had to be a real physical relationship between the angular velocity (w) and the momentum (p) of the electron and its energy (E) as evidenced spectroscopically by its wavelength in angstroms or wave frequency. He well knew from Bohr's studies of Balmer lines that electron energies "jumped" by clear separations and that this jump in energy between orbits had a fundamental ratio from the lowest state to the next higher energy state in terms of integer numbers. An important assumption also was that this change in different energy levels between orbits had to reflected in similar changes in the angular velocity of the electron around the nucleus and its distance from the center. Heisenberg's own explanation for the famous Zeeman effect three years earlier in 1922 was given mathematically by the equation below:

$$\text{Energy Difference} = (n + 1/2)\frac{h\omega}{2\pi} \;. = (n + 1/2)\, h\omega\,)$$

The addition of the one half fraction was a bold innovation, although he had discussed this possibility with student friends at college and had been encouraged by Wolfgang Pauli. Notice that the jumps in energy levels from shell to shell for the electron still remain integers however.

n = l/2 to 3/2, or 3/2 to 5/2, or 5/2 to 7/2, all equal one.

Later on the recognition of spin and Fermions confirmed his judgment.

Reverting back to the picture of an electron as a material object in motion and still abiding with the older classical physics, Heisenberg depended upon another mathematical relationship familiar to High School students shown below[3]:

3. See the derivation of this equation in the Appendix at the back of this book.

Kinetic Energy = K.E. =
When: p = Momentum.
m_0 = The rest mass of a particle (in this case the electron).
ω = Angular velocity of revolution.
$\hbar = h / 2\pi$
$k = 2\pi / \lambda$ or 2π divided by the wave length selected.

It was these innocuous function relationships which ultimately became the starting points for the formulation of a mechanical oscillator type of equation for a force free particle in quantum mechanics. Heisenberg set about grouping the appropriate quantum numbers and wave length functions in a table of columns and rows or matrix with each group having a rational and physical relationship, keeping in mind all the while his objective of expressing their correlation in some reasonable wave form equation. Very roughly speaking, momentum expressions had a relationship with the frequency of the oscillator equation and the position in space of the particle or electron was a function of the wave amplitude. His exact thought processes and logic in the arrangement for such a mathematical simile at the time of invention remain somewhat obscure to this day. One reason is that his own description was contained to a large extent in personal letters to friends and associates written in German. Another reason is that his spoken lectures and memoirs in later years had the benefit of hindsight and refinements subsequently offered by other scientists. Mental intuition, as well as reason, played a large part. At any rate, he came to the conclusion that he could make sense out of all this confusion by selecting the correct numbers from his table and manipulating them properly, in some cases multiplying them and in others dividing them. He emphasized carefully in his publication that the order or sequence of multiplication of some functions was critically important; and that a reverse order could result in a wrong answer. This puzzled him; and it is possible that Heisenberg at the time was unfamiliar with the rules for matrix algebra and multiplication. This was in spite of the fact that professional mathematicians had already played with it.

An example of a very rudimentary matrix table which is familiar to most readers is the common chart printed on many road maps for a given State which list in a table the mileage distances between various cities or towns. More complicated and useful matrices may be constructed for such purposes as a tally of basketball or tennis games with scores and player performance or even the solution of three simultaneous equations with three unknowns. In all of these there are very specific rules for the multiplication of numbers in different cells, with the consequence that matrix algebra became a known skill. Much has been made of this known peculiarity in some books about quantum mechanics with the statement that "p x" is *not* equal to "x p," as though it were some mysterious aberration of quantum mechanics alone.

Thus Heisenberg introduced the concept of an electron visualized as some type of oscillating wave packet of very limited longitudinal direction at a location "q" in three dimensional space and which had a fixed frequency and a wave amplitude varying from a maximum at the center to zero at its ends. His goal was to find an expression for a "virtual anharmonic" linear oscillator compatible with this assumption.

We need here to pay a little closer attention to the technical meaning for a few of the words used in the previous sentence. "Virtual" is a token concession to the fact that many people still prefer to think of the electron as a unique corpuscle and that the wave function expression which best describes its behavior inside the atom is only an alternate representation in accordance with Bohr's "Principle of Complementarity." Dirac, on the other hand, preferred to use the bald word "fictitious."

The simplest definition for "anharmonic" might be "Non-Harmonic," or a compound vibration of distorted wave form. It is really a technical term in mathematics which refers to a wave form which could be described as a curve which intersects a straight line at four ordered points with the lengths of the segments in between these node points being unequal or even plus or minus in direction. In other words, perturbations from other outside influences may complicate the original wave so that it does not represent a pure sine wave or tone simulation in music. Its graphical representation can be very complex with overtones.

"Linear oscillator" refers to a vibration to and fro, or up and down, type of wave motion, which can be an alternately compressed and then attenuated frequency along some longitudinal axis in the direction of motion. A physical demonstration of such an oscillation could be the motion of a ball weight at the bottom end of a vertical spring which is attached at its upper end to a fixed and unyielding support and which has been set into motion by a downward external force upon the ball.

Heisenberg eventually arrived at a long and complicated expression for the wave mechanics which conformed to his selected observed facts; but it is not our intention to inflict upon the reader the perplexity of an equation which was subsequently superseded by a better one. Not having complete confidence in what he knew to be most unconventional mathematics, he hesitated to send his dissertation for publication after he had returned to the mainland. Instead he mailed one copy of the manuscript to his friend Pauli in Hamburg for comments and submitted another copy to his mentor, Max Born, in Gottingen.

Pauli enthusiastically urged its publication in the German *Journal of Physics*, which transpired in July of 1925. Born, at first, turned it over to a twenty–two year old student of his named Pascual Jordan; but in September of the same year turned his personal attention to the manuscript. Both these people quickly recognized that Heisenberg's idiosyncratic rules for multiplication were nothing more than the little known, but already established, rules for matrix arrays. Thus some three months after Heisenberg had published his paper Born and Jordan took up the subject with him seriously. Essentially they endorsed his idea of a wave equation based upon harmonic oscillators with a few modifications and detailed a reasonable solution for electron behavior by means of conventional matrix algebra. A joint Born-Jordan-Heisenberg paper was then published on the sixteenth of November, 1925.

In the meantime, another Austrian physics major named Erwin Schrodinger (1887 to 1961) had independently taken up the same challenge of atomic behavior. Born an only child to a cultivated and prosperous family in Vienna, his early education was primarily entrusted to private tutors and his parents. He did attend a local public elementary school in Innsbruck for a few weeks. His father had inherited a factory which made glazed "oil cloth" for popular consumption; but had become interested in chemistry as a young man before indulging in other scientific subjects. At age eleven the child entered the prestigious Gymnasium of Vienna for the requisite academic preparation for the University of Vienna which he was expected to attend. It was not until the age of twenty that Erwin Schrodinger became seriously intrigued with theoretical physics and began attending lectures on the subject. He eventually graduated from the University of Vienna in 1910 with a doctorate under the sponsorship of Friedrich Hasenorhl, who was a renowned scientist. He subsequently became an assistant and apprentice to Egon von Schweidler in the laboratory of the Second Physics Institute.

Upon the outbreak of World War I he became an officer in the German artillery and served at various isolated fortresses in comparative safety from combat until 1916 at the Italian front. By 1918 the Austro-Hungarian monarchy had collapsed and also with it his dreams of becoming a full professor in Chernovtsy in the Ukraine. During this chaotic period he also became immersed in the writings of philosophers such as Spinoza, Kant, Schopenhauer, and others, as any good German intellectual might. He soon returned back home to the University of Vienna, and married in 1920. He ultimately obtained a full professorship at the University of Breslau in Switzerland.

The defense of his dissertation by Louis de Broglie in November of 1924 at the Sorbonne in Paris, in which he treated electrons as a wave phenomenon, made a profound impression on Schrodinger. This was reinforced by Albert Einstein's ideas as contained in his paper titled *Quantum Theory on Atoms in Ideal Gases*, which was published in February of 1925. Schrodinger became an enthusiastic and convinced convert to the extent of expounding upon de Broglie's ideas in a lecture at the University of Zurich in November of the same year. He proceeded to throw himself into the possible mathematics of the riddle of hydrogen atom spectroscopy, as well as helium, which then–current physics was unable to explain. Essentially, he unequivocally adopted the notion that an electron could be treated as a kind of wave packet or wave amplitude crest against a background of other wave harmonics. With respect to the "wave packet" idea, the reader should be aware that there is one big problem which led to its eventual lack of popularity. What is to prevent it from radiating and dissipating its energy into surrounding space and eventually disappearing altogether? This assumption was in spite of the prevailing wave -versus- particle argument and genuine contradictory objections based upon the dissipation of electromagnetic wave energy required by Maxwell's equations. He also borrowed heavily from his previous five years of work at the University of Breslau on gas kinematics and vibrations. Basically, he decided the only way to explain the electromagnetic behavior of the atom was by a statistical analysis of its quantum response to certain key conditions. He consciously rejected the matrix approach of Bohr and Heisenberg with their materialistic "tiny marbles" and astronomical similes.

At this point in our story the patient reader should be warned that he stands metaphorically on the brink of a plunge into very deep intellectual waters to which our synopsis cannot really do justice.

The eminent mathematician, Max Born, whose role during the evolution of quantum mechanics was not entirely dissimilar from that of an obstetrician presiding at the birth of a new baby, stated the problem both honestly and succinctly. In his book *Atomic Physics*, the first English–language edition of which was published in 1935, he states: "There is, naturally, no way of deducing the wave equation by strict logic; the formal steps which lead to it are merely matters of clever guessing." This remark was not intended as a snide deprecation of Schrodinger's work, as it might first appear; but was a candid statement of the truth that a rigorous classical derivation of the final accepted equation is simply not possible. It was actually patched together by a number of people over time as an empirical accomodation of observed laboratory facts. The joint efforts over subsequent years of improvements ultimately resulted in a very abstract specialty in higher mathematics using arcane symbols and requiring graduate degree training for full comprehension and dexterity in actual use. This fact rapidly became a barrier which tended to isolate quantum physicists from the ordinary crowd. With the much later advent of World War II and the atomic bomb its practitioners became overnight almost a priesthood.

The accepted equation did not occur to Schrodinger in a sudden and brilliant flash of inspiration. He was working away at the problem over a period of time while at the University of Zurich with frequent communications with other theoreticians in the new science. By his own admission he needed to investigate the more intricate aspects of Maxwell's wave theory as already developed by Richard Courant and David Hilbert, the inventor of "Hilbert Space." He also appealed to Hermann Weyl for assistance at one point.

The entire theory starts with the adoption of the concepts of both de Broglie and Dirac in which we imagine a wave form and not an electron corpuscle revolving around the nucleus of an atom. As the case of Schrodinger, the structure of this wave function must conform with that of a mechanical harmonic oscillator subject to perturbations, such as the already familiar picture of a weight vibrating through a limited distance on the end of a coil spring. It was further assumed that the total energy of the system must equal the sum of the kinetic and potential energies and that the final wave equation must also reflect that stricture. When mathematically expressed as a Hamiltonian function, this requirement became that shown below:

$$H = \frac{p^2}{2m_o} + \frac{m_o w^2 x^2}{2}$$

Inasmuch as this book is not a textbook, there is little reason to attempt any detailed explanation here; but both Heisenberg and Schrodinger agreed on the need for this basic relationship. There exist numerous variations of Schrodinger's famous equation which were all contrived to be a simplification adapted to a certain special and limiting circumstance under study. For the insatiably curious reader, however, a most general and popular version of his final equation has the appearance shown below and applies to a time dependent rate at which some quantum system changes:

$$i\hbar \frac{\partial \Psi_{r,t}}{\partial t} = (-\frac{\hbar^2}{2m_o} \nabla^2 + V_r)\Psi_{r,t}$$

To most people this expression in complex variables has all the significance of enigmatic cryptography, which, in a sense, it is. A superficial definition of the various terms might give us a comfortable illusion of comprehension.

One may recall that the letter "i" designates the imaginary square root of minus one. Squaring or multiplying by itself makes it become one.

"ħ" = Planck's constant divided by 2π, which is a commonly used energy conversion factor which is equal to $1.0545887 \times 10^{-34}$ joule seconds.

r = A distance of electron from proton is a defined three dimensional space.

m_o = The rest mass of the electron.

Ψ = The wave function or state vector for a particle (usually an electron) which determines its most probable condition. Mathematically, it is a harmonic vibration amplitude of energy in a three dimensional space.

V = The potential energy in an electric voltage field.

In addition a La Place transformation operator for enabling to calculus solution of three double derivative function equations in calculating involving electron position.

The fact that Schrodinger's equations could be expressed in calculus derivatives was welcomed by all physicists of his day, who were mostly unfamiliar with matrix algebra and found Heisenberg's approach rather awkward and time consuming. As Schrodinger himself put it so forthrightly, "I...feel intimidated, not to say repelled, by what seem to me the very difficult methods of matrix mechanics and by the lack of clarity." Although the two methods seemed utterly different, they proved actually in the end to be not all that far apart in basic theory and quite equivalent in their final results. Quoting Schrodinger once again to reveal his attitude: "The ordinary quantization rule can be replaced by another condition in which the term 'integral number' no longer appears. Rather, the integrality occurs in the same natural way as, say, the integrality in modal numbers of a vibrating string." Thus particles become merely a wave crest on the background of numerous lesser waves. Using the actual words of the originator again, "it is really a vibration (or, as the case may be, an interference) process which occurs with the same frequency as the one which we observe in the spectroscope." To the reader, this all seems equivalent to reducing what happens inside an atom to wave mechanics. What happens to the substance of matter?

Schrodinger found the courage of his convictions and published his ideas somewhat deferentially in a series of papers in the German *Annals of Physics*, starting in December of 1925 and continuing through January to February 23rd in 1926. He followed up on June 21st of the later year with a supplementary paper giving a non-relativistic time dependent equation which we have seen.

Schrodinger's memorable opus was received by his peers with admiration and respect; but there was also a mixture of bafflement and healthy skepticism in some quarters. As we already explained, it was not classical mathematics of a conventional sort. His old Gottingen school friends, however, seized upon his publications with enthusiastic avidity and proceeded to analyze his theories in professional detail. They disagreed with him that the equation strictly represented a "wave packet;" and Max Born lost little time in demolishing that interpretation also - much to the inventor's personal distress. Much more important, Born introduced the notion of statistical probability and came to the conclusion that the doubling of amplitudes represented by the quantity Ψ2 was a kind of "ghost field" pilot wave which determined the behavior of the quantum particle. Specifically, he reasoned that it was necessary anyhow to take the square of the modulus of the wave function "Ψ" in order to get rid of the complex and imaginary coefficient "i." Physicists today accept this trick as the linear superposition of two wave form amplitudes one on top of the other. The probability attribute is quite similar to that encountered with throwing two gambling dice on the table. There is a probability of getting six times six or six squared different possibilities for the total sum of dots on the upturned faces of the dice. This viewpoint has prevailed, although the question of exactly when such a calculated probability becomes a physical fact in the form of a particle in the laboratory remains the subject of an on-going debate in science to this day. Schrodinger himself stated later that if he had foreseen this consequence of his work he would rather have not published his papers at all! That is the natural reaction of a man who had been educated and worked for years in the classical science tradition when confronted with what is now commonly called quantum weirdness. Nevertheless, his fame spread rapidly, and he was invited to assume the chair of theoretical physics at the University of Berlin in 1927. In spite of his great love for Switzerland and hiking in the Alps mountains with his friends, he accepted and stayed there until 1933.

The reader should be aware there are many complications and peculiarities to the quantum mechanics equations which we have avoided in this very abbreviated synopsis. Quantum mechanics calculations are definitely not for amateurs and sometimes defy our natural intuitions, just as do the results of passing light rays through two slits in an opaque screen. One strange aspect of the general wave equation which is fairly obvious is the multiplication of the fundamental wave function by the square root of minus one. This "imaginary" number conspires to make two possible versions of the quantum equation, one the mirror image of the other with the algebraic signs reversed. Theoretically, this makes it possible for time to go either forward or backward in the calculations. It also means that an independent particle may go off in different directions and have a spin of either plus or minus one! Such an idea, of course, does violence to our own conscious experience and intuition about the nature of our world reality, as well as contradicting the evidence of entropy in the entire cosmos which Sir Arthur Eddington called the "Arrow of Time." Thus for many sensible reasons the physicists of all eras have chosen to ignore backward time. General relativity between different inertial systems, as we know, makes strange things possible; and even Richard Feynman in his diagrams for particle reactions such as photons and electrons talks about particles "going backward in time," using general relativity as his excuse.

More important is the propensity for the general wave function equation, as it stands alone, to yield infinities for answers. In order to avoid this frustration it becomes necessary to factor in very specific limits or parameter numbers before going through the calculus derivation processes. These stipulated limits must reflect observed and measured physically significant numbers found in the laboratory. Such empirical factors might include such things as displacement or boundary limits, nearby coulomb charges, any electromagnetic barriers, and the steps or wave length "jumps" which can be expected for different particle energy levels. Such mathematical restrictions or "operators" were first christened "eigen" values or "eigen" states by the Germans and are essential for any successful calculus to result in non-trivial results.

There was yet another esoteric problem to overcome. It was known that the electron, especially when it is free and unbound, has the capacity to constantly absorb and then instantly emit photons of energy, whether real or virtual. As is also known from the photo-electric effect of light rays on certain chemicals or metals, even a "bound" electron can absorb enough energy in this manner to shake loose from its atomic capture. This extra energy from movement above its initial and lowest rest mass value is called "self energy" and can theoretically become infinite. The terms of Special Relativity also demand that the particle mass may increase. Obviously, this kind of situation can make the quantum calculations an amorphous mess. Thus the scientists had little choice left to them but to assign a firm value to the electron's (or any other particle's) rest mass (m_O) and to either subtract entirely or else give a fixed value to the "self energy" quantity.

All this mathematical fudging of the starting numbers was dignified by the name "renormalization" and became the accepted practice. If you happen to be a person who likes to believe in causality and not in a world of mostly probabilities, or are trained to believe in a predictable end to a given calculation, then this "doctoring" approach may very well be intellectually unsettling. In fact, the successful application of the renormalization procedure requires considerable skill. Some people still think that renormalization is somehow cheating. The argument in the defense of both renormalization and eigenvalues is quite ingenuously simple. In the lack of any other knowledge the scientists have no other alternative — at present anyway. Besides, repeated experience has demonstrated to their satisfaction that, if the quantum calculations with all their complications are performed correctly, the answers are "realistic" and have a repeatable accuracy to an astounding degree. In spite of many ingenious attempts at explanations, the wave equation actually works well, even though very learned experts are not exactly certain why it should. That is its incontestable claim to wide acceptance.

In the meantime, Erwin Heisenberg had not been idle. He had been in continuous and detailed correspondence with Wolfgang Pauli, with one of his last letters running to fourteen pages. They were both preoccupied with the necessity to choose specific and continuous Hamiltonian variables for the momentum (designated by the letter "p") and the general space coordinate (usually designated for all three dimensions by the letter "q," but sometimes in limited cases by "x") in calculations for the electron which forms part of an atom. The momentum had to be exact; but the position or path of the electron could not be known exactly. Thus it happened that, scarcely a month after the publication of Schrodinger's last paper, Heisenberg appeared in the Journal of Physics with a paper which identified and gave the mathematical proof for one of the most important quandaries in quantum mechanics. Although it already had been deduced by others before him, Heisenberg's paper was published in March of 1927 under the title of On the content of *Quantum Theoretical Kinematics and Mechanics*. It enunciated for all history what soon became known as "The Principle of Indeterminancy" and since has been elevated to the staus of an incontrovertible doctrine. In simple mathematical abbreviation he showed that:

Δ p times Δ q must be equal to or greater than $h/2\pi$

when "h" is Planck's Constant.

There is a corresponding statement involving time. When one remembers, after a moment of reflection, that momentum represents mass times a velocity, then it becomes obvious that Heisenberg's principle must apply to time measurements as well. The correct expression for this becomes:

Δ E times Δ t = or $\geq$ h

"E" in this case represents total energy.

In particle accelerator laboratories and in the elastic scattering of particles by a target this last expression becomes very important. The time, and consequently the length of its track, that a fragment can be seen in a cloud chamber or scintillation screen is a form of measurement of its kinetic energy. Sometimes that information alone identifies the particle.

The appearance of Planck's Constant in any calculation always means there is a finite practical limit to how well one may predict the position of a particle under given physical conditions. Since momentum involves mass times velocity and the latter is merely a translation through space in a certain unit of time, it is inevitable that the position of a particle is also subject to this uncertainty. In experiments one must decide which of these two properties is of the greatest immediate importance and then make a long series of tediously exact measurements in order to obtain the best average.

Moreover, precision in the location of a particle will automatically exclude equal or simultaneous precision in determining momentum. Not only are all atomic and sub-atomic particles dancing around frenetically; but occasionally they can turn up where they are not even reasonably supposed to be. To the layman living essentially in a Newtonian macro–cosmos this sounds like scientific heresy, but it has become clear that Nature abhors absolute shackles for a unique case and bows only to mean statistical probabilities for the mass behavior of the millions of particles, whether electrons or photons. There remains always the secret potential for the miraculous—or at least, the surprising. Heisenberg and others elevated this basic observation to the status of academic probity and doctrine by naming it the "Uncertainty Principle," thus tacitly admitting the demise of the clockwork machine world of predetermined fate.

Richard Feynman, who persisted during his entire professional life in searching for "real life" analogies or similes to explain quantum mechanics and its relationships to other physical phenomena, once semi-facetiously compared the Uncertainty Principle to standing near a waterfall watching a cascade of water plunge over a ledge and dash upon rocks beneath. If you stand near it long enough, the probability is that a drop of water may land on your nose. Actually, the uncertainty problem is a great deal more profound than that and involves how we measure things and the possible effect of our decisions on such measurement on the laboratory results. It happens that the uncertainty relationship is intimately connected with the minimum size of an atom and just how close in radius about the nucleus an electron is allowed to orbit. Given certain simple and basic assumptions, however approximate, physicists have toyed with the centifugal forces involved and resulting momentum to discover that _p times _x does indeed calculate out very close to the value of Planck's Constant. Those assumptions are:

Size of the Atom = 10^{-8} centimeter in diameter

Mass of the Electron = 9×10^{-28} gram

Velocity of Electron in Circular Orbit = 10^{-8} centimeters per second

Approx. Momentum of Electron = 10^{-20} gram centimeter per second

Abraham Pais, who was an early nuclear physicist of distinction and who became a writer and historian, objected with considerable vehemence and emphasis in his book Inward Bound to the term "uncertainty relations" being applied to these two statements. Such a description he regards as inappropriate and misleading. He claims there is nothing uncertain about it. If one learns from careful measurement and observation one of the two parameters, then it becomes impossible to know anything about the other with any exactitude. One can only make guesses. Nature will allow us to be precise about one of her secrets or the other; but not both at once. Thus one must decide which of the two variables is of the greatest immediate importance.

It is said that Heisenberg also believed at the time that his paper would demonstrate a critical flaw in Schrodinger's "unreal" or non-materialistic wave equation, which he resented. Whether it did or not, the "uncertainty" paper had as much influence as any of his other treatises to earn for Heisenberg a share in a Nobel Prize award in 1933, along with Schrodinger and Dirac.

It now remained only for the youngest member of this group of luminaries, one Paul Adrien Maurice Dirac (1902 to 1984), to clean up some details and add the finishing touches to the mathematics for the new Quantum Theory. In spite of the French name, Dirac was born in Bristol,

England, as one of three children of the daughter of a ship captain and an immigrant Swiss-Frenchman who taught the French language and history at the local Merchant Venturer's Technical College. His father, it would seem, was a rather eccentric misanthrope and martinet who would not permit any other language but French in his home. Young Dirac, who was a genuine intellectual genius, grew up quite socially inhibited and all his life tended to be quite monosyllabic and even to stutter badly whenever he attempted to speak English to his associates. Nevertheless, he quickly excelled at mathematics and science in school and at the age of only sixteen entered Bristol University with electrical engineering as his intended major. Although he had technically completed courses for graduation in 1922, he quickly accepted an offer of free tuition there in order to continue his studies for another two years. Employment opportunities were scarce immediately after World War I. The big change in his life was his transfer in 1923 to Saint John's College at the University of Cambridge with the help of a scholarship grant. There his interests permanently shifted over to mathematics and theoretical physics. This is scarcely surprising, since it was Dirac who once wrote that "God is a mathematician of a very high order; and He used very advanced mathematics in constructing the universe." Dirac became a Fellow in the same college at Cambridge at the age of only twenty–five.

Fortunately for nuclear physics his assigned professorial advisor at Cambridge was Ralph Fowler, who was heavily involved in laboratory experiments in particle physics and became the first expert in England on the American originated cyclotron accelerator. This new development, which came later, speeded up charged particles in their circular tracks inside two evacuated "Dee" shaped chambers placed between the poles of a huge electro-magnet by means of timed radio frequency voltage pulses. Although the early development of this machine, which was invented by Ernest Orlando Lawrence, was largely paid for by medical grants in an effort to find a cheaper source of X-rays than radium, it eventually became an important tool in nuclear research in England. Fowler became a leader for using this machine for research. Dirac was also strongly influenced by reading *Mathematical Theory of Relativity* by Sir Arthur Eddington, which had been published in 1923. It was Fowler, however, who first introduced his student to the first published papers of Werner Heisenberg in 1926. Dirac immediately became fascinated and made them the subject of his doctoral dissertation, which he completed without delay in June of 1926. He then traveled to Norway in 1927 in order to confer with Neils Bohr and to organize his own thoughts about the duality question on corpuscles versus electromagnetic waves. He also became familiar with the recently published papers of Erwin Schrodinger.

Among many other of his contributions to the discipline, Dirac introduced a method for accounting for relativity effects on the behavior of electrons in their orbits around the nucleus and also the quantization for radiation interaction with the atom using Maxwell's formulae. All this study culminated in 1930 with the publication of his *The Principles of Quantum Mechanics*, which immediately became a textbook for the subject.

This brings us to the end of the very brief historical outline of the beginnings of quantum mechanics which we have offered in this book. It has undoubtedly tried the patience of many of you readers, but it still has not done justice to the complications of the subject or to the contributions of many scientists other than the ten or so singled out. All of them literally changed the scientific world forever and simultaneously introduced profound philosophical implications for the nature of reality as we perceive it. Many books and discussions by brilliant people are still proceeding sixty–five or more years later in attempts to explain its mysteries. No longer can the universe be regarded as a potentially predictable "clockwork" mechanism and the notion of predetermined "Fate" for the individual is challenged. On the contrary, small probabilities introduce the element of chance into all things and open a window to what we might call miraculous. The author urges every reader to pursue his or her reading on the subject elsewhere in the available literature. As we stated previously, the great majority of practicing physicists tend to ignore or even deride any speculations upon the philosophical or metaphysical implications of quantum mechanics, feeling that such mental activity is an occupational "dead end street." It would be similar to asking "What is Time really?" or "Whence did the energy come from in the beginning of our own universe?" Nonetheless, many thoughtful people of acknowledged intellectual and scientific credentials continue to ponder the subject and even write books about it. But be forewarned. There still exists a curtain behind which we are unable to peer.

In his book *Dreams of a Final Theory*[4] Steven Weinberg relates an anecdotal conversation which he had with a fellow physics professor in Texas one time. He casually asked his friend what had become of a specially promising graduate student who had disappeared from his classes. The wry answer was, "He tried to understand quantum mechanics."

---

4. Stephen Weinberg, *Dreams of a Final Theory*, (New York, Pantheon Books, 1992).

Chapter 8
Cosmic Rays and the Meson Showers

The planet earth is constantly being bombarded with a host of different particles and a wide spectrum of electro-magnetic radiation frequencies as it wheels around in its orbit about the sun. Much of this bath of energy comes from our central sun, of course, but a surprising amount emanates from distant sources far out in space. The visual evidence of this bombardment is the Aurora Borealis, commonly known as the "Northern Lights," above the Arctic Circle latitudes. The real nature of these cosmic "rays" took many decades to discover.

Possibly the first person to produce a reliable visual record of the arrival of cosmic rays on the surface of the earth was a Russian in his early thirties named Dimitri Skobeltzyn working in Lenigrad. He was experimenting with a small circular cloud chamber, which had just been invented, between the years of 1924 and 1928. He was the first to conceive of the idea of placing this detector in a strong magnetic field and photographing the resulting tracks. Normal electrons curved to one side, but the cosmic particles having very high energies made what appeared to be straight line tracks. The source was not specifically identified. It remained for Victor Franz Hess (1883 to 1964) to demonstrate by means of scintillation counters on the ground that the inaccurately named "cosmic rays" were extra-terrestrial in origin and actually came from the sky. Hess was the son of the Chief Forester in the service of a prince and landowner in Austria, and had obtained all his education in the City of Graz, culminating with a PhD degree from the University there in 1906.

With the help of the Austrian Academy of Sciences and a local balloon club Hess made ten ascents in a balloon basket between the years of 1911 and 1912 to heights approaching 5000 meters (16,400 feet approximately) while carrying his scintillation counters with him. He noted that the intensity of the particle radiation steadily increased as his balloon rose higher above 500 feet. His originality and personal daring won him a Nobel prize in 1936. When the Nazi fascists occupied Austria in 1938, Hess fled to Fordham University in New York City to continue his scientific career.

The term "cosmic rays" is now customarily applied to baryons, or heavy particles, coming from all directions from outer space and impinging upon the earth's outer atmosphere at the rate of one thousand particles per square meter per second. The quantity and variety astonished the scientists and their exact nature became the subject of intense curiosity and a widespread scientific interest. In the absence of today's large accelerator machines for the purpose of laboratory simulation, this natural rain of unknown particles from the sky became temporarily the best hope for understanding the atom. We know today that these misnamed "rays" are actually for the most part ionized atomic nuclei consisting of ninety percent protons, nine percent alpha particles and the remaining one percent being different heavier elements. A great many come in a constant shower from our own sun, some originate from within our own "Milky Way" galaxy and some few are thought to originate in super-novae or neutron stars in outer space well beyond the limits of our own galaxy. All of these particles achieve relativistic or Einsteinian velocities approaching that of light in varying degrees, although the greatest velocities come from outside our galaxy and not the sun. A few of the latter approach a kinetic energy in the order of 10^{20} electron volts (or 500 TeV), thus achieving an effective mass eleven times that of the stationary proton and a colossal impact momentum.

Our own sun emits a plasma wind which, in combination with the sun's own magnetic field, concentrates the abundance of cosmic particles into a doughnut–shaped zone called the heliosphere. This cloud extends out from the sun's surface a distance exceeding that of the earth's orbit. Intermittent and cataclysmic solar gas flares, however, intermittently upset a fragile equilibrium with gravity and permit huge streams of protons, electrons and neutrinos to reach the earth 's troposphere. The result of such disturbances is intense radio wave interference in certain frequencies and fluctuations in electric power line voltages for our workaday existence. The shower of millions of tiny and probably massless neutrinos coming our way pass through the earth and our own bodies easily with minimal or no effect.

Confusing the entire picture is the fact that many of these unknown particles which experimenters were eagerly detecting were secondary results of the high velocity protons colliding with molecules of air or water vapor in the upper atmosphere and producing fragments. For instance, there is a great abundance of neutrinos at our ground or sea level, but they do not come from any cosmic source. Instead, they result from the decay of these secondary fragments or mesons such as pions or kaons as they travel earthward from the thin atmosphere at higher altitudes. These secondary subatomic fragments from Nature's own accelerator are more easily detectable and provide indirect evidence of an unremitting bombardment from space. Their study produced a long collection of such lower mass particles which utterly surprised and confounded the physicists of the early 1900's era. The atom was no longer simple.

Section 8.1 Mesons

The name "meson" is somewhat misleading and was invented by the Indian physicist H. J. Bhaba in 1939. In the very first discoveries of cosmic ray research almost any particle which had a mass greater than the electron and less than the proton was called a meson, which was derived from the Greek word for "middle" or "intermediate." This definition was quite basic and applied for quite a while, but subsequent discoveries revealed many exceptions to the rule. Particles which had a mass equal to or greater than the proton appeared, so they became classified as "baryons," which are discussed elsewhere in this text.

Mesons are considered as being very susceptible to connection with the strong force inside the nucleus of the atom, but in their free and uncaptured state they have extremely short lives of varying degrees of time before decaying. A mean life of only 10^{-6} seconds would be normal. The first statement, however, does not mean that the weak force cannot operate within mesons. As research progressed, this group of particles ultimately proliferated into a bewildering array of particles with different masses and charges. The first result among scientists was utter confusion and dismay, since the appealing simplicity of Bohr's picture of the atom was being shattered. Eventually it was realized that all these cosmic created particles were ephemeral clues to the structure of all matter. Their study by hundreds of physicists under the most scrupulous and exacting laboratory conditions yielded a compendium of raw data which enabled the theorists to venture their explanations and construct a speculative order to the data. The eventual achievement of the theorists in being able to predict the discovery and properties of newer and unknown particles cannot be lightly dismissed.

The heavier mesons often spontaneously decay into a cascade series of lighter fragments which exist briefly and then themselves decay into a shower of electrons, positrons and neutrinos. The 1994 edition of *The Review of Particle Properties* handbook, as published by the Particle Data Group with their headquarters at the Lawrence Berkeley Laboratory in California, lists at least eighty–four different varieties of mesons or observed atomic nucleus fragments. These all differ from one another in their mass, electric charge (positive, negative or zero), mean lifetimes, and decay modes.

The entire study of mesons began in the 1920 to 1930 decade with the detection and measurement of the strange ionized particles produced in our atmosphere by cosmic rays. This was a time when scientists climbed to the top of mountains, flew balloons, or rode airplanes in order to pursue their work. Many years later the emphasis shifted to particle accelerator machines located on the ground and costing hundreds of thousands of dollars. Mountain climbing opportunities became scarce. The public knows these new machines as "atom smashers"; and this vividly picturesque language is crudely justifiable. It is important to realize, however, that many of the products resulting from either cosmic rays or accelerators are not necessarily elementary divisions of the atom. The Special Theory of Relativity is always operating and some of the kinetic energy of particles which have been accelerated to speeds approaching that of light is converted upon collision into a material mass for brief duration. Thus the total mass of fragments resulting from a high velocity collision can actually exceed the original rest mass of the projectiles used.

Although best known for his exact measurements of the magnitude of the electrical charge carried by the electron in 1909, for which he received a Nobel prize in 1923, Robert A. Millikan began his own serious study of cosmic rays in 1920 in order to establish their real origin. He had been raised as a boy in a very small town in Iowa in typical midwestern fashion and remained a church member and of conservative political persuasion all his life. His mother was a graduate of Oberlin College in Ohio and insisted he enter that college when he became eighteen. He graduated in 1891 from an essentially classical and academic curriculum; but, having been an honor student, he was asked to teach physics there to preparatory status students. He became more interested in the subject and two years later decided to embark on graduate studies in physics at Columbia University in New York City on a fellowship grant. He later moved to the University of Chicago which was closer to home and graduated there with a PhD in 1895. After a celebratory trip to Paris and Berlin, he took a position on the faculty of the University of Chicago as a teaching assistant. By 1920 his scientific reputation had been established.

Millikan was a personality of tremendous vitality and determination. Once interested in the riddle of cosmic rays he pursued measurements all over the globe with the conviction that they were indeed cosmic in origin and not terrestrial. He began with un-manned sounding balloons carrying a scintillation screen and motion picture camera to a height of 15 kilometers. Subsequently he took lead–shielded electroscopes to the top of Pike's Peak in Colorado in 1923. He also packed his detector equipment onto the backs of mules and climbed on foot with his son and friends to an altitude of 14,260 feet at the top of Mount Whitney in California. He then conceived the idea of submerging his ionization meters under water in Lakes Arrowhead and Muir in California and later Lake Titicaca in Peru. Science can be fun! Part of his idea was to protect the apparatus from random radiation in the air. Allowing for the differences in elevation above sea level between the lakes, the readings were all remarkably similar. This helped to convince Millikan that the ionized particles all came from space beyond our planet. More than that, he calculated that the water depth and earth's atmosphere above both represented a barrier equivalent to six feet of lead, which made the so-called cosmic rays more energetic than anything previously known. By 1928 he had become convinced that the particles were high speed photons. In that he was totally wrong and was compelled to admit it by 1933. In all this activity he recruited and started many younger men on the path of scientific discovery and also stimulated world wide interest.

8.2 Positrons and Anti-Matter

As mentioned previously, the 1930 to 1940 decade was a time of prolific discovery in particle physics. Moreover most of the successful experiments were accomplished with ridiculously inexpensive apparatus as compared to the costs which became normal during the 1960's. The invention of the Geiger counter in 1928 by Hans Geiger and Walther Muller of the Physics Institute in Kiel, Germany, gave a huge technical boost to the art of particle ray detection. Their new detector was quite simple in theory and consisted essentially of a positively charged wire carrying a high voltage direct current which ran longitudinally down the center of a thin metal tube. The metal tube was connected to the negative side of a bank of storage batteries or a rectifier. The tube was evacuated of air and then filled with a gas which was easily susceptible to ionization. Whenever a stray photon or proton penetrated the tube shell and hit a molecule of the gas it triggered an electrical discharge or momentary current spark. This current pulse could be made to flash a light or make a clicking noise by means of an electronic amplifier, as well as activating some type of counter.

Another detector which became a vastly more important tool for the identification of the properties of ionized particles was the cloud chamber. The first of these devices was invented by C. T. R. Wilson in 1911 for the purpose of detecting the debris from radioactivity and it immediately became popular. The basic principle was to enclose a volume of gas saturated with ethyl alcohol or some other volatile vapor in a shallow cylinder having a glass window at one end. The trick was to exactly time the sudden withdrawal of a piston from the cylinder in order to create a sudden adiabatic expansion of the gas. The vapor condenses around any ionized particle inside the chamber, thus making a visible trail which could be photographed. Such tracks made temporarily visible could be immensely valuable in judging the electric charge polarity, the particle's velocity or kinetic energy and, in some cases, its mean lifetime. In order to discern the nature of the electric charge, if any, it was necessary to install the cloud chamber between the poles of a strong electromagnet. The direction of the resultant curvature of the particle, whether to the left or to the right, was a clue as to whether it carried a positive or negative charge.

This new ability to identify the electric charge and relative energy content of an unknown particle was a significant leap forward in the technique of detection and a welcomed improvement on the old method of photosensitive emulsions on glass plates. With Nature now furnishing these unknown particles in great abundance it was small wonder that the study of cosmic rays became the new playground of physicists. No expensive accelerator machines were required and there was no dearth of perplexing phenomena to justify a professional paper.

One of the smartest moves made by Robert Millikan, aside from moving himself to the California Institute of Technology and its laboratory facilities, was to take under his tutelage a very gifted young graduate student named Carl David Anderson, who was born in New York City in 1905. Starting in the summer of 1930 these two men designed and built a vertical cloud chamber of fourteen centimeters in diameter which they placed between the poles of a large magnet which produced a field intensity of 15,000 gauss. On a warm August day about two years later they photographed a particle track which totally confounded all their most reasonable expectations. Not only did the particle, entering from the bottom, have enough energy to pass through a 6 millimeter thick lead plate barrier, but it curved to the left, thus making the convex side of the curve face to the right in the photograph. This was a clear indication of a positively charged particle, which would lead one to the conclusion that it was probably a proton. The only trouble was that the energy calculations and length of the track in air with little curvature or collision contradicted such an assumption. They found that out of 1300 cloud chamber photographs there were 15 cases that entirely duplicated such a behavior.

In a world famous paper Anderson published his conclusions in February of 1933 in the *Physical Review*. He identified the observed strange particle as a positively charged electron (now called positron) having a mass identical with the normal electron, but carrying a charge polarity of opposite sign. At this time Anderson was quite aware that Dirac had already predicted the possible existence of such a particle from his theoretical calculations. Anderson received the Nobel prize for his discovery in 1936. The big question immediately suggested was: "Who needs a positive electron anyway, since the negative electron does such a good job of creating chemistry and the material stuff of our environment?"

Guiseppe Occhialini and Patrick Blackett confirmed the production of positrons by cosmic rays in the Cavendish Laboratory at Cambridge, England, in 1932, using a similar cloud chamber. Most of the electron or positron tracks photographed originated from the collision of an outside particle arriving at tremendous speed and impinging upon the copper or brass walls of the cloud chamber itself. It soon became customary to regard cosmic rays as having two different components: (1) "soft" rays found around sea level and consisting mostly of electrons, positrons, and gamma ray photons and (2) "hard" rays permeating the very upper reaches of the atmosphere and consisting mostly of high velocity protons and heavy mesons. Although considered to be numerous in the gases surrounding stars, positrons usually escape detection in our ordinary habitat by being immediately absorbed and canceling each other out in a minute flash of gamma energy by collision with any old free electron around. Their life is very brief, being only a fraction of a second.

The usual symbol encountered in textbooks for the positron is "e^{+}" although recent publications are tending towards the use of the symbol "e," which is compatible with the idea of the existence of anti-matter. The positron is a lepton, just as the normal electron is.

This novel notion of anti-matter needs an explanation. As we already mentioned in both Chapters 7 and 8, Paul Dirac was responsible for it. In 1928 he had written down his equation for the energy of a particle of finite size under Einstein's conditions for Special Relativity, as shown below :

$$E^2 = c^2 p^2 + m^2 c^4$$

For a given momentum "p" and mass "m" and solving for the energy it becomes obvious by simple algebra that there must be two possible root answers, either plus or minus. Dirac, with a single minded and tenacious adherence to the implications of his mathematics, argued there might be the possibility of what he called anti-particles of opposite sign, whether electrons or protons or anything else. When challenged on how negative energy solutions could have any real meaning in a physical sense, he countered with an imaginative abstract explanation. He claimed that the empty vacuum of space as we know it is actually a "sea" of negative energy which is fully occupied by our normal and negatively charged electrons. However, if a normal electron is excited with the input of enough positive energy to become a positron, it would leave a hole in the hypothetical sea of negative energy. Space thus is conceived as having the possibility of both positive and negative energy with a firm boundary in between. As Tony Hey puts it in his book *The Quantum Universe*, vacuum space suddenly becomes a bubbling soup of virtual particles and particle-antiparticle pairs winking in and out of existence. There were justifiable objections to such a picture, but, nevertheless, Dirac adamantly stuck with the courage of his convictions and stated in May of 1931, speaking of electrons, that "A hole, if there were one, would be a new kind of particle unknown to experimental physics and having the same mass and opposite charge of the electron." Carl Anderson proved him correct.

If one were to presume to gauge the initial reception of Dirac's "sea" concept by many physicists of his time, it probably was "Well, it certainly is an interesting idea; but let's not take it too seriously right now." The evolution of events later compelled them to confront the problem more soberly. Moreover, the cosmologists began to take his proposition seriously when they got to talking among themselves about the "Big Bang" theory of the origin of our universe. There happens to be an extraordinarily powerful source of gamma ray radiation located close to the center of our particular galaxy. The area is called "The Great Attractor" and it is speculated that it might be an enormous "black hole" or singularity. Radio receiver dish antennae have measured sporadic bursts of energy at a level of 511×10^3 electron volts. This happens to coincide with the calculations for the annihilation of positrons by electrons.

As one might expect, the discovery of the positron anti-particle created a world sensation. Science fiction writers had a field day writing about anti-matter galaxies encountering ours or even anti-matter aliens invading our planet with explosive results. Nonetheless, the fundamental question remains: "Why?."

8.3 Mesotrons and Muons

Showers from cosmic rays also produced evidence for a new and unknown particle which behaved somewhat like an electron, but having a significantly greater mass. This was first called the "mesotron" for middle particle between the electron and the proton, but it later became permanently known as the "muon." Roughly three quarters of all the secondary particles resulting from cosmic rays or their fragments which are detected at the surface of the earth are muons.

In 1937 Seth Neddermeyer and Carl D. Anderson, the same man at the California Institute of Technology who received the Nobel prize for discovering the positron, announced a new discovery of an unknown particle which behaved like a high velocity electron, but which had a calculated mass some two hundred times greater. Their method this time was to insert a one centimeter thick platinum plate inside the cloud chamber and then measure the energy loss as this unknown particle emerged from the other side. Many were slowed down, but some kept right on going, which was an indication of a heavier mass than the electron. Later and more precise measurements indicated a mass of 207 times that of the electron or 105.7 MeV. No one knew what it could be.

It happened that two years earlier a highly respected Japanese theoretician named Hideki Yukawa had published a paper in which he proposed that the strong force which held the protons and neutrons together in the nucleus at very close ranges might become evidenced as a still undiscovered particle, which he estimated might have a mass of two hundred times that of the electron. In later and more developed language he was predicting a boson or virtual particle which skips back and forth transferring charge and thus holding the nucleons together. A neutron changes into a proton with a negatively charged meson and the proton changes into a neutron with a positively charged boson. Some people have raised the crude analogy of two people passing a medicine ball back and forth between them; but this picture does not really appeal to the author. At any rate, it is not at all surprising that other physicists immediately associated Yukawa's theory with the heavy electron when it was first announced.

Unfortunately, the laboratory results did not jibe at all with the supposed boson carrier of the strong force passing through the platinum plate unhindered with no reaction whatsoever with the barrier of neutrons and protons it represented. This was no way for the carrier of the strong force to behave at all. Later research developments compelled the scientific fraternity to conclude that the Neddermeyer and Anderson particle was unique and most peculiar.

This particle's earlier name of the "Mu" meson was ultimately abandoned for the name "Muon," which was derived from the Greek letter "μ" The custom of designating the rapidly proliferating mesons by Greek letters had begun. Moreover, the fact that its mass was still so close to that of an electron certainly did not justify the category of "middle," although it is sometimes called the "fat" electron. It is now considered a lepton.

The muon has a spin of 1/2 h, which makes it a fermion and prohibited from occupying the same space in the same state as another fermion.

Why the muon should exist at all is an unsolved puzzle and a clue to the structure of the universe that we really do not fully understand. The very famous and frequently repeated remark in many textbooks of Isadore Rabi of Harvard when told of the discovery of the muon was "Who ordered that?." His involuntary remark was not merely a flippant or facetious witticism, but most probably an honest reaction of surprise and dismay to an completely unexpected complication to existing theory. The muon is a peculiar redundancy to the electron, and it also introduced the new idea of "families" of certain particles having similar properties, but different masses. The muon helped to shatter the innocent assumptions made prior to 1940 that the atom was simple and that the basic elemental component particles were few in number and inherently stable.

The normal decay mode of the muon is into an electron and two neutrinos, as shown below:

$$\mu \Rightarrow e^- + \nu_e + \nu_\mu$$

The scientists had expected it to decay into an electron and emit a photon of gamma ray energy but this mode seems to be forbidden. Again, the reason is obscure. All we know is that the electron is stable for eternity, so far as we can visualize it, but the muon is entirely transitory. Does its existence hint at the possibility of another world or is the reason merely that its mass conflicts with some basic law ? The great majority of muons which appear naturally on the earth are the by-products of the decay of heavier mesons such as pions and kaons.

The mean life of a muon as an observable entity by itself and moving relatively slowly, if not at rest, is a quite precise and observable value and averages around 2.197×10^{-6} seconds. Knowing this and realizing it had an electrical charge, the physicists at the CERN facility in Switzerland decided to accelerate a beam of them electromagnetically in their large ring to velocities approaching that of light and then to measure their duration of life in a cloud chamber before instantaneous decay. They found, not unexpectedly, that the mean life span had increased, thus proving Einstein right once more.

8.4 The Pi - Meson or Pion

"Pi" stands for the Greek letter "π" .

The pions are the next to lightest of the entire group of mesons, the muon just discussed having the least mass. Pions have the strange property that they come in three different varieties with different electric charges: positive, negative, and neutral or zero. Although their masses are not precisely identical, they average roughly 270 times that of the normal electron. This amounts to one ninth the mass of the proton. All varieties of pions have a spin of zero, which places them in the boson category.

Much of the research which confirmed the existence of pions was done by using the energy of cosmic rays to disintegrate individual atoms in a special thick emulsion coating on photographic plates carried by airplanes or balloons. Moreover, high velocity particles of heavy nucleii or baryons from outer space, when they encounter an atom of nitrogen in our atmosphere, may cause it to fragment and yield a free proton and a neutron thus leaving an isotope of nitrogen behind. The pion does not stay around very long. In a period from 1947 to 1948 a collaborative group of scientists headed by Cecil F. Powell at the University of Bristol in England published the results of their extended research into the nature of cosmic rays. They had been placing high stacks of photographic plates with silver nitrite emulsion on the tops of mountains, notably Pic du Midi in the French Alps and Kilimanjaro in Tanzania. At the latter the altitude reached 19,000 feet.

The pion satisfies quite nicely the precise predictions made by the Japanese theorist Hideki Yukawa back in 1935 long before its experimental discovery. The pion is supposed to be the carrier of the strong force between the protons and neutrons in the nucleus of the atom binding them together. Pions could be emitted or absorbed, as the case might be, by either protons or neutrons. Yukawa, at first, also believed that the pion might have something to do with the Beta decay of the neutron, but subsequent theory demanded a boson with a much greater mass.

The discovery of the pion removed all doubt about the muon being entirely unique unto itself. The discovery of these new pion particles at the end of World War II marked a very much more complicated stage in nuclear physics and quantum mechanics. Mesons of all kinds proliferated. Speculation was rife and the notion that mesons represented the carriers of the strong force became popular. Proton to proton attraction gives rise to a neutrally charged pion. Proton to neutron yields a positively charged pion. Neutron to proton strong force emits a negatively charged pion. The force thus exerted by these virtual pions can only be exerted over a minute distance of 10^{-13} centimeters or one half Fermi.

It is worth noting that the free pion, being a boson of the correct charge, can react with any available protons with which it may collide to produce an entirely new reaction which may transform the inadvertent proton into a neutron. Thus:

$$\pi^+ + p = \pi^+ + \pi^+ + n$$

In 1951 a team consisting of W. K. H. (" Pief ") Panofsky, R. L. Aamodt, and J. Hadley conducted experiments in California shooting negatively charged pions at a hydrogen target and getting a resultant neutron in combination with either a gamma ray photon or else a zero charged pion. This work helped to confirm pion energy levels. Thus pions gained respectability and official acceptance at the conference of "high energy" physicists at Rochester, New York, in January of 1952. However, a really coherent explanation for the three possible forms of the pion still bewildered the scientists and instigated a great deal of thought and conjecture. The ultimate result many years later was the radical idea of yet another list of sub-elementary particles inside the proton and neutron, but we will reach that story later.

The three common forms of the Pi- meson are:

(1) The positively charged pion having a mass of 139.6 MeV and a mean lifetime of 2.60×10^{-8} seconds. In more recent years it has become regarded as the anti-particle of the negatively charged pion.

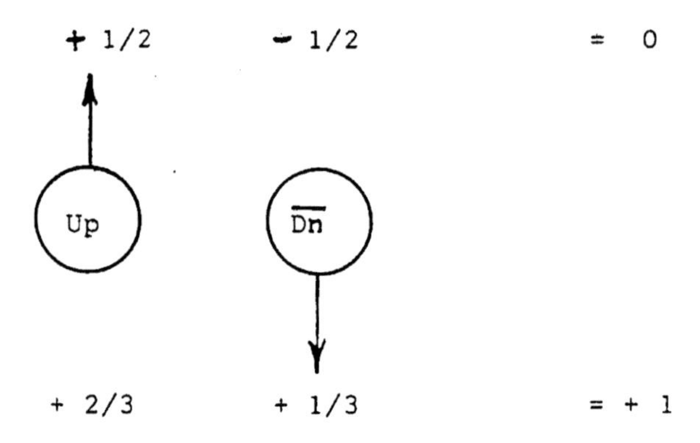

The negatively charged pion also has the same 139.6 MeV mass and a similar mean lifetime of 2.60×10^{-8} seconds.

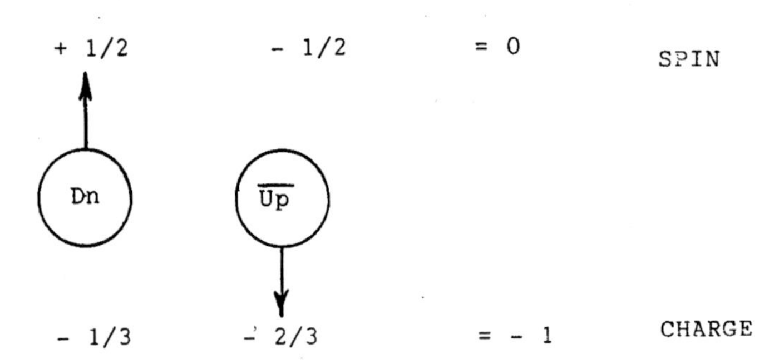

The neutral pion with a zero charge has a mass of 134.97 MeV and a mean lifetime of only 0.84×10^{-16} seconds or less than half that of its two siblings. This extraordinarily short life means that the neutral pion almost instantaneously decays into two gamma ray photons. If the result of cosmic rays, it usually decays before even reaching our atmosphere. Since it posseses no charge, it is almost impossible to track in standard detectors. Physicists consequently hit upon the scheme of interrupting its presumed path with lead plates or tantalum foil and then photographing the showers of electrons and positrons produced by the high energy photons.

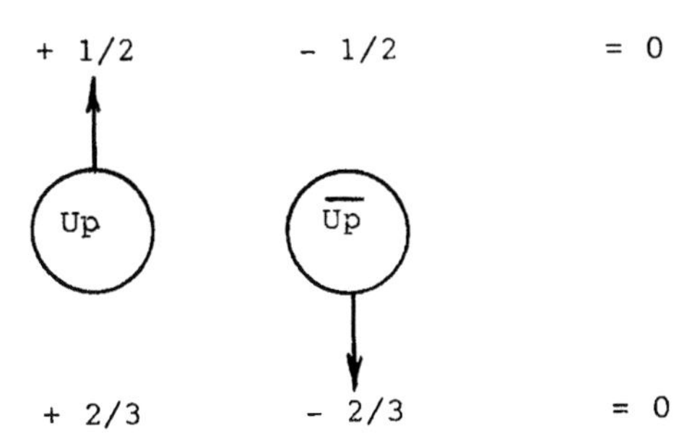

The thoughtful reader may have deduced that all these mesons described are a combination of only two quarks, with one of them being an "anti" quark with a reversed electric charge opposite to the normal.

The slightly heavier and electrically charged pion particles, when isolated by themselves, spontaneously decay into a mu lepton plus a neutrino. The lighter and neutral pion usually decays into two photons with the mass being converted into pure energy. Thus:

$$\pi^+ = \mu^+ + \nu_\mu$$

$$\pi^- = \pi^- + \nu_\mu$$

$$\pi^0 = \gamma + \gamma$$

In all these decay reactions material particles shed energy in various forms in order to return, allegorically speaking, to their original form in the heart of stars. It is important to realize that the author has a general policy of giving only the most common method of disintegration known; and many deviations to reach the same ultimate end are often possible.

The "K" Mesons or Kaons

The study of the many and various secondary particles produced by high intensity cosmic rays continued apace after World War II at many universities. During the late 1940 decade a great number of exotic and quite unexpected particles were identified. Among them was a group called the "K" mesons, all of which had a mass of roughly 1000 times that of the normal electron.

The first identification of the kaon particles is generally attributed to George Rochester and Clifford Butler, who were both graduate student assistants to Professor Patrick Blackett in Manchester, England. Blackett was of a very practical research turn of mind, when he moved to the college in Manchester in 1937, he had brought with him a large eleven ton electromagnet. The energy restrictions during World War II postponed any real research operations employing this magnet until 1946. Butler, in the meantime, had devised an ingenious method of projecting the particle tracks obtained in photographs of a cloud chamber onto a screen and then measuring their exact degree of curvature under the magnetic field by means of a glass prism. His associate, Rochester, built a target apparatus of a thin lead plate with Geiger counter detectors located on both the top and bottom of the cloud chamber. This entire apparatus was inserted between the poles of the large magnet so that the observers could thus determine the direction from which the high energy particles came and their intensity. A third thin lead plate remained inside the cloud chamber at is center in order to slow the cosmic particles down and to produce secondary mesons by collision. Tracks thus photographed in the cloud chamber obviously represented very energetic outside particles.

By October of 1946 they were startled to see the appearance of a forked or "vee" shaped track below the lead sheet in the cloud chamber which had no obvious cause or preliminary track. They quickly conjectured that the vee had to represent the spontaneous decay of some hitherto unknown particle with a mass of at least 800 times that of the normal electron which resulted from a collision which had taken place in the lead plate below. During some 1500 hours of equipment operation they obtained about 5000 photographs in all. They selected only two for their utmost clarity in their official and joint publication in the December of 1947 issue of *Nature* magazine.

The laboratory staff then decided to move the entire cumbersome apparatus to the site of an astronomical observatory 2850 meters high on the top of a mountain called Pic du Midi de Bigorre in the Pyrenees range of Southern France. Their intent was to obtain more and better pictures of the vee or "K" phenomenon; but the physical transfer and set–up of all the equipment took more than two years.

In the meantime Carl Anderson and Eugene Cowan entered the game in White Mountain, California, and confirmed the existence of the same vee tracks with twenty eight pictures. Eventually, Rochester and Butler came back in with some 10,000 cloud chamber photographs taken between July of 1950 and March of 1951. Although some few of the prints indicated the disintegration of a particle having a mass greater than that of the proton, most were kaons. The latter had become an established fact and were new and different from anything discovered previously. Although similar to their cousins, the pions, in coming in three different varieties of charge, these new particles had a mass three times greater than the pion. The electric charges may be positive, negative or zero (neutral). Their spin was zero or identical also.

The positively charged kaon has a mass of 493.7 MeV and a mean lifetime of 1.24×10^{-8} seconds.

The negatively charged kaon has a mass of 493.7 MeV and the same mean lifetime duration as its brother.

The neutral kaon with no charge whatsoever has a mass of 497.7 MeV and a mean lifetime of 0.89×10^{-10} seconds, which is significantly shorter than that of the two charged versions.

All the kaons were eventually categorized by physicists as "strange," along with other particles which we have not yet discussed. The reason for this nomenclature originally came about because they were expected to have average lifetimes in the order of 10^{-23} seconds, which would be representative of strong force decay reactions. Instead, as we learned, they actually have a relatively long lifetime in the realm of 10^{-10}, which would normally indicate a weak force reaction. This was a perplexing anomaly, but the scientists decided to give the feature a "strangeness" quantum number of one (1) and go on with their work in hope of a good explanation later.

In 1954 a giant stack of 250 different sheets of photographic emulsion was flown for six hours at an altitude of more than sixteen miles above the earth's surface. The result was not merely a clear identification of the existence of kaons, but also a statistical knowledge of their methods of decay.

The positive kaon most usually decays into a muon and a muon neutrino. The scientists had rather expected it to produce an electron.

$$K^+ = \mu^+ + \nu_\mu$$

A second favorite way for the positively charged pion to break down is into two pions, one positively charged and the other neutral.

$$K^+ = \pi^+ + \pi^0$$

As we stated previously, mesons have the quixotic capacity to disintegrate back down to energy in a multitude of different paths, and the positive kaon is a prime example. It occasionally displays some twenty–five other modes of decay of much less probability of occurrence. Obviously, this would complicate their detection and identification. Remembering that the muon has both an electrical charge and a polarization spin of one half, the topmost and most common decay behavior later became a valuable research tool for the physicists to verify their mathematical theories with particular respect to something called "parity" or charge conjugation in Nature.

The decay of the negatively charged pion is quite similar to its positive sibling, except that the charges of the by-products of lesser mass have conjugate or opposite electrical charges.

The neutral kaon seems to be quite ambivalent, one half of the time going one way and the other half of the time flopping differently. As a general expectation, the two most common modes are shown below:

$$K_o = \pi^- + \pi^+$$
$$K_o = \pi^o + \pi^o$$

The reader will undoubtedly find all this information both confusing and superfluous, but he or she should be aware these behaviors were not only typical for mesons and other nuclear fragments, but also provided the scientists with clues as to the internal construction of nucleons themselves. Constant study of hundreds of similar cascades provided them with a long and growing list of ephemeral particles to catalogue, but few answers on why they happened. Thus the 1935 to 1945 era produced a genuine intellectual ferment of ideas and hundreds of mathematical papers to justify them.

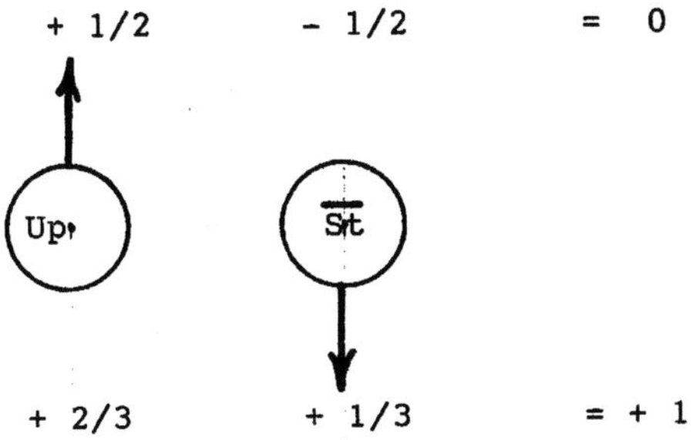

+ 1/2   - 1/2   = 0

Up $\overline{St}$

+ 2/3   + 1/3   = + 1

THE POSITIVE KAON

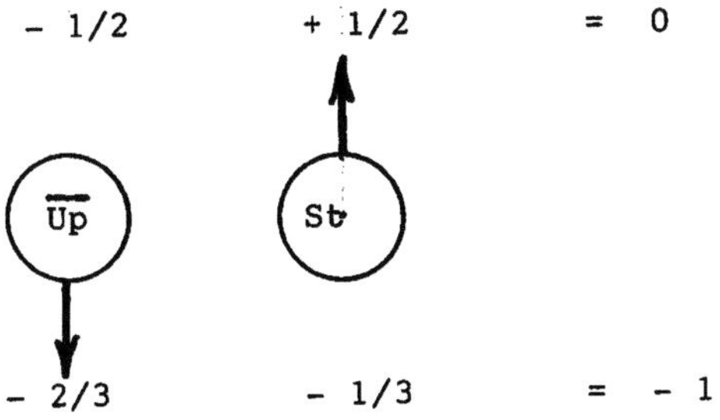

- 1/2   + 1/2   = 0

$\overline{Up}$ St

- 2/3   - 1/3   = - 1

THE NEGATIVE KAON

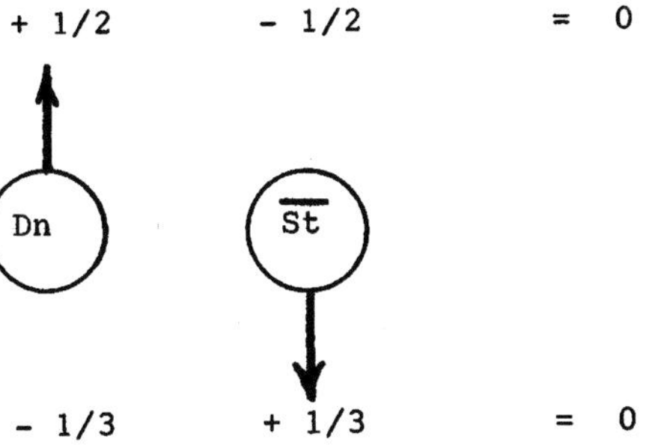

+ 1/2   - 1/2   = 0

Dn $\overline{St}$

- 1/3   + 1/3   = 0

THE NEUTRAL KAON

Chapter 9
The Insubstantial and the Ephemeral

Section 9.1

As the reader learned from Chapter 6, the neutrino became necessary as an intellectual subterfuge for physics theoreticians in order to balance their reaction equations for compliance with the Law of Conservation of Energy. However, proving what they assumed to be actually real and not merely conjectural turned out to be quite difficult.

The symbol for the neutrino is the Greek letter "ν" or "Nu."

While a professor at the University of Zurich in Switzerland, Wolfgang Pauli became interested in the phenomenon of " Beta Decay " of the neutron. In 1930 he argued with the eminent Neils Bohr that the absence of a complete energy balance in its apparent decay was not an excuse to abandon the Conservation of Energy principle and that, in addition to the electron, there had to be some other as yet unknown particle. Enrico Fermi was responsible for suggesting the name "neutrino" or "little neutral one" from the Italian point of view. By Pauli's own admission he is quoted as saying: "I have committed the ultimate sin, for I have predicted the existence of a particle that can never be observed." By that he meant it would have no charge and an infinitesimal mass and hence be difficult to detect.

In addition to its other peculiarities the normal electron neutrino also has a left-handed spin of 1/2 (h/2π), which means that, when looking at it from its direction of motion, it is rotating clockwise (or conversely, looking in the direction in which it is moving, it would appear to be rotating counter-clockwise). This makes it a fermion. The anti-neutrino is right-handed and thus exactly the reverse. They cannot have the same spin direction, which once more confounded the physicist's expectations until it was later demonstrated that non-conservation of parity could make this possible.

For a long time the actual mass of the electron neutrino (ν_e) was the subject of furious debate; and there were ardent arguments that it had to be zero. However, the *Physical Review D*, volume 50, in 1994 listed its mass as

5.1 electron volts maximum, which is so minuscule as to be almost nothing. In order to reach some agreement, a series of eight or more expensive laboratory tests were made at different universities all over the world. The results ranged from sixty electron volts down to zero before the current official value was adopted. Just how a particle which has very little or no mass and no discernible radius can have a spin or magnetic moment is not explained, but it is merely accepted as one of its quantum numbers necessary to identify its properties. The same thing is true of its momentum, which makes the measurement of its energy possible. They must certainly be a first cousin to the photon because many of them move at close to the speed of light. Being so extremely tiny and having no charge, neutrinos are totally indifferent to electromagnetic forces or even the strong force. As a result they can pass through matter, including atoms or molecules, as if they were mostly empty space. Floods of millions of neutrinos from the sun pass entirely through the earth every day. Only at very rare intervals, when they have been slowed down by dense barriers and when they come close enough to an electron or other particle, will they become captured by the weak force.

Consequently, the actual laboratory proof of the neutrino's real existence did not occur until 1956. Two physicists, Frederick Reines and Clyde L. Cowan, Jr., who were still working at the Los Alamos National Laboratory in New Mexico after World War II, decided to attack the impossible. Their very first notion was to use the program then current in 1952 for testing atomic bombs as a high flux source of neutrinos and look for the reaction where a neutrino collision changes a proton into a neutron plus an electron. This is the inverse of Beta decay. They planned to suspend their detector in an ancillary excavated shaft adjacent to the test well, but this scheme was abandoned as being too dangerous and a bureaucratic nuisance. Instead they went to the United States government breeder reactor at Hanford, Washington, to work with their detectors in an unheated building for a time during the winter of 1953. The background radiation to which they were exposed there also made their results too confused, so they moved to the government's Savannah River complex near Aiken, South Carolina, instead. The plutonium fueled reactor there at that time produced 700 megawatts of power with an electron-antineutrino flux of about 10^{13} per square centimeter per second! Just as important, the researchers were provided with a small room deep underground and only eleven meters from the center of the reactor. They then went to work in their cramped cell on "Project Poltergeist." Their detector was two tanks of 400 liters of pure water containing dissolved calcium chloride placed between three liquid scintillation detectors. For cheap shielding they adopted wood sawdust wetted down with water. Their theory was that when the positron created by the impact of an anti-neutrino on a proton became shortly annihilated by an electron, the resulting two photons of energy produced a flash of light which registered on the scintillation counters. The usual newly created neutron might possibly also be slowed down and merge with a cadmium nucleus seconds later and produce a second flash. By June of 1956 they became convinced that they had observed the expected neutrino reactions and sent a telegram to Wolfgang Pauli. His answer was: "Everything comes to him who knows how to wait."

Thirty five years later the Nobel prize committee became convinced and awarded Reines his honor, although shared by Milton L. Perl for his discovery of the Tau lepton.

Neutrino research became big business in physics with many nations subsidizing the experiments. The sheer number of expensive facilities for the measurement of the neutrino barrage from space which eventually appeared over the years in many countries of the world is impressive. In addition to the fascination of theoretical physicists with the Weak Force, Beta decay and the question of "mirror image" parity, a very considerable impetus for these expenditures came from astronomers and cosmologists. Astrophysicists were enthusiastic about measuring the neutrino radiation from our sun and stars beyond for two major reasons: (1) the intensity of the neutrino emission would give them a good clue as to the exact nature of the fusion energy reaction in the heart of the sun and other stars and, moreover, might explain the sudden explosion and then gravity imploding of super-novae; and (2) they had become convinced that there had to be a very great deal of "dark matter" in the universe which they could not see in their telescopes. This conviction arose from studies of the expansion rate and movements of our own galaxy and the recession of other galaxies from one another in the entire universe. The rotation of nebulae, furthermore, did not conform in the least to their calculations for gravitational coherence. It had become very popular, at first, to believe that this unknown dark matter could be explained by clouds of neutrinos circling at the outer fringes of these nebulae, assuming that neutrinos had mass. Neutrino research was ultimately to prove discouraging to all these speculations and lead to sober reconsideration.

A pioneer on the subject of neutrino detection was Dr. Raymond Davis, Jr., who was publishing papers on the feasibility of measuring solar neutrinos as early as 1955. He had originally become interested in the difference, if any, between a neutrino and its anti-neutrino. He also had read a much earlier suggestion that the heavy isotope of chlorine (designated number 37), which occurs naturally and has 17 protons and 22 neutrons, might make a good neutrino detector. The theory was that an energetic neutrino impacting on one of the neutrons in this particular atom might change it into a proton and thus transform the atom into ordinary chlorine, which is chemically distinct and identifiable. Davis selected carbon tetrachloride ($C\,Cl_4$) for tests at the

Brookhaven National Laboratory on Long Island in New York; but the results were not encouraging.

Subsequently, in 1962 Davis was recruited by Willy Fowler of the Kellogg Radiation Laboratory at the California Institute of Technology for studies in astro-physics. He immediately there formed a friendly and fruitful partnership with John Bahcall from the University of Indiana. Both their fortunes rose when the Kellogg group received a large grant for the construction of a solar neutrino detector. During a solicitation for proposals the Homestake Mining Company made a financially attractive offer to excavate a large cavern 1500 meters deep underground near their gold mine at Lead in the Black Hills of South Dakota. The construction of this cave began in 1965 and took only two months to complete. The purpose of the subterranean location was to minimize any background radiation or stray mesons from the earth surface, a precaution which became standard in all research for neutrinos. The cavern was designed to house a large and gas–tight tank made of special non-radioactive steel by the Chicago Bridge and Iron Company. The tank itself measured 47.6 feet long by 6 feet in diameter and was intended for the storage of 10,034 gallons of perchloro-ethylene fluid ($C_2 Cl_4$) which took the Frontier Chemical Company from Wichita, Kansas, five weeks to fill with such a rare fluid.

After the installation of pumps and piping and liquid helium storage, the facility went into operation in 1966. During actual operation without shut-downs the entire cavern was often filled with ordinary water as an additional radiation shield. The actual chemistry was quite intricate, which caused Bahcall serious worries, but not Davis. He expected the gradual accumulation in the tank of an isotope of argon (number 37) in gaseous form over a period of two to three months due to neutrino collisions. The tank was then flushed out, using helium gas as the solvent for argon. The gas mixture was then passed through a charcoal trap and cooled there by liquid nitrogen to a temperature where the argon liquefied and was thus separated out. The amount of argon atoms were then measured by means of an electrical ionization counter. The amazing thing is that Davis and Bahcall made the operation work almost continuously for twenty–five years with their constant cooperation and and they soon became confident of a consistent yield of one atom of argon 37 appearing every two or three days. Their published results first appeared in *Physical Review Letters* in 1968. Although the experiment worked satisfactorily in principle, the actual quantity of neutrino events measured was sharply lower than expected by the astro-physicists by some 50 to 60 percent. This was disheartening to the cosmologists and triggered some serious questions about solar fusion dogma. It also encouraged a trend toward the construction of gallium detectors.

The Japanese universities and their National Laboratory for High Energy Physics, in the meantime, had already devised an elaborate experiment which was set up 1000 meters underground in a mine in the mountains 300 kilometers west of Tokyo. The facility had originally been designated for an investigation of proton decay; but as the reader has already learned, this possibility was so statistically remote as to discourage even the most patient scientist from sitting around for his lifetime in order to witness it. In 1984 the Japanese invited a group of physicists from the University of Pennsylvania to discuss the modification of their detector to measure neutrino flux densities instead. The cheapest alternative decided upon was a large tank of very pure water and the new effort was christened Kamiokande II. The tank was absolutely huge, being 15.6 meters in diameter and 16 meters high. The top, bottom and circular sides were festooned with hundreds of 19 1/2 inch diameter glass portholes housing light sensitive photoelectric sensors.

After learning how to recognize and discard many flashes of light which were spurious signals from cosmic ray muons or gamma rays, the facility started serious operation in 1986 with some confidence that they were actually detecting neutrino collisions. It happened that on February 23, 1987, a large blue giant star of 12th magnitude named Sanduleak -69° 202 in the Greater Magellanic Cloud in the southern hemisphere skies suddenly blew up and turned into the most recently observed supernova of unusual brightness witnessed in 400 years. It had been acting in a peculiar manner previously and astronomers in Australia were in a privileged position to observe the progress of the gigantic explosion. To the joy and gratification of all astro-physicists and experimenters alike a very unusually strong burst of neutrinos was recorded then by both the Kamiokande II detector and a similar installation belonging jointly to Irvine at the University of California, the University of Michigan, and the Brookhaven National Laboratory, abbreviated to I. M. B. The latter large tank of purified water was located 600 meters below ground in a salt mine belonging to the Morton-Thiokol Corporation in Fairport, Ohio.

With respect to the neutrino bombardment from the sun, however, the delight was moderated to chagrin. Published reports from the Japanese in 1990 showed the number of luminescent reactions caused by neutrinos in their tank was less than one half that predicted by the current theories on solar fusion. Undeterred, the Japanese initiated plans to build a new giant water detector having thirty times the volume!

By this time the Europeans had become interested. Till Kirsten of the Max Planck Institute for Nuclear Studies managed to persuade the West German government to advance twenty million deutchmarks or almost one half the estimated cost for 30 tonnes of rare Gallium metal, or more than the world production at the time. A Russian named Vladimir Kuzman was responsible for convincing the German consortium that Gallium-71 with 31 protons and 30 neutrons would be a better detector for low energy neutrinos such as expected in proton combinations at the center of the sun. It also has the rare property of turning into a liquid at any temperatures higher than 86 degrees Fahrenheit (30 degrees Centigrade). The Italian government did its part in 1981 by agreeing to excavate a cavern out of solid rock off to the side of a tunnel for automobiles being constructed right through a range of mountains named the Gran Sasso Masif about 75 miles north of Rome. This would place the new laboratory under 1200 meters of rock for shielding. The cylindrical steel tank itself was 9 1/2 meters long with a diameter of about 2.715 meters made to hold 55 cubic meters of a very concentrated solution of gallium chloride. When an atom of gallium number 71 absorbs a neutrino, it converts to Germanium, so the task was to count the number of germanium chloride molecules created over a period of time.

A consortium of international scientists was assembled in 1987 in order to operate the detector facility, which was then named "Gallex" as an abbreviation of Gallium Experiment. They began taking data in 1991. An amusing sidelight to this story is that the Italian physicists read of lead ingots being salvaged from the wreck of a Roman cargo vessel which sank around 250 B.C. off the island of Mal de Ventre near Sardinia. The scientists plotted to get the 73 pound ingots from the archaeologists to use around their apparatus as shielding from gamma rays. The reason was that such old lead would be almost inert from any self-contained radioactivity. At any rate, apparent results thus far seem to indicate a neutrino emission from the sun which is roughly less than two thirds of the quantity required by conventional fusion theory taught to students. The discomfiture of the astro-physicists was not significantly allayed.

In spite of the remnants of the "cold war," scientists from the Los Alamos Laboratory, the University of Pennsylvania, and Louisiana State and Princeton Universities in the United States were in close correspondence with the Institute of Nuclear Research in Moscow on the possibility of another gallium detector. Thus "SAGE" or the Soviet American Gallium Experiment was born with the Russians V. N. Gavrin and George Zatsepin as Directors. The Baksan Neutrino Observatory was unbelievably grand and expensive and not merely because of the excessive cost of the gallium metal. Located not far from Mount Elbrus in the Caucus Mountains, the Russians excavated a tunnel 4 kilometers long, starting in Baksan gorge or valley at the foot of Mount Andyrchi, and extending to a vertical depth of 2 kilometers beneath the mountain peak. The various laboratory caverns along the length of the tunnel were almost as huge as that for the eight detector reactor tanks themselves, four of which contained 30 tons of gallium used in the initial start-up. The installation began serious data collection in January of 1990 and the first report of the team was presented to a conference at CERN in Geneva in June of the same year. The results were even more disappointing for the astro-physicists than all the other laboratories previously in operation, since it confirmed the great discrepancy with predictions. Later the gallium itself became vulnerable to stealing by criminals.

Meanwhile, Canada came to the rescue with an entirely different form of detector which was optimistically believed to be much more sensitive than any other. Herbert Chen of the University of California had proposed the use of heavy water or deuterium oxide. Canada was attractive because it had a surplus of heavy water made for nuclear reactors and, moreover, the Creighton Mine at Sudbury, Ontario, was available free of charge with an underground site at the bottom of a 2070 meter deep shaft at the center of an ancient meteor crater 2700 meters below ground. George Ewan from Queen's University in Ontario then became the new Director of the Sudbury Neutrino Observatory (abbreviated to S. N. O.).

The 1000 tons of heavy water were contained in a transparent spherical vessel made of acrylic plastic. The exterior surface of this sphere was then entirely surrounded by 6400 photo-electric light sensor tubes. This entire apparatus was then immersed in another 7300 tons of ordinary water, which not only provided an external radiation shield, but also added a compressive pressure envelope to structurally strengthen the plastic sphere inside.

Great hopes, which were later confirmed, were entertained for the Sudbury detector to have a much greater sensitivity to neutrinos, since three different atomic reactions involving weak neutral currents were considered possible.

(1) An electron neutrino could impact with a deuteron nucleus consisting of one proton and one neutron , thus involving a W minus boson and causing the neutron to change into a proton and have another ordinary electron be ejected from the nucleus. If this secondary electron had an energy in excess of 5 MeV or higher, it would produce a cone of violet Cerenkov radiation which would be visible to the photon sensitive tubes.

(2) Neutrinos resulting from the decay of Boron 8 into Beryllium 8 in the core of the sun which reached the earth with an energy of in excess of 14 MeV might impact with a shell electron in a deuterium atom and with the energy of a W plus boson knock it out or scatter it with a velocity great enough to register a light flash bright enough to be detected.

(3) A less probable reaction might be that a slower solar neutrino might simply be absorbed by a deuteron and thus break apart the bond between the proton anD neutron by what is called a "Neutral Currents Reaction" with the ejection of an entirely new neutrino in the process. If detected, such a reaction involving a Z zero boson would yield important clues about the

masses of different types of neutrinos which were then postulated. Although it had been hoped that this Sudbury facility would be in actual operation by year 1995, the schedule slipped to May of 1998. By the year 2001, however, the experimental results confirmed the scientists' suspicion that many of the electron neutrinos entering from space were indeed being metamorphosed into either muon or taon neutrinos or both. This evidence was also interpreted as proof that all neutrinos had to have some mass, however tiny or undetectable.

Yet another imaginative venture which started construction in 1992 is the second phase of DUMAND, which is an acronym for Deep Underwater Muon and Neutrino Detector. This is an array of optically sensitive spheres located some 4700 meters or two miles down in the black deep of the Pacific Ocean about 18 miles off Keahole Point on the west coast of the large island of Hawaii. The nine strings of photo-electric sensors are all connected to a monitoring station on the mainland by a fiber–optics cable. One must admire the bold daring in coping with the anchorage of such an array and arranging for the cable wiring in the face of heavy seas and susceptibility to storms.

By this time the French decided to enter the arena. Never have so many nations spent so much money on the study of so little a material subject, all to the great envy of their other scientific brethren who were not physicists. The new neutrino detector was given the name of "Gargamelle," which was derived from satirical stories by a 16th century writer named Francois Rabelais about a giant called Gargantua and his mother, Gargamelle.

The person who was most active and instrumental in urging the building of a huge liquid bubble chamber to be used as a neutrino and gamma ray detector in conjunction with the just constructed CERN ring accelerator in Switzerland was Andre Laggarique of the Ecole Polytechnique in Paris. The actual agreement to undertake his ambitius plan was signed by the European consortium with the French Atomic Energy Commission in December of 1995. The device was truly grandiose in concept and the designing of it required the combined brains of a large team of both physicists and engineers.

The bubble chamber itself was a large partially flattened cylindrical chamber measuring 4.8 meters long by 1.85 meters wide which would hold 12,000 liters or 18 tonnes of liquid Freon (CF_3 Br). This is the same synthetic compound of Carbon, Fluorine, and Bromine originally devised for use in refrigeration machines and which earned opprobrium as the destroyer of the ozone layer over the earth's poles. The tank was very intricately manufactured with a row of four large circular port holes or windows near its top on both sides equipped with "fish eye" lenses for observation by humans in addition to a myriad of other round and smaller openings for photo-electric sensors and motion picture cameras or floodlights. The entire tank assembly had to be fitted within the poles of a huge electromagnet above and below at the CERN site. Being a regular bubble chamber in operation, there also had to be machinery for suddenly reducing the pressure inside the cylinder mechanically whenever high velocity particles were anticipated to be streaming through the liquid contents. Five years elapsed before the detector was operating on line with a beam from the accelerator and the typical curling "ram's horn" tracks of electrons and positrons were photographed successfully. The bubble chamber did little to add anything new to the already available data about the density of the neutrino bombardment from the sun.

With respect to the quandary of the astro-physicists over the much lower neutrino flux density indicated by the majority of the detectors in operation from their original predictions, the problem remains under discussion still. The scientists are understandably reluctant to discard their carefully derived theories for the mechanics of the fusion energy in the sun and stars, but the discrepancy is real. There are many conjectural explanations, such as the neutrinos being absorbed somehow by gas clouds at or near the surface of the sun or the neutrinos somehow "oscillating" and changing "flavor"or mass as they enter the earth's magnetic field, or even being converted to muons as they enter our atmosphere.This second idea is gaining current favor. The jury has not reached a consensus on the reason quite yet, although the efficiency of the detectors does allow for some doubts about the degree of accuracy of the observational statistics on the actual distribution quantities of the transformed neutinos.

With respect to the missing "dark" mass in the universe required for a steady state or gravitational behavior of nebulae, the neutrino cloud theory is no longer quite as fashionable, since an absolutely prodigious number of them would be required. The astronomers now talk about non-baryonic Weakly Interacting Massive Particles (WIMPS) or Massive Compact Halo Objects (MACHOS), ostensibly dead asteroids or fragments of matter. It is reassuring to note that the scientists do allow themselves a sense of humor at times.

One should be aware that, quite apart from the problems of the astrophysicists with the neutrino measurements, this strange particle has other peculiarities which attracted attention. The reader may recall that back in Chapter 8 we mentioned without explanation that one may derive either electron neutrinos or muon neutrinos, depending upon the nature of the nuclear reaction by which they were produced. With the invention and development of narrowly focused beams of protons circulating between the quadrupole magnets in series it became feasible to build higher energy accelerator machines reaching to 28 GeV and beyond. In 1960 these included the Proton Synchrotron at Geneva, Switzerland, and the Alternating Gradient Synchrotron at the Brookhaven Laboratory on Long Island. Two years later at the latter facility a team of experimenters from Columbia University led by three distinguished physicists (Jack Steinberger, Leon Lederman, and Melvin

Schwartz) directed a beam of protons at a beryllium metal target enclosed in the barrel of a scrapped naval cannon. The resultant particle fragments are occasionally K mesons, but usually pions. The kaons soon break down into pions anyway. If left to themselves, the pions would probably travel about 50 meters before going into their normal decay transformation. However, placing a forty foot thick barrier of steel plates salvaged from the scrap metal from an obsolete battleship (waste not, want not) slows down and captures about 90 percent of the pions and other debris. Typically, the pions then disintegrate into muons and some form of neutrinos. The only particles which would emerge from the shielding wall were expected to be neutrinos or anti-neutrinos, and so it happened.

The neutrino detector used in this installation was quite innovative and consisted of a long series of four foot square aluminum plates one inch thick spaced at only one half an inch apart. The space between the plates was filled with an ionizing gas. The aluminum plates themselves were subjected to a very high direct current voltage difference, thus creating a gigantic spark chamber to track the neutrinos, whether an electron anti-neutrino or a muon anti-neutrino.

The question arises of how one can distinguish in the laboratory between a muon neutrino and the more ordinary electron neutrino? A glib answer would be "by the company they keep." One method is the difference in results when a slow neutrino merges with a neutron for the following reactions:

$$\text{Neutron} + \nu_{\mu} = \mu^{-} + \text{Proton}$$
$$\text{Neutron} + \nu_{e} = e^{-} + \text{Proton}$$

The neutrinos are not interchangeable. No muon has ever been seen to decay into a normal negatively charged electron and a photon of gamma ray energy. The reasons for all of this is obscure, but it has something to do with the conservation of properties or distinctive attributes. The fact that a particle so tiny and having so trifling mass could have two or more "personalities" or behavior properties is no small wonder. Their study, however, has added to a better understanding of the weak force interactions and the fundamental theory of the atom. Yet perplexities remain. If neutrinos are to fit within the restrictions of an elaborately constructed theory which physicists call "The Standard Model," there should be at least three different varieties of different masses. What bothers the researchers most is the possibility that these slightly different neutrinos may indeed exhibit a propensity to transpose or change into one another under certain circumstances and thus not be stable.This idea was named "oscillation"; but it cannot happen without mass. That is most disturbing to their first fundamental assumptions about the sun neutrino radiation, but does indicate that all varieties possess some mass, however tiny it may be, and this needs to be verified.

The mass of the muon neutrino is expected to be 270 electron volts or slightly more than its electron cousin. It also has a spin of one half. The early discovery of the positron or "anti-electron" quite logically encouraged the idea of other "anti" particles, or mirror reflections of the more commonplace and stable particles. Since the neutrino has no charge whatsoever, the requisite anti-neutrino for symmetry requirements must have a spin in the opposite direction. More specifically, its rotation by the old-fashioned hand and fingers rule for the direction of an electrical current in a wire should indicate a counter-clockwise rotation with the "thumb" pointing in the direction of its motion of translation.

The symbol for the anti-neutrino is ν .

Much more spectacular was the decision by the Japanese to build a new "Super Kamiokande" of gigantic proportions. Actual construction started in December of 1991. It was located in a deep cylindrical pit excavated into rock at the bottom of a zinc mine.The walls and bottom of this huge forty meter diameter hole are lined with stainless steel plates in order to retain 50,000 cubic meters of purified water. Contained within this outer cylinder is a second cylindrical cage of metal which is separated from the outer walls by approximately three meters all around. This framework, which includes the bottom and top of the pit as well, is intended to support 11,000 photo-multiplier light detector tubes with lenses twenty inches in diameter. Test operation began in 1996. After eighteen months of ensuing experimental operation, Yoji Totsuka, the Director of the Kamiokande team, announced their tentative results to an enthusiastic crowd. Muon neutrinos produced in the earth's atmosphere and, in particular, those from the bottom side of the planet from Japan indeed show by their numbers evidence that the incoming neutrinos must have a tiny mass and some were changing into either muon or taon neutrinos along the way to our detectors..

9.2 The Anti-proton

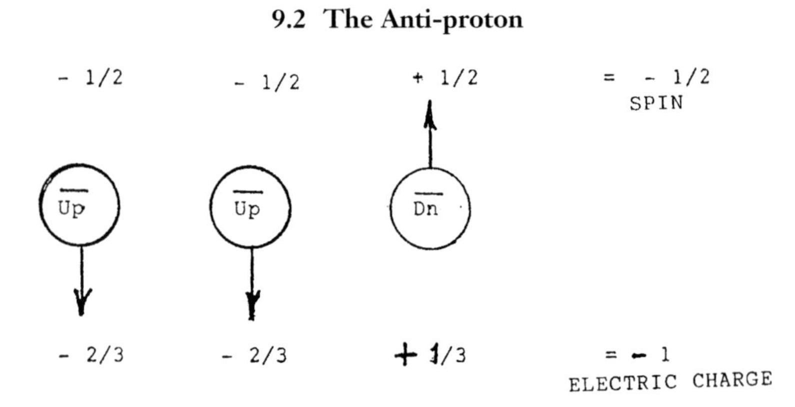

The symbol for the anti-proton or negative proton is "p̄."

By 1951 the conviction that the proton should have its corresponding anti-particle, however ephemeral its existence might be, was gaining credence. The proven existence of the positron back in 1932, as well as the positively and negatively charged muons and pions and the prevailing notion of symmetry in all the particle grouping, conspired to encourage a determined search. The reader may remember that Paul Dirac initiated this idea. When challenged to justify his complete faith in the mathematical plus and minus signs with a physical picture, he came up with the notion that what we call the vacuum of space is really a "sea" of negative energy. This clearly contradicts the older classical assumption that the vacuum of empty space is a featureless, passive void of nothing whatsoever.[1] Although many scientists remained dubious of his explanation, the idea nevertheless achieved a wider acceptance in the scientific community over time. Thus the new assumption being circulated today is that vacuum represents an arbitrary barrier in the ground state of minimum energy for which permanent material particles in our "real" world can exist, so far as we can experience or sense it. Notwithstanding our intuition to the contrary, it is asserted that lower "negative" energies are possible and that quantum oscillations come into existence at random by the mechanism of the Uncertainty Principle and thus produce virtual particles and anti-particles in the wink of an eye. The supposed vacuum thus becomes a teeming and invisible "sea" of anti-particles created simultaneously with the normal particle in pairs on the other side of the barrier which constantly annihilate each other to revert to ground zero energy almost instantaneously. If this explanation seems weird to the reader, be assured that you are not alone. There are other hypotheses with which you might feel more compatible.

The search for the anti-proton was very much in the minds of the designers of the new "Bevatron" accelerator machine at the Lawrence Berkeley Radiation Laboratory in California, which became operational in 1954. The theorists had calculated that a minimum beam energy of 5.7 giga electron volts would be required for that purpose. The actual search for the anti-proton began in 1955 with beam energies ultimately approaching 6.2 GeV. The solution for getting there was an ingenious harnessing of machines in a series.

The initial protons to be accelerated were obtained from an old-fashioned Van de Graaff electro-static belt generator at the small kinetic energy of three mega electron volts. They were then fed into the "doughnut" synchrotron ring of the so-called "Cosmotron" accelerator, which was to be abandoned later in 1966. There the protons were raised to an energy of 3 Giga electron volts and, in turn, fed into the newly constructed and larger

1. G. D. Coughlin and I.E. Dodd, The Ideas of Particle Physics, (Cambridge University Press, 1984).

"Bevatron" ring designed for this purpose and where they achieved a kinetic energy of around 6 GeV. This pulsed beam of high velocity protons was then directed onto a copper target, where the collision with stationary protons or neutrons created a veritable shower of particles consisting primarily of other protons, a cloud of negative pions and a few anti-protons. Clearly, the detection and reliable identification of the tracks of the anti-protons became a crucial task with only a one in forty–four thousand chance of finding the anti-proton among the cloud of pions. The success of the effort thus depended upon weeding out the tracks of the pions with a mass lighter than that of a proton and also any kaons with a mass greater than that of the proton. Calculating momentum and velocity differences from different tracks of the fragments became the basic identification technique. A series of magnetic spectrometers, scintillation detectors, and Cerenkov radiation detectors were necessary. One technique for determining velocities was the placement of two Cerenkov ultra-violet radiation counters thirty–seven feet apart and measuring the time differences between arrival and departure of the track. As another back-up check the scientists measured the emission angles as the particles passed through an organic liquid such as carbon fluoride. It was necessary to measure the velocities and length by several methods for absolute confirmation and then calculate the statistical mean from the bell curve of data thus obtained for the mass of the particle observed to check whether it agreed with that of the normal proton. The fact that the lifetime for the decay of the anti-proton is approximately 10.2×10^{-8} seconds, which is a longer life than most mesons, afforded another plausible argument for its identification.

The painstaking care and tedious repetition of meticulously conducted measurements in this experiment is quite typical of the vast numbers of hours of dedication required from good scientists in order to establish evidence to the satisfaction of their peers. By September of 1955 Emilio Segre and Owen Chamberlain, along with four other associate experimenters, were confident that they had sorted out the evidence for the creation of at least sixty anti-protons. This was judged enough to justify an announcement in the *Physical Review* journal. Their joint paper with Claude Weigand and Thomas Ypilantis of the University of California at Berkeley Laboratories appeared in October of that year. They received a Nobel prize in 1959 for their achievement in demonstrating that the negative proton or anti-particle could actually exist as a reality, however rare and ephemeral; and that it was not merely the figment of some theoretician's imagination. Their discovery was confirmed by Italian scientists in Rome in 1956.

In time and with the advent of very high powered circular synchrotron particle accelerators using narrowly constricted and focused beams made possible by huge electro-magnets chilled with liquid helium or nitrogen the technique for manufacturing anti-protons improved tremendously. In particular, the Fermi National Accelerator Laboratory outside of Chicago was financed by the United States Department of Energy with the deliberate intention of building the "Tevatron" machine capable of colliding protons with anti-protons. This was quite a daring idea. After more than eighteen years of design development and construction work on this one and one quarter mile diameter synchrotron ring, Leon M. Lederman and his large group of associates succeeded in starting up the world's first super-conducting magnet synchrotron accelerator on July 3, 1983.

Negative hydrogen single ions were first generated in what is called a small Cockcroft-Walton accelerator in which a magnetic field propelled ions (protons with accompanying surplus electrons) up to the minuscule energy of 0.75 MeV. These were then led into a short electro-magnetic linear accelerator only 492 feet long which boosted the ions to 200 MeV energy, at which point they were directed into a carbon target. This block was dimensioned to pass the protons, but absorb the electrons. The continuing protons were then fed into an older and smaller synchrotron ring 522 feet in diameter, where they were again boosted to a kinetic energy of 8 GeV. Here they were injected into the next stage, which was known as the "Main Ring" of the newly constructed machine, and were accelerated by one thousand conventional copper wire coiled magnets to an energy of 120 GeV to 150 GeV before being allowed to strike a copper target. As the reader by now may expect, this is where huge numbers of negative anti-protons were generated, along with some normal positive protons.

These separated anti-protons were focused by a cylinder of liquid lithium and then directed to a small and irregularly shaped toroid ring called a "De-buncher," where they were cooled and collected before injection into the new and large Tevatron ring having 1000 extra strength superconducting magnets. Although the early operational trials reached only 512 GeV, this colossal machine can accelerate both the normal protons and anti-protons to kinetic energies potentially approaching 900 GeV. Because the two particles have exactly opposite electric charges, they circulate in the accelerator ring in entirely opposite directions in diffuse clouds which prevent most actual collisions with each other inside the toroid. Ultimately the two beams are magnetically constricted down to a diameter of only 0.1 millimeter, which is comparable to the diameter of a human hair.The two different particle beams are then allowed to collide with each other. The total of the two beam energies in collision has been claimed to be 1.8 trillion electron volts or 1.8 TeV. The impact results in a veritable cascade of exotic sub-particles, including pions, protons, neutrons, kaons, neutrinos and some others as yet undiscovered. It thus fully justified the physicist's expectations and the great amount of federal money expended on it. Leon M. Lederman, as the Project Manager, became justly famous and received a Nobel prize in 1988.

The Tevatron machine was a direct assault upon key questions about the structural formation of the nucleons of the atom and the nature of the strong force by United States physicists in competition with the European CERN complex. With full respect for the incredibly complicated design engineering and fine machine manufacturing tolerances achieved in creating the two companion rings, the startling success of this accelerator was due to two fairly simple physics concepts. These were (1) the collision of two particle beams coming from opposite directions, thus doubling the momentum available, and (2) recognition of the truth of Albert Einstein's Special Relativity theory and the fact that the mass of a piece of matter increases with its velocity of translation as it nears the speed of light. Both these circumstances account for the seemingly impossible laboratory result that the point impact of a proton and anti-proton yields a shower of meson particles, as well as photons, electrons and muons, which in total may add up to a rest mass exceeding that of the original two colliding particles. In short, mass has been created from pure energy. The kinds of secondary particles or fragments obtained from such a collision vary in size, mass, and velocity and are not always the same in a series of identical experiments. Whether the particles hit directly head-on or only glance off in a deflection is often the important factor. Ultimately, however, all these strange particles decompose into the fundamental and stable ones with which we are already familiar.

Although the very existence of anti-protons still seems improbable to most of the world's population, a "post-doc" teacher at the University of Washington in 1984 named Gerald Gabrielse had total faith and a dream of collecting them. This imaginative and boldly daring young man, then in his thirties, (now a professor in the Graduate School of Harvard) decided he would capture a swarm of anti-protons in a bottle and store them for a time and eventually did so.

Meaningful experiments to study the behavior of anti-protons requires they be slowed down to energies less than one thousandths of an electron volt, essentially freezing them. Gabrielse conceived a method of accomplishing this task by cold storage after they left a low energy LEAR accelerator ring at a velocity approaching ten percent that of light. It would become necessary to retain them in the core of a large electro-magnet, where the anti-protons were to be frozen still at a temperature of 4 degrees above absolute zero (4 degrees Kelvin) for observation.

The technical difficulties in achieving this goal were so daunting that many physicists were dubious of success. Yet another complication was that the proposed experiment of Gabrielese demanded approval for the interruption of scheduled operation of one of the only two accelerator facilities then known which could manufacture the anti-protons for him in the first place. Fermilab in Chicago disassembled their small storage ring for anti-protons in 1984, thus eliminating that possibility. That left CERN in Switzerland. Not surprisingly, he had difficulty in securing grant money for the experiment; but eventually the team of Gerald Gabrielse and three other physicists persuaded the United States National Science Foundation and the West German State Ministry for Research. In 1986 he and Hartmut Kalinowsky from the University of Mainz, young Thomas Trainor from the University of Washington, and William Kells from Fermilab went to work building their apparatus at the CERN complex near Geneva. Workers there had scavenged parts from earlier storage rings to construct the Low Energy Antiproton Ring, or LEAR for short, in 1982. This accelerator had a modest circumference of only 79 meters, but was adequate for this job. In May of 1986 his group was given permission to access the LEAR in order to demonstrate the feasibility of his proposal. They had borrowed a super-conducting solenoid magnet which contained about 25 miles of copper wire and which was cooled by liquid helium. The great advantage of a super-conducting magnet is that a momentary current burst of 37 amperes of electricity will cause the circulation of the electrons through the coils almost indefinitely in an approach to perpetual motion. This not only saves a prodigious electric energy bill from the local power utility company but also minimizes the amount of heat losses from electrical resistance which must inevitably occur in the solenoid wires.

The heart of the entire apparatus was the anti-proton trap itself, which was the invention of Gabrielse and his friends. This was a small glass cylinder or Dewar flask scarcely fifteen centimeters long which was constructed of alternate layers of gold plated thick copper electrodes separated by glass ring spacers. Each burst of the anti-protons from the Lear accelerator was directed towards the bottom of the trap and encouraged to pass through the copper barrier by a modest positive direct current voltage of 100 volts and pass into the interior of the trap. A negative voltage of - 3000 volts was then quickly applied to the copper disc barriers at top and bottom at just the right milli-second instant. The anti-protons then started to oscillate up and down vertically in confusion and encountered a cloud of normal negative electrons and a chilled surrounding atmosphere all inside the intense focusing effect of the powerful torus magnet. They thus tended to accumulate in the center of the tubular vacuum trap and gradually lose their kinetic energy. The anti-protons and electrons are mutually indifferent and repel each other since they have the same negative charge, although they continue to interact with each other as a gas mixture. Eventually, about one out of every five thousand anti-protons which started out at a high energy of almost 6 MeV were expected to be slowed down to about one ten billionth of that figure and become virtually motionless in the middle of what Gabrielse called a Dewar flask. Those not so captured were mostly annihilated upon striking normal protons in the solid copper electrodes at the top and bottom. Because of all the dangerously radioactive byproducts thus generated, the entire apparatus with its' magnet had to be enclosed by concrete wall shields. The entire experimental equipment stood about twenty-three feet high above the second floor of the building.

After many difficulties and crises, including the arrival from the United States of a glass thermos bottle trap cracked and damaged by transport in a truck instead of the expected airplane, and internal electrical short circuits caused by the condensation of moisture, the apparatus went on line for tests on Friday, July 17th, 1986. The experiment was a success and attracted the enthusiastic attention of all international physicists. Even the attitude of the CERN administrative complex became more benevolent and allowed time for the continuation of the experiments and improvement of the apparatus. By 1989 Gabrielse and his team of seven others were able to confine 10,000 anti-protons from a single burst from the LEAR ring accelerator and hold them frozen for as long as two months. They immediately set about making measurements to determine the mass and electric charge magnitudes of the anti-proton and were thus able to confirm within an extremely small margin of error that these quantities were quite identical to the "normal" positively charged protons found inside the atoms of our world. Symmetry had again been preserved.

The confirmation of the rare existence of both positrons and negative anti-protons was a hugely important event in nuclear physics and instigated the idea of symmetry for all particles in what ultimately became known as the "Standard Model" despite their extremely brief life in our own local space and time frame. The theories of Paul Dirac and Erwin Heisenberg on the subject had been thoroughly vindicated yet a bothersome practical question still remains. If the existence of anti-particles is actually feasible, then why have they all effectively disappeared in our cosmos? Could not matter in the universe in its creation have remained a vast soup of normal and anti-particles winking in and out of their material entities ceaselessly? What broke the symmetry which might have existed in the beginning hours of the "Big Bang"?

In an entirely different vein of thought, it occurred to some bright students that atoms of anti-matter might also be possible. Specifically, would it not be simple to arrange for a positron or positive electron to bond and orbit around a negative anti-proton? Charles Munger, who was working at the Stanford Linear Accelerator in California became preoccupied with this possibility in 1992. In cooperation with a fellow graduate student at SLAC named Stanley Brodsky and in consultation with other friends at other universities, such as Ivan Schmidt in Valaparaiso in Chile, he submitted a proposal for a feasible experiment. The European institutions in charge of the

CERN facility in Switzerland became interested and assigned them some time and equipment in 1996. A series of sophisticated detectors were devised to verify such an anti-hydrogen atom after it impacted with a silicon counter and there shed its positron satellite. They claimed success the same year and published their results. Verification by others still awaits; although some scientists still regard it as a mere circus trick of only ephemeral plausibility.

CHAPTER 10
THE ADVENT OF PARTICLE ACCELERATIONS

"I have heard it said that the finder of a new elementary particle used to be rewarded by a Nobel prize; but such a discovery now ought to be punished by a $ 10,000 fine." — Willis E. Lamb in a Nobel Lecture on December 12, 1955.

The advent of the particle accelerator machines for use in an enclosed laboratory on the ground liberated the physicists for a time from the astronomers and cosmic rays on mountain tops. In their delight with these new machines they even began talking about their work as "high energy" physics, thus making the forces unleashed by Nature in the stars sound a trifle minimal.

One of the very first successful proton accelerator machines was devised by Cockcroft and Walton in Cambridge, England, around 1930. It was an electrostatic generator which built up a one and one half million direct current voltage charge on a large hollow copper ball by means of the accumulation of electrons by brushes scraping over a vertically moving insulated belt over a period of time. A sudden discharge of electrons in an artificial lightning bolt down an evacuated glass tube propelled hydrogen ions onto metal targets of lithium, beryllium, or even lead. Although reasonably effective, this type of accelerator design was eventually supplanted by other more controllable and less limited machines, although the Van De Graff generator, as it was originally called, continued to be the source of relatively low energy protons for feeding into the bigger machines for many years.

The idea of accelerating electrons or other charged ions artificially by means of machines instead of depending upon rare and very expensive natural radio-active elements received a tremendous boost from a young physicist turned inventor named Ernest Orlando Lawrence (1901 to 1958). Both his parents were teachers of Norwegian ancestry who had settled in South Dakota. His father became the Superintendent of Education for that state for a time before undertaking the presidency of a small college. As the elder of two brothers, Lawrence was sent to a small Lutheran college in Minnesota for one year after graduation from public High School, but soon transferred to the University of South Dakota in Pierre. To a large extent his educational career depended upon scholarships which Lawrence had little trouble earning as a bright student. Graduating from the State University in 1922, he followed a friend of similar scientific interests to the University of Minnesota for graduate studies. There he came under the tutelage of Professor W. F. G. Swann, an early scientist of considerable brilliance, and there completed his studies for a master's degree. Swann was quite a peripatetic personality and had achieved some eminence in the field of cosmic ray particles and the oscillating magnetic fields which surround our planet. In his later years Swann was conspicuous for a head of bushy white hair and was in great demand for popular lectures on science, frequently appearing at the Franklin Institute in Philadelphia. He also became a Director of the Bartol Research Foundation at suburban Swarthmore College. The author was once treated to a pessimistic dissertation by Professor Swann in the company of a group of power engineers on the subject of the unsatisfactorily low efficiency of gas–fired turbines for driving electric generators. Although he was quite correct in his opinion of the gas turbine's heat efficiency, this did not deter natural gas producers in the southwest of United States from employing such turbines for the purpose of driving the pumps on their transcontinental pipe lines nor did it slow down the development of jet engines for airplanes. Specialization in technological applications sometimes makes an exception to the general rule.

At any rate, Lawrence accompanied Swann when the latter moved to the University of Chicago and later to Yale University in New Haven in 1924. Lawrence completed his PhD dissertation there on photoelectric effects on potassium vapor (a cousin to the now familiar sodium vapor street lamps). He then remained at Yale as an Assistant Professor and research Fellow on photoelectric phenomenon. However, he and Swann both became keenly interested in the possibility of accelerating electrons or other ions to velocities high enough to act as "atom–smashing" projectiles. Swann was not optimistic, fearing any such charged particles, and especially electrons, would still be too slow as produced by the known accelerators of the day, or else would be soon absorbed by stray gas molecules in the evacuated chamber, or even by the walls of the tube itself. Typically, Swann's enthusiasms were for abstract ideas but experimenting with actual hardware was not to his liking. Fortunately, Lawrence was not deterred. Furthermore, his reputation had grown to the extent that the administrators for the relatively new State University in southern California decided in 1926 that they wanted him for their physics laboratory at Berkeley, offering several enticements. After World War I Berkeley's experimental physics group had become the beneficiaries of money from the newly formed National Research Council and also from W. K. Kellogg of the cornflakes fortune. Two years later at age 27 Lawrence decided to accept their offer and left the east coast for the west. Within two years of arriving he became a full Professor at the astonishingly handsome salary for those years of around forty–five hundred dollars per year.

The beginning speculations about an electron accelerator were not original with Lawrence, since many other scientists of the day had indulged in suggestions of how it might be accomplished. In particular, an electrical engineer in Germany named Rolf Wideroe came up with a circular orbit in glass tubes in 1927 but his apparatus failed. The publication of Wideroe's attempts inspired Lawrence with the idea of series of electrical voltage boosts for electrons (or protons) in a circular path instead of the straight linear cascade methods which had already been tried. More than that, however, Lawrence realized the particles could be kept bunched in a horizontal plane by placing the evacuated chamber between the poles of a powerful electro-magnet. Furthermore, his real inspiration was that they could be circulated between two semi-circular electrodes called "dees" (after the alphabet letter suggested by their shape) which alternated their electric polarity at regular timed intervals in order to provide an accelerating voltage boost as the particles crossed the gap between one dee and another. The hidden secret of the success of his machine was his perceptive insight that if the ions were introduced at the center of the of radius of the two air-evacuated "dees" they would travel in an outward spiraling path at circles of ever increasing radius due to the magnetic field and ultimately leave the flat canisters near their outer circumference. At each voltage kick given to the charged particles to increase their kinetic energy the centrifugal force would cause the radius of their trajectory to increase. For any given time interval and circumferential velocity the centrifugal force must be balanced by the restraining or counter-balancing force of the magnetic field in order to contain the ions within the flat cylinder of the cyclotron.[1]

While working through the physics equations which express these relationships, Lawrence suddenly discerned in a flash of inspiration that for a given time interval of complete circuit of a particle inside the chamber, the radius did not matter. The "r's" in his equation canceled each other out. The time that it would take for any given particle to complete its full circumference depends only upon the strength of the magnetic field and the mass and electric charge of the particle, whet-her electron or proton or alpha particle. As it gains energy at each jump over the electrode gap it simply moves out into a longer trajectory having a greater radius. The time interval remains essentially the same, disregarding velocities which involve relativity. This meant that the frequency of the oscillating voltage change between dees, when carefully adjusted, could remain fairly constant.

---

1. For a lucid explanation of the simple mathematics of the cyclotron see the book *From Atoms to Quarks* by James S. Trefil.

When Lawrence first discovered this important fact, his enthusiasm could not be contained and he went around the laboratory buttonholing his associates and allegedly exclaiming "R cancels R," much like Archimedes when he discovered the secret of buoyancy in water.

Putting all theory aside, a second reason for the success of the new "cyclotron," as this type of magnetic resonance accelerator was christened, was hard and very practical work exerted in its development and manufacture. Lawrence was the antithesis of the journalistic myth of the lone, introverted genius removed from the common everyday world of people. He was a happily married man with a wife whom he adored, if occasionally neglected for his work, and who presented him with five children. He was socially gregarious and popular as a public speaker on matters scientific. On top of all that he was also an astute business man who dealt with a parsimonious college administration, as well as successfully wheedling money from foundations, or wealthy individuals, or donations of free and valuable equipment from manufacturing corporations. One of his early bi-polar electromagnets was begged from the Federal Telegraph Company. He inspired great loyalty and prodigious work hours from his graduate students, and won their admiration with his inventiveness in finding expedient practical solutions to problems in the proper operation of the accelerator. As an instance, when it became difficult to obtain a perfectly uniform magnetic field in the horizontal plane between the magnet poles, he devised the ingeniously simple solution of fashioning and inserting flat teardrop shaped wedges made from soft iron at strategic spots around the electrodes canister. Success was achieved by trial manual adjustments.

Although his graduate students were numerous and helpful, the two most important assistants that he ever recruited were unquestionably David Sloan and M. Stanley Livingston. The latter came from Dartmouth College with a Master's degree in physics in 1930. Sloan, at first, concentrated his efforts on an X-ray machine for medical applications. Livingston brought superior ability in the design and making of equipment combined with a sound theoretical knowledge. During the early development of the first cyclotrons his remarkable devotion to the task drove him to work long and exhausting hours continuously, to the endangerment of his health. Although he never received the same degree of fame or public credit that he may have deserved, Livingston's contribution to their ultimate success was probably just as important on a technological level as that of his boss, Lawrence.

It must be acknowledged that the goal of this laboratory group was not entirely pure scientific research for theoretical purposes only during the starting years. It happened that the younger brother of Lawrence had chosen a career in medical biological research and ultimately became a physician also associated with the University of California. Consequently, some of the justification for striving to build an ion accelerator became legitimately associated with finding radioactive isotopes of metals or other machine–made substitutes for natural, and expensive, radium or polonium from which X-rays were derived in that day. Much later the cyclotron was used to generate a beam of neutrons for the radiation of malignant tumors in patients. Many of the organizations or benevolent foundations which Lawrence solicited for financial aid were medically oriented and often responded generously, with applications as hospital equipment in mind .

During the early development of the cyclotron the working conditions under which the graduate students labored were quite primitive and would be considered dangerously unhealthy by today's standards for atomic energy plants. They were frequently exposed to unmeasured and random X-ray and neutron radiation despite attempts at shielding around the machine with fifty–five gallon barrels filled with water. While the machine was operating or being tested the danger of accidental electrocution from touching or coming too near to the bare high voltage leads or terminals was real. This was realized once by an assistant. Careless and prolonged exposure to the beam exiting from the machine caused a permanent burn scar on the hand of one student and caused him to wear a black glove thereafter. All of them occasionally breathed the fumes of natural gas when the outer canister around the dees was being tested for pin hole leaks by smelling the odor of escaping gas. Merely turning on the electric power to the magnet would cause all the lights in surrounding rooms or buildings to dim. All watches carelessly worn on bystanders standing near the cyclotron machine would become magnetized and stop, and all loose tools or wrenches made of iron could fly through the air like lethal projectiles. The oscillating frequency of the direct current voltage on the dee electrodes was in the vicinity of 20 megacycles, thus interfering with short wave radio reception for miles around the neighborhood. Getting the first cyclotrons to work properly at best efficiency always took some tinkering and fine tuning, being a radio ham as Lawrence was in his youth, was always helpful. A trick which never failed to impress visitors was to hold an ordinary electric lamp bulb to some metal surface in the room and have it glow with light without any wire connections.

Notwithstanding, their team succeeded remarkably. The first pilot model, which was essentially put together by Livingston, measured roughly only 5 1/2 inches in overall diameter and could be held in one hand. In December of 1930 this device produced evidence of hydrogen ions (protons) circulating around in spirals ten times. This indicated that the basic concept was feasible. The group immediately set about building an 11 inch diameter cyclotron. They also acquired a larger magnet and a 500 watt short wave radio energy oscillator from Federal Telegraph Company. By July of 1931 they were rewarded by the acceleration of ions to 900,000 volt levels.

Lawrence was ecstatic. In the summer of that same year he announced his engagement to his childhood sweetheart, Molly. Moreover, he was awarded space by the Civil Engineering Department in their old material testing laboratory which had a solid and reinforced concrete floor slab suitable for heavy loads such as giant magnets. This was the nascence of what was to become the Lawrence Radiation Laboratory, which became a separate facility by itself in 1936. They immediately moved from a room in the Chemistry Department, thus significantly demonstrating the liberation of nuclear physics from domination by that field in academic organization.

The Radiation Laboratory at Berkely went on to build a 27 inch diameter cyclotron, which reached an energy level of 3.6 MeV by September of 1936, and a 37 1/2 inch diameter machine in 1938. The latter was the most successful of all, since it incorporated numerous technological improvements. It used deuterons instead of protons; and it could operate reliably for almost eight hours a day for a seven day week at particle energies of 8.5 MeV.

By this time Lawrence had suddenly become world famous. The University of California would have been more happy if he obtained legal patents for everything that he did. Lawrence never really abandoned his dual purposes of scientific research for ultimately transmuting elements by bombardment with heavy nucleons and obtaining breeder machines for the altruistic purpose of making artificial radioactive materials which might be useful for the treatment of people with cancer. Moreover, he was generously free with advice and suggestions to other research laboratories at universities in other foreign countries as well as his own. Many of these later hired his own graduate students. He became the Director of the Radiation Laboratory in 1936 and received the Nobel prize in 1938 at the age of thirty–eight. He was the first scientist working at a state controlled and financed university to do so. The largest cyclotron he ever constructed was a giant 60 inches in diameter in 1939. With the advent of World War II government funding for physics research had become very much easier to obtain. Lawrence at Berkeley Laboratories found himself in competition with the Carnegie Institution in Washington and, coincidentally, with his old boyhood friend from South Dakota, Merle A. Tuve.

The shift to artificially produced particle beams was clearly emphasized during an international conference in July of 1953 at Bagneres in France. By that time almost two dozen copies of the cyclotron had proliferated all over the world. However, there were real practical limits to the size and particle velocities which could be achieved by the cyclotron method. This was quite well recognized by its inventors, as well as other developers. The reason is that, as the positively charged ions are accelerated to velocities approaching that of light, Special Relativity enters the picture. The mass of the particles increases significantly and the simple equations used by Lawrence for calculating equilibrium between centrifugal force and magnetic constraint begin to break down. The practical energy limit for protons in a cyclotron was estimated to be around 22 MeV. Other means needed to be discovered for higher beam energies. In spite of this electro-mechanical limitation, scientists everywhere became fascinated with the idea that particles restrained within a circular and air–evacuated tube and having their kinetic energy boosted by a series of successive and continuous voltage charge kicks had a fantastic promise.

The practicability of circular accelerators was enhanced by a curious and fortuitous circumstance in the action of Newtonian physics. The timing of the voltage boost to the moving particles does not need to be absolutely precise. The faster or sooner one particle reaches the electrode gap, the weaker the kinetic energy kick that is furnished to that particular ion, whereas a slower moving particle lagging behind gets the benefit of a larger boost and greater acceleration from the electric magnetic voltage difference. Thus the latter tends to catch up with the first and the average velocity of the beam tends to even out somewhat.

The post World War II interval between the years of 1950 and 1965 was a ferment of activity and transition from preoccupation with atomic bombs among nuclear physicists. Many were returning to teaching at universities or laboratory research once more. The importance of nuclear physics had been amply and spectacularly demonstrated to the people by the "Bomb." Competent people were readily finding jobs and, equally important, governments of many nationalities were suddenly amenable to lavishing money for research grants. Politicians became acquiescent to voting large funds for ambitious research facilities. This flood of taxpayer's money, often justified on the basis of either national defense or else the advertised promise of ultra-cheap electric power from uranium fuel, resulted in particle accelerator machines springing up all over the world. As they came into operation these new machines quickly supplanted the older cosmic ray studies and the Lawrence "cyclotron" as well. Scientists were discovering that smashing high velocity protons into the proton or neutron nucleus of metallic atoms produced a wonderful shower of fragments. Some of these particles they already knew about; but dozens were completely novel and unexpected and had brief lifetimes measured in a hundred thousandth of a second or less.

This plethora of strange new particles was greeted by many physicists with mixed feelings of wonder and regret. Exciting as the new discoveries were, there was no doubt that their previous ideas of the atom being relatively simple and comprehensible had been shattered. Clearly there were things going on inside the nucleus of an atom which they had only dimly perceived, and they were confused. The theoreticians were pitched into a turmoil of speculation on how to explain the creation of all these fragments. No major university worth its salt could afford not to have access to some particle accelerator of whatever design for precise study with reliable results.

A hybrid synchrotron-plus-cyclotron of 400 MeV capacity, which was really a large and modified cyclotron, was built between 1947 and 1949 for Columbia University at a former DuPont estate known as *Ben Nevis* at Irvington near the Hudson River in New York State. It was dedicated by General Dwight Eisenhower, then President, in 1950, although the machine was not quite in operation at the time. It was originally intended to generate pions in quantities by directing a beam of protons at a target of graphite, carbon, or copper. This marked the beginning of the transition over to the synchrotron idea.

One of the first really large "atom smashers," as the journalists soon fixed them in the minds of the public, was the Stanford University Linear Collider or "S.L.C." near Palo Alto in California. In this case the accelerator was actually a tube housed in a straight "shot" two mile long underground tunnel which by necessity passed beneath the suburban countryside and one major concrete highway. Construction began in 1962, and by 1966 it was in operation, colliding high velocity electrons into metal targets. The straight line tube acted as a kind of magnetic gun barrel with microwave radio energy boosters placed in series in order to propel the electrons at ever increasing velocities. The design intentionally departed from the circular track trend because the energy losses inherent in magnetically bending the ionized particle beam into a curved trajectory were thus avoided and, as a result, electric power costs were also reduced.

Over a period of years many ingenious engineering innovations and new equipment were added to increase the power of the beam. These modifications ultimately resulted in two new synchrotron rings being added at the target end of the tunnel. The research facility became known as the Standard Linear Accelerator Center or "S.L.A.C." By 1988 it was colliding electrons with positrons head on from opposing directions, thus doubling the momentum of the impact and presaging the existence of what later became known as quarks. It thus became a stunningly successful research tool.

Most subsequent accelerator machines, however, adhered to the circular track, using a series of very large electro-magnets to constrict the charged particle beam during its many transits of the ring circumference. The international race for learning the secrets of nuclear power had started. With the endorsement and financing by the newly formed United States Atomic Energy Commission the "Cosmotron" ring accelerator started into operation at the Brookhaven National Laboratory in north eastern Long Island in late 1952. It had a rated power of 3 GeV and was intended for the circulation of protons. This machine was ultimately replaced in 1960 by the larger 24 GeV "Alternating Gradient Synchrotron" or "A. G. S."

The same Commission also entrusted the design and construction of the 6.2 GeV "Bevatron" accelerator to the Radiation Laboratory at the University of California at Berkeley. This synchrotron started operation in 1954 also using ionized hydrogen nucleons.

The Cornell University "Electron Storage Ring" or "C.E.S.R." first started operation at Ithaca, New York, in 1967. It eventually reached a power level of 12 GeV and later was converted from electron to positron beam in 1977.

The largest accelerator facility presently existing in the United States is the "Tevatron" synchrotron with its name derived from the design energy of 1 TeV. It is located at the Fermi National Accelerator Laboratory (or Fermilab for short) at Batavia, Illinois, some thirty miles south of Chicago. The first constructed ring was a four inch diameter stainless steel tube having the air evacuated to a very high vacuum. It was housed in a six and one half foot wide tunnel with a roof elevation thiry feet below the ground surface. This tunnel had a diameter of one and one quarter miles and traced a four mile circumferential path underground. A series of powerful magnets kept the beam of protons centered within the tube. At regular intervals the protons being accelerated passed through specially shaped copper cylinders called "cavities." These were subjected to pulses of high voltage at radio frequencies to create a field of standing waves of high energy which, upon transfer to the protons, gave a boost to their velocity.

The first synchrotron ring was completed in 1972. Later on in 1983 a second circular ring was added below the original, but in this case the electromagnets were of the super-conducting type chilled by liquid helium. Either protons or anti-protons, since they both carried electric charges, could be circulated in opposite directions and make 470,000 complete circuits. Their velocities approached that of the speed of light with the result that their added energies, when made to collide, reached 1000 GeV (1 TeV). The electricity costs incurred in the operation of this facility could be as much as $ 16,000,000 per year.

The Fermilab was in deliberate competition with the syndicate of European scientists, who had intelligently pooled their national resources into the *Conseil European pour la Recherche Nucleaire*, commonly known as "CERN." This huge accelerator ring of today started out with what was modestly called a Proton Synchrotron (P. S.) near Geneva, Switzerland, with a design energy of only 28 GeV. It was completed for operation in 1959. Subsequently the consortium decided to aim for a 400 to 500 GeV ring four and one third miles in circumference, which they called the "Super Proton Accelerator" or "S. P. S." This machine began a very successful operation in 1976. However, even this energy level no longer satisfied the physicists. The theories of nuclear particles had become even more complex and mathematical in nature, and they became convinced that higher particle velocities would confirm their deductions and speculations. Concurrently with this drive for more knowledge was the helpful fact that their particle detectors had reached a very sophisticated level of engineering with the use of computers for recording a mass of observational measurements. CERN then undertook to build a fantastic underground tunnel having a 16 1/4 mile circumference, which was large enough to require that part of the ring arc pass through southern France. Quite justifiably, this one was christened the "Large Electron Proton Collider" or "L. E. P." It was scheduled for completion in 1989 with a deceptively low energy rating of 100 GeV, which doubles when one collides those two different particles with each other coming from opposite directions. The extremely large diameter of the ring was decided by the gradual curve requiring a smaller and less expensive investment in less powerful super-conducting controlling magnets. However, the ultimate goal was really dictated by compatibility with the intent to demonstrate the existence of quarks inside the nucleons - a totally new idea which we have not yet discussed.

Earlier in this race West Germany decided to keep up with the crowd independently and not be left in ignorance, as they certainly had been (the world's good fortune) during Hitler's regime and the United States secret Manhattan Project. The government authorized the 7 GeV "Deutches Elektron Synchrotron" or "D.E.S.Y." near Hamburg. This machine started operation in 1974 and consisted of two rings, one on top of the other. For this reason it became known among physicists as the "Double Ring Storage Facility" or "D.O.R.I.S." It was intended to collide electrons with positrons. However, modification plans in 1977 placed the two rings in series in order to raise the energy level. This revision caused the name to become "Positron Electron Tandem Ring Accelerator" or otherwise "P.E.T.R.A." when it was completed in 1978. Another up-grade improvement called "H.E.R.A." followed later.

Not to be outdone, the Union of Soviet Republics started to build their "Synchro-phasotron" laboratory in total secrecy at Dubna, just outside Moscow.

Although the list just given is by no means complete, it nevertheless should give the reader some an idea of the tremendous amount of money, time, and engineering technology which nations suddenly concentrated upon theoretical research in nuclear or "High Energy" physics. Only space exploration has become a major competitor in the United States for similar amounts of money, but it has the advantage of being highly popular.

It is a curious and little known fact that during the beginning days of the excitement over synchrotron accelerating machines, which were then utterly new and untried, many qualified physicists became alarmed and considered them possibly to be terribly dangerous. Their private worries were real and derived essentially from astronomy. This genuine anxiety arose as early as the plans for the "Bevalac" machine at the Lawrence Berkeley Laboratory in the mid 1950 decade. It was resurrected into a serious and crucial debate during the planning sessions for the "Relativistic Heavy Ion Accelerator Collider" or "R.H.I.C." as late as 1983 at the Brookhaven National Laboratory in New York state.

The fear was that in the course of the experiments they might inadvertently simulate on the earth the action of some collapsing stars and create "strange matter" or abnormally compressed and dense matter. These condensed particles, it was conjectured, might attract and trigger a cascade of adjacent "normal" matter into a similar agglutination and destroy the entire world around it. This fantastic idea was borrowed from the astronomers who had been observing and puzzling over the formation of nebulae and neutron stars. The Nobel laureate Tsung Dao Lee and the late Gian Carlo Wick raised this scary possibility together. The more energetic the collisions of ever more massive particles become in the coming new accelerators, the more this worry returns.

Not surprisingly, discussions about the remote possibility of such a dire catastrophe took place in the utmost secrecy, since all the physicists involved feared "bad publicity" for their projects! Ultimately the majority of prominent theorists assured them that such a possibility would not happen at the puny earthly energies expected and with the small amounts of particles with which the colliders were operating. Another one of their arguments is that, in spite of the continuous barrage by comets, asteroids and meteors in the past millions of years, the moons and planets in our solar system were still intact, though pock-marked. Perhaps it is just as well that we humans are not expected to be able to mimic the huge energies of the stars.

CHAPTER 11
THE PARTICLE ZOO

Section 11.1 - The Lambda Zero

Meanwhile, whether by using natural cosmic rays or by employing early model particle accelerators, the discovery of new and strange nucleon fragments resulting from collisions continued like snow in a "winter of discontent." This development was both exciting and challenging to the scientists, but it cannot be denied as a serious threat to their complacency and their philosophy of simplicity in nature.

One of these was the Lambda Zero, which was first discovered by Clifford Butler and G. D. Rochester at the University of Manchester in England in March of 1951 and later confirmed by the "Cosmotron" machine at Brookhaven National Laboratory in the United States in 1953. Originally called the "Vee" particles because of the tracks made in cloud chamber detectors by their decomposition, their symbol ultimately became the Greek capital letter Lambda or "Λ."

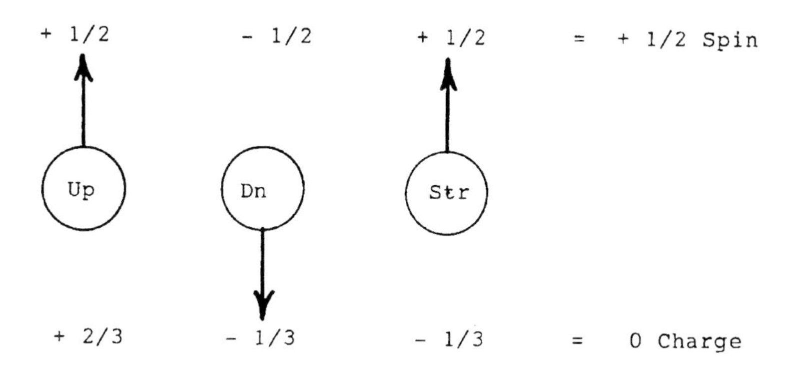

The mass of the Lambda Zero is 1115.68 MeV, which the alert reader will have noticed is significantly larger than that of the proton. In fact, the Lambda Zero is regarded as behaving somewhat like a heavy neutron. That fact also makes it a baryon. Its electrical charge is zero, of course, and its spin is one half. Its mean lifetime is 2.63 x 10^{-10} seconds, which is long enough to place it in the same category as kaons as "strange." Sometimes both of these mentioned particles are produced together at the same time.The reason for such a simultaneous appearance happening is that if a high energy collision of a proton and pion should produce a neutral kaon having a strangeness of + 1, then a second particle must also appear with a strangeness of - 1 in order to balance the books.

The cosmic ray laboratory built by Manchester University under the supervision of Patrick Blackett on the top of Pic du Midi de Bigorre mountain in the French Pyrenees was an especially fertile discovery source for many previously unknown particles. The Lambda was one of them. In a single period of from July of 1950 to March of 1951 some sixty–seven cloud chamber photographs were obtained by Carl Anderson and Eugene Cowan which showed a mysterious Vee shaped track. These had also been reported by the White Mountain observatory in California. The pictures puzzled the scientists, at first, since the lack of any electrical charge on the particle yielded no previous track or warning of its existence in the usual cloud chamber detector. It also usually appeared suddenly at some distance away in a shower produced by a proton versus proton collision and had its characteristic vee signature.

For apparatus the experimenters at the University of Manchester were using a lead target placed inside an air and alcohol cloud chamber in order to slow down the velocity of a stream of high energy cosmic particles. The vee shaped track would appear on the far or down side of the lead foil. Further investigation and measurement of the two diverging decay particles suggested that they consisted of a proton and a negatively charged pion (π). However, there were two quite unusual and startling facts about the reaction which did not fit their preconceived ideas at all. First, the then unknown particle had a mass greater than that of the proton; and, secondly, the distance that it traveled after emerging from the metal target calculated out to a lifetime of around 2.6 x 10^{-6} seconds. This time was vastly longer than the 10^{-24} seconds considered normal with strong force reactions originating from a nucleus.

Approximately three years later the accelerator machine at the Brookhaven National Laboratory had been completed and became operational with a beam of protons. Using an eighty inch diameter bubble chamber filled with liquid hydrogen as a detector, the neutral Lambda particle was again confirmed in conjunction with a negative kaon. The results were published in an article titled "Production of Heavy Unstable Particles by Negative Pions" in 1954 by W. B. Fowler, R. P. Shutt, A.M. Thorndike, and W. L. Whittemore. Thus the new age of teamwork was emphasized.

The two most typical decay modes for the Lambda were disintegration into either a proton and negative pion or else into a neutron and a neutral pion.

$$\Lambda^0 \rightarrow p + \pi^-$$
$$\Lambda^0 \rightarrow n + \pi^0$$

In both cases the neutral electric charge is conserved, as expected. Note also that the daughter products do not share the strangeness property.

The weak force, such as evidenced by the Beta decay of the neutron into fragments smaller and lighter than the proton, was already familiar to the physicists. However, the Lambda was not only more massive than the proton to start with; but it hung around for a mean lifetime about a million billion times longer than expected for a decay involving strong force products. The pion was the clue to a presumed collapse of the strong force that must have been involved. This was indeed a new and strange phenomenon to the experimenters and was genuinely bewildering. Eventually in 1953 Murray Gell-Mann of the United States and T. Nakone of Japan, both leading theoretical physicists, together proposed that some new and poorly understood property called " strangeness " needed to be added to their list of the properties of matter which could be designated by a new quantum number.

Section 11.2 - The Xi Particle

The symbol for this particle is the Greek capital letter "Ξ," which is occasionally printed tilted somewhat to the right.

There exist two common varieties of this particular particle, depending upon what type of electric charge they carry, if any. A third and very rare possibility, which had a positive electric charge, was detected several years later after the first two; but it is a weird aberration having an infinitesimally short lifetime and belonging to what was eventually called a "charmed" group. They were not all identified simultaneously. The first to be discovered was the negatively charged Xi, which has a mass of 1321.3 MeV and a "J" spin of one half.

A particle of such an extreme mass was initially called a "hyper" fragment, meaning unusually large; but later it was included in a special category known as "hyperons" when it was discovered that there were others.

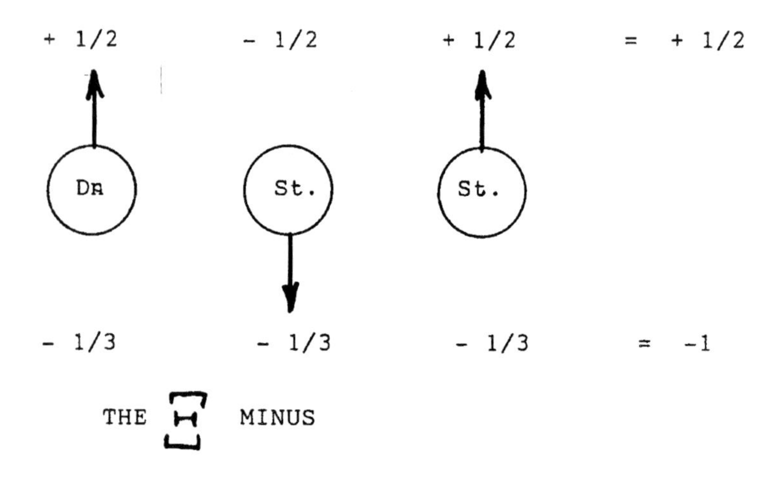

The peculiarity of the Xi was that it is doubly strange with a strangeness number of – 2, and for a long time no one could figure out why it existed at all.

Although the early experimenters in 1952 did not really know what they were looking at in their cloud chamber, the track of the negatively charged Xi particle is quite distinctive and identifiable. Upon entering the detector cloud chamber it suddenly takes a sharp bend or "kink" where it sends off a negative pion at an angle. It also sends off a Lambda particle simultaneously in a different direction at the kink. However, that decay fragment, as we have explained previously, leaves no visible track for a short distance until the Lambda itself breaks down into a proton and negatively charged pion. The result is a disconnected vee with its apex pointing backward to the site of the original Xi decay kink.

The existence of the negative Xi was confirmed by E. W. Cowan while working at the California Institute of Technology; and he published his announcement in the *Physical Review* in 1954.

The mean lifetime of the negatively charged Xi is 1.64×10^{-10} seconds. Normally the first decay products are a neutral Lambda and a negative pion, as we just explained above. Once this particle "decides" to split up, it does so rapidly in a chain of sequential separate disintegrations proceeding downward in individual masses like a string of ignited firecrackers. This effect is called a "cascade" and is similar to a series of stepped waterfalls. The entire process is a veritable shower of fragments and may be represented by the diagram shown below:

$\Xi^- = \Lambda^0 + \pi^-$

$\mu^- + \nu_\mu$

$n + \pi^0$ $\quad e^- + \nu_\mu + \nu_e$

$\gamma + \gamma$

$$p + e + \nu_e$$

Interpreting the scientist's short-hand diagram into words:

- The negative Xi decays into a neutral Lambda and a negative pion.
- The lambda Zero decays into a neutron plus a neutral pion.
- The negative pion decays into a negative muon and a muon neutrino.
- The neutron further decays into a proton plus an electron plus an electron anti-neutrino.
- The neutral pion decays into two photons of gamma ray radiant energy.
- he negative muon decays into an electron plus a muon neutrino plus an electron anti- neutrino.

The writer does not intend to inflict this sort of thing upon the reader too often; but it serves to illustrate just how intricate the decay tracks of a particle can become and why people, sometimes women with the patience, are hired to study the cloud chamber photographs meticulously with the purpose of clarifying the observable paths. It becomes a fascinating puzzle.

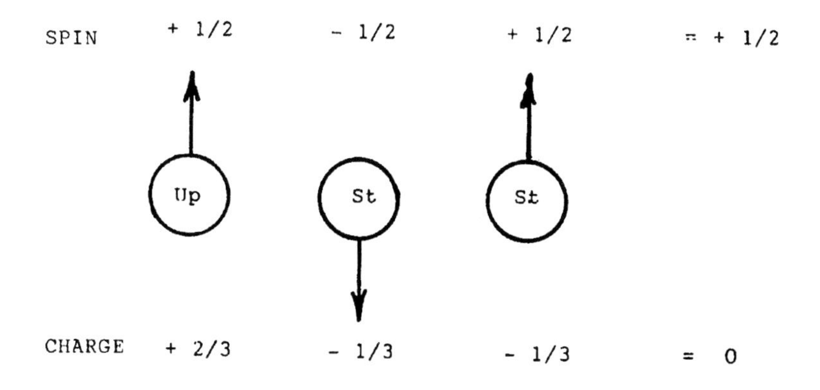

The Neutral Xi

The neutral variety was not discovered (or, at least, recognized) until seen years after the negative type. Its production by high energy pions is very infrequent and the early experiments using only cosmic rays did not allow very specific searches among the chance collisions provided by nature.

Louis Alverez and his associates at the Lawrence Berkeley Laboratory started a program in 1958 which used a beam of high energy negative kaon mesons obtained from the recently built "Bevatron" synchrotron in California. The kaons were separated out from other particles by means of their masses as evidenced by their tracks inside a fifteen inch diameter bubble chamber. This detector had only recently been invented by Donald

Glaser in 1953. The experiment was not the result of a mere happy accident, since two prominent theorists has already predicted the probable existence of the neutral Xi by this date. The learning curve among the physicists was well on its way with the advantages of controlled and regular beams obtained by means of the new accelerators. The confirmation of the Xi Zero particle was duly recorded in 1959 by a letter from Louis Alverez to the *Physical Review* titled *Neutral Cascade Hyperon Event.*

The neutral Xi has a mass of 1314.9 MeV and a mean lifetime of 2.90 x 10^{-10} seconds. It also has a spin of one half and a strangeness of number of -2 like its brother the Xi Minus. The usual decay mode is by changing into a Lambda zero particle and a neutral with a zero electrical charge also. Since both of these fragments possess no charge, they become extraordinarily difficult to detect in any bubble chamber. Their tracks are quite invisible until these daughter products fall apart themselves subsequently. That makes it a real detective story to identify the scene of the original "crime."

Section 11.3 - The Sigma Particle

The symbol for these baryons is the Greek letter "Sigma" or "Σ." The letter "s" in the Sigma name was supposed to suggest the idea of a "super-proton" because of this particle's commanding mass. As with the Kaons we have studied previously, they come in three different varieties: negatively charged, positively charged and neutral. All have a spin of one half.

All three varieties carry one unit of negative strangeness, although how or why this should be so was unknown at the time of discovery and remained so for several years.

The persevering reader by this time should have deduced that the little circles shown in our prematurely displayed pictographs denote what are now regarded as the fundamental sub-particles of baryons and mesons called "quarks." Their derivation comes later in our story and remains to be explained in subsequent chapters.

1953 was a banner year for the scientists operating the "Cosmotron" accelerator machine at the Brookhaven National Laboratory on Long Island. Four Italians in Milan, Italy, named A. Bonetti, R. Levi, M. Panetti, and G. Tomasini had demonstrated evidence for the positively charged Sigma by using cosmic rays and photographic emulsion plates. Their results were published in that year. The Brookhaven team took up the search and confirmed the existence of two hyperons known as Sigma Plus and Sigma Minus later in the same year. Their article published in 1954 bore the perfectly factual and banal title of *Production of Heavy Unstable Particles by Negative Pions*. It would be hard to argue with that. The truth was that the nuclear physics community was in a ferment of activity

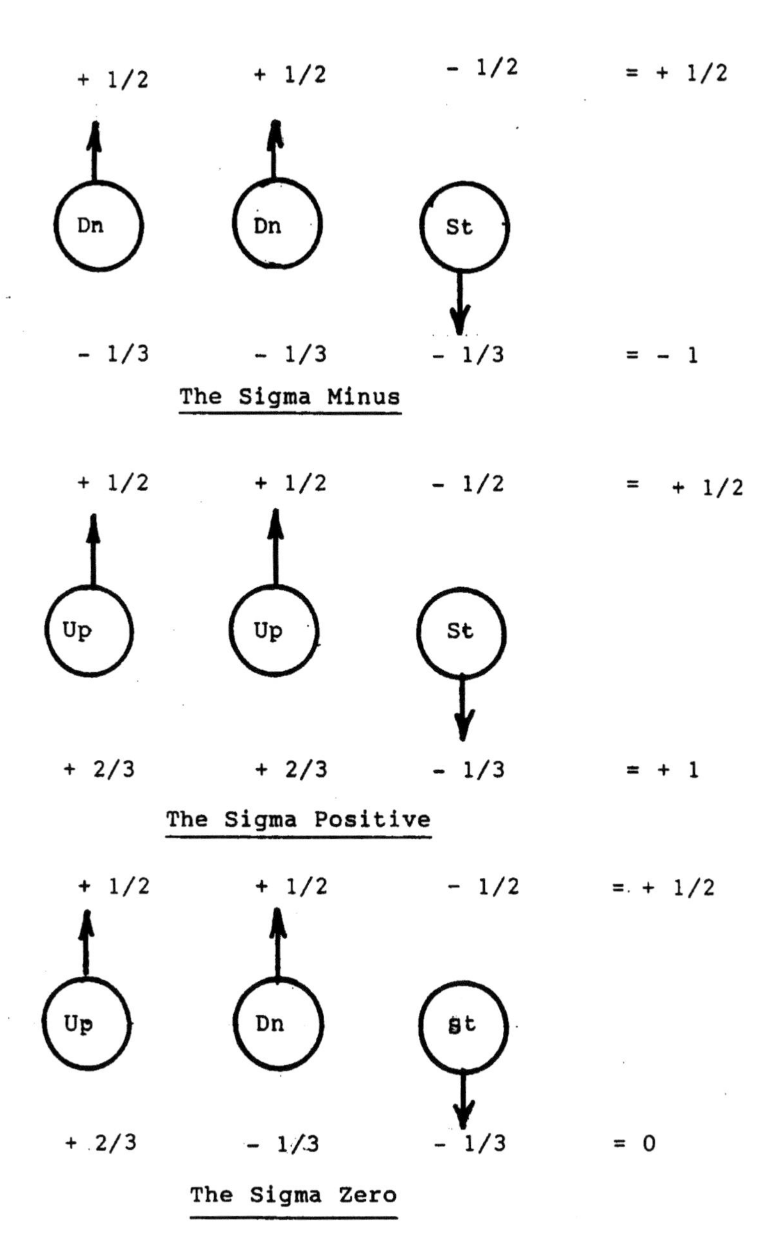

The Sigma Minus

The Sigma Positive

The Sigma Zero

and speculation upon the interactions of the strong and weak forces and the puzzling "long life" characteristic of "strangeness." Such top echelon theorists as Abraham Pais, Murray Gell-Mann, as well as T. Nakano and Kazahiko Nishijima of Japan, were all conferring and working out a logical system.

The cosmic ray scientists were not to be outdone. They flew an airplane at the high altitude of 27,000 feet for six hours carrying a giant 15 liter stack of photographic emulsion plates thick enough to record all almost all possible decay products to their rest point. As you already know, the existence of Σ^+ and Σ^- particles had been thereby deduced, but other tracks suggested the possibility of the Sigma Zero. W. D.Walker from Bristol, England, wrote the report for the latter in *Physical Review* in 1955; but the third Sigma was not confirmed until 1956 by R. Plano at the Brookhaven Laboratory as a follow up.

The positive Sigma has a mass of 1189.4 MeV and a mean lifetime of 0.799×10^{-10} seconds. Notice the accuracy of the numbers to which the scientists believe that they can adhere. It has two common decay modes:

$$\Sigma^+ = p + \pi^0$$
$$\Sigma^+ = n + \pi^0$$

The pions, of course, eventually decay themselves or else attach themselves to other atoms, depending somewhat on their velocity.

The negatively charged Sigma has a mass of 1197.4 MeV and a mean lifetime of 7.4×10^{-10} seconds. Ninety percent of the time it decays into a neutron and a pion as shown above for the alternate on the positive particle.

The neutral Sigma has a mass of 1192.6 MeV and a mean lifetime of 7.4×10^{-20} seconds or 700 billionths of a second. It is thus no wonder that this particle eluded detection for three years or more. One marvels that the physicists even gave credibility to such a fleeting existence as a real particle at all. The zero charge Sigma decays into a Lambda particle and a photon of radiant energy. The Lambda, of course, follows our expected cascade disintegration.

Section 11.4 - Classification

All of the atomic or nuclear sub-particles presented to the patient reader thus far are quite reproducible and material fragments which the knowledgeable amateur might cheerfully call "real." Experimental physicists believe implicitly in their existence, however brief and elusive. Their measurements can be duplicated in a well equipped laboratory. However, their numbers represent only the tip of the iceberg, figuratively speaking. There are tens of dozens more listed in the *Physical Review Manual D*, published every two years by the American Physical Society in Berkeley, California. Many would severely tax our tolerance to admit them as "real" by any macroscopic definition which the ordinary citizen might understand and thus are of significance only to the physicists themselves.

The proliferation of the strange and unexpected particles of different masses and electric charges utterly startled the scientists in their initial stages of research. The reader's own bewilderment at all the particles described was entirely shared by the experimenters themselves up to the year of 1959. The famed Enrico Fermi is alleged to have exclaimed "If I could remember the names of all these particles, I'd have become a botanist instead of a physicist!" True, quantum mechanics had given them a powerful tool towards understanding what was happening; but their first reaction to the confusion caused by the rapid pace of new announcements was to establish some sort of organized method of classification.

Initially, the physicists were compelled by their quandary to follow the example of the much older sciences of natural history and zoology. Far back in 1753 a Swedish botanist named Carolus Linnaeus made the first attempt to classify all the known living things in the world, starting with the simple division of plants versus animals. The variety of organisms of all kinds being discovered by the naturalists during the ensuing centuries became so numerous and intricate that a better system had to be adopted. Present day taxonomy follows a revised system devised by R. H. Whittaker in 1950 in which he basically divides the non- vegetation world of animals into the following arbitrary categories:

Phylum, such as Flat Worms or Ribbon Worms
Chordate, such as vertebrates like reptiles
Class, such as mammals or egg laying
Order, such as primates
Family, such as Hominidae or Equidae
Genus, such as a Lynx or Leopard
Species, such as Homo Sapiens.

Any analogy of the problem which confronted physicists with zoological taxonomy is a poor one; but it does serve to illustrate their mental effort to summon order out of chaos. From their point of view a corresponding list of salient characteristics one might look for in nuclear particles might be that given below:

Rest Mass
Baryon Number
Electric Charge
Angular Momentum or Isotropic Spin

Mean Lifetime
Strangeness Number
Observation of Parity

An imaginative very general classification table for the various different types of particles encountered in ascending order of masses could be:

Photons
Leptons
Quarks
Nucleons
Hadrons (including mesons and hyperons)
Atoms

It becomes evident from our illustrative exercise why the facetious term "particle zoo" arose as a whimsical jest. Physicists frequently afford themselves a sense of humor; and it certainly was needed during the proliferation of particles. At this point in our story It would seem a good idea to arm the innocent reader with an abbreviated dictionary of terms constantly used by the professionals in this discipline with some descriptive explanation.

Section 11.4 Leptons

This name is taken from the Greek word "leptos," meaning small, fine or light.

The reader is already familiar with the electron, which belongs to this category. The electron has extremely little mass, a negative electric charge and a spin of one half. As far as its size is concerned, modern thought regards the lepton as being so insignificantly tiny as to be regarded as only a point in space.

A second lepton which we have described already is the muon, which has a slightly heavier mass than the electron. However, there is a third variety. Later research revealed the Tauon, which the general chronology of our story has not permitted us to mention yet.

Leptons are immune and quite indifferent to the strong force which holds protons and neutrons together in the nucleus. They feel electric charge polarity only. Their masses go in ascending order of magnitude.

Section 11.5 - Baryons

Baryons, meaning "heavy ones," is merely a convenient and arbitrary category given to nuclear particles having a mass roughly equal to or greater than that of the proton. It has little to do with charge or spin. The need for such a classification never appeared until the "particle zoo" proliferation appeared. A baryon may also be a hadron.

The protons and neutrons are the most stable of all baryons and give some reassuring substance to our visible and tactile environment, whether it be earth, trees, rocks, or the stars. We feel from school age that such matter is the stuff of reality, including the organic chemistry of our own bodies. It may be disturbing to our complacency that the cosmologists hold the opinion that, even if the "Big Bang" theory is only approximately correct, baryons account for as little as ten percent of the mass of matter in space. Obviously there are quantities of hydrogen and helium gases whirling around, as well as meteorites, asteroids, planets, and comets. Nevertheless, the composition of the bulk of the missing "dark" mass is still totally unknown.

Since the definition of the upper limit to the mass was left somewhat vague, it was formerly possible to talk about short-lived and unstable baryons having masses greater than the proton or neutron. Conversely, baryons which come into violent collision with one another under conditions of high velocity may break apart and produce mesons with significantly lower mass.We will discover why later. Conservation of the "baryon number" in any given reaction implies the involvement of the strong force.

Section 11.6 - Hyperons

The name is derived from the Greek word "iper," meaning "above" or "beyond." It was devised during the 1950's decade as a very loosely used term for any of the baryons which were discovered to be heavier in mass than the stable proton or neutron and which were unstable and had short lifetimes. Today, starting with the Lambda and going up in mass through the Sigma and Xi, the term has been extended to include other heavy resonances, such as the Delta or Omega. In contemporary theory all hyperons must be composed of three quarks.

Section 11.7 - Hadrons

The term comes from the Greek root "hadrys" for strong or, alternatively, "hadros" for thick or bulky. In an early attempt to classify the welter of newly discovered particles into separate groups it occurred to a Russian physicist to assign one term to any of those particles which appeared to be held together by the strong force. Although still little understood, it was evident that the strong force was needed to overcome both the repulsion of like electrical charges and also neutral indifference. Obviously this category must include the two nucleons; but it also had to embrace the heavier fragments produced by cosmic rays or accelerator machines and the lighter mesons, including the Kaons. Although the hadron classification has survived even today, it was quite arbitrary and has led to some confusion among non-physicists. The advent of quarks on the research scene years later, which we will discuss in subsequent chapters, brought some logic and justification for the use of the word. Today hadrons are generally understood as a combination of two or three quarks all held together for a period of time by the strong force.

Section 11.8 - Fermions

Invented by Enrico Fermi, this term has already been discussed superficially in our Chapter 3, and more seriously in Chapter 7. We reiterate it here only for the purpose of emphasis, since the concept is of vital importance for the formation of atoms and, consequently, all chemically bound molecules which comprise the matter of our entire world.

The fermion is any particle which has a spin (or total basic unit of angular momentum) of one half or half integer multiples of one half such as 1/2, 3 / 2, 5 / 2 and so on. The reader will recall that these numbers are always the spin constant multiplied by the expression $h / 2\pi$. This category includes all leptons, the proton and neutron and also, as we will shortly discover, all quarks. Any fermion by definition must obey the Pauli Exclusion Principle in that no two of them can ever occupy the same space in the same quantum state at the same instant of time. Reverting briefly to the quantum mechanics theory for the benefit of the amateur scientist, the reader should remember that the symbol "psi" or "ψ" in Schrodinger's famous equation has become traditionally associated with a wave function or state vector description for a wave traveling in time along an assumed direction. It is usually supposed to represent the state of motion of a particle when the proper operative factors are included in the equation. Fermions mathematically have the peculiarity that when one definite and identifiable particle is interchanged or replaced at a specific dimensional position in space the algebraic sign for its spin must become minus or opposite. Plus ψ cancels out minus ψ and the equation becomes anti-symmetric. The net wave function becomes zero and the wave function description collapses, confirming the Exclusion Principle

Section 11.9 - Strangeness

"Strange" is an attempted classification of certain particles which surfaced somewhat later in the study of mesons when the new accelerator machines began supplementing cosmic rays. A somewhat complicated theoretical concept in quantum mechanics was endowed with this name by Murray Gell-Mann, although a fellow theoretician named Kazahiko Nishijima in Japan would have preferred to categorize them as having the "eta" or "η" factor. The puzzle arose when it became apparent that some of the heavier mesons having nearly the same masses could come in a family of positive, negative, and neutral or zero electric charges. More important, however, was the discovery made possible by measuring the length of their tracks in a cloud chamber or photographic emulsion that they stayed around much longer than expected for their mean lifetime before decaying. By that it is meant around 10^{-8} or 10^{-10} seconds instead of the incredibly shorter 10^{-23}. Mesons such as the Kaon, Lambda, Epsilon and Xi are all examples of strange particles.

Arbitrary numbers were assigned to this strange factor for any given particle ranging from zero for non-strange through -1 to -3 for extremely strange. They all decay by the weak force action. Degree of strangeness became yet another quantum number, although any elucidation as to the reason why had to wait for another ten years or so. The situation precipitated a great deal of thought and collaboration between physicists.

CHAPTER 12
RESONANCES AND BOSONS

Section 12.1 - Resonances

As the cyclotrons and linear accelerators came on line and into operation, it became increasingly obvious strange and new particles were not only becoming more numerous, but their classification and qualitative evaluation of mass, spin, and electric charge was also becoming more difficult. Things were getting much more complicated than anticipated. This was particularly true with the new technique of barraging proton targets with charged pions or kaons or observing their reactions with protons in the bubble chamber detector invented by Donald Glaser of the University of Michigan in 1952. His chamber contains a volatile liquid, whether it be liquid hydrogen, freon, propane, or ethyl ether, kept just below its boiling point. This chamber is placed between the poles of a large magnet. At the exact instant when a beam of particles is directed into the chamber the pressure inside is abruptly lowered, with the result that the paths of ionized particles become visible as a stream of bubbles of vaporized liquid. This type of detector became very popular and larger ones were constantly being fabricated. One of the biggest was constructed by Louis Alverez at the Berkeley campus in California in 1959.

By this time the scientists had developed a technique known as "measuring the cross section" as particles such as pions or kaons were scattered as they impacted a heavier particle. Many of the new and unknown particles had a mean lifetime so fantastically short (in the order of 10^{-23} seconds) that they were virtually invisible. Many repetitious measurements had to be made and a graph constructed of the number of observed events as an ordinate and the calculated or estimated energy on the abscissa. A statistical mean then became the adopted mass for the particle at the absolute peak of the curve. These ephemeral and numerous new particles became known as "resonances"; and they represented a new and serious puzzle. They were more like ghosts or shadows of those particles which we are prone to call "real" or relatively stable.

Max Born acknowledged their existence in his 1969 textbook on *Atomic Physics* with the statement that they could not be elementary particle states. The reaction of this genius founder of quantum mechanics to resonances is philosophically quite trenchant. To quote him verbatim "If, even in inanimate nature, the physicist comes up against absolute limits at which strict causal connection ceases and must be replaced by statistics, then we should be prepared in the realm of living things, and emphatically so in processes connected with consciousness and will, to meet insurmountable barriers where the mechanistic explanations and goal of older natural philosophy becomes entirely meaningless."

Resonances are a temporary affair in which a "bump" in the cross section measurements is centered on a variable mass of some unstable and intermediate particle which has been formed, only to soon disintegrate. Because of Heisenberg's Uncertainty Principle it is impossible to establish any perfectly exact measurement of mass. We only know that the "bump" in the graph of detector results spreads out to become wider at the base in direct proportion to the shortness of the particle's lifetime. The Uncertainty Principle of quantum mechanics allows these unimaginably brief lifetimes by a relationship with Planck's constant shown below:

$$\Delta t . \Delta E = \hbar = h / 2\pi$$

Thus the incremental ΔE for the energy uncertainty allows a band width or ambiguity in any measurement of mass for a resonance ranging from 10 to 200 MeV. It might be said there is no great difference, other than mean lifetime, between a resonance and other unstable particles in nature. The difference is that resonances are usually bosons. Physicists were reluctant to admit this for a time, since they did not really care for a plethora of weird particles whose existence they could not really explain. At first it was thought that resonances might be classified separately as ultra-strange particles which vigorously split apart almost instantly. Even this distinction became blurred with the discovery that a sharp definition of a mass peak was frequently impossible. In such cases the selection of a nominal mass was left to the arbitrary judgment of a committee. The end result was to abandon their handbook classification as spurious aberrations and to include them in the list of mesons and hadrons with their various possible masses.

If resonances are so elusive and insubstantial, then why bother with them at all? Perhaps one reason might be that there are so many of them that there must be some importance by the weight of sheer numbers, amounting to hundreds even, as a hint that something is going on of significance. One example is the Sigma particle which we introduced in Chapter 11. Louis Alvarez with three co-workers was experimenting in 1960 with separating out the minus kaon particles from a beam when they accidentally detected a resonance pair having a mass of 1385 MeV. This fleeting coupling decayed quickly back into Sigma and Kaon particles again and is now called the Sigma (1385), although its more exact mass actually varies from 1382.8 for the Σ^+ to 1387.2 for the Σ^-. If that were not bad enough, no less than eight additional other resonances were subsequently discovered for the Sigma and listed in the *Physical Review, Volume "D."*

The real reason for their study by scientists was that resonances afforded a beginning clue as to why and how all the strange particles formed at all. Although resonances complicated the life of experimental physicists, their ultimate contribution was undoubtedly that they instigated a great deal of theoretical thought and thus built a receptive situation for the advent of quarks as a more comprehensible explanation.

If one happens to be a mathematician quite accustomed to thinking abstractly in symbols of logic and, moreover, has become conversant in the superposition of sine and cosine functions in order to create a picture of complicated wave forms with energy content plotted against a locus in time, then it may become easy to think of resonances as merely a peculiar aberration or superposition of a wave form. Thus we can attempt to describe the phenomenon of resonance as the sympathetic combined wave packet forms represented by two merging and formerly separate particles which in a fleeting marriage causes their separate entities to represent a seemingly new particle. None of these mergers last very long and soon split up, much like the random promiscuity of some of the current younger generation of people. A resonance can thus be viewed by some as a kind of discordance or harmonic "blip" in an otherwise smoother wave form.

If, on the other hand, a person happens to be comfortably immersed in the normal world of our five senses and prefers to deal with pictures which are compatible with everyday experiences having tangible and material objects, then the analogy of two droplets of water on the surface of a sloping glass window pane has been suggested. One smaller droplet could happen to meet with very moderate velocity with a slightly larger drop and temporarily merge with it into a larger globule of water. The resultant bigger droplet may oscillate or quiver like jelly for a moment until the surface tension on the containing envelope is somehow broken. At that time the enlarged drop of water may suddenly split up into new parts again. Such an accidental encounter could, on occasion, be actually observed, but its significance in the larger macroscopic world of our own environment may safely be ignored.

It is possible that in some reactions which involve the collisions of two constituent nucleons, the incident energy of the projectile particle becomes transformed upon the impact on the target atom and it becomes an indistinguishable part of what becomes a "compound nucleus." Often a heavier isotope of the original target atom may be created, such as the conversion of Indium 115 to Indium 116. This is a form of fusion.

Section 12.2 - The Delta Particle

The symbol for the Delta is the Greek letter "Δ."

Early experiments at the University of Chicago in 1952 with the scattering of positive pions in hydrogen gas with the use of a 480 MeV capacity cyclotron machine gave tentative evidence that it was possible for a positively charged pion to briefly coalesce with a proton. In the great majority of cases the pions would ricochet off in various directions, but the positive kaon having the same charge as the proton was much more likely than the negative version to form a resonance peak in the scattering measurements. Enrico Fermi, who was in charge of the team at the time, suggested that these were resonances and evidence of a "phase shift" in isospins. Resonances usually exhibit spins of 3/2, 5/2 or even 7/2.

Subsequently very careful work by J. Askin with a cyclotron accelerator at the University of Rochester verified that there was indeed a resonance having a mass of 1232 MeV. It appeared to be the result of a head-on collision of a proton pion with a proton at relatively low velocities. All experimental evidence indicated that at a certain definite energy level a pion particle with a positive charge will be attracted to and for a fleeting instant combine with that proton to form a new and separate entity having the unusual attribute of two positive electric charges. The latter has been designated the Delta plus plus or Δ^{++} and it quickly decays back into its separate original identities. Simple addition proves such a happening is feasible, and that the two particles may coalesce into a unity for 10^{-23} second or less. Considering only the masses:

938	+	139	+	155	=	1232
(for the proton)		(for the pion)		(for kinetic energy)		

There is some justification for objecting to the Delta double plus as a genuine particle at all. As some wit who had a sense of humor expressed it, that particle is like an automobile which falls apart before it even leaves the assembly line at the factory!

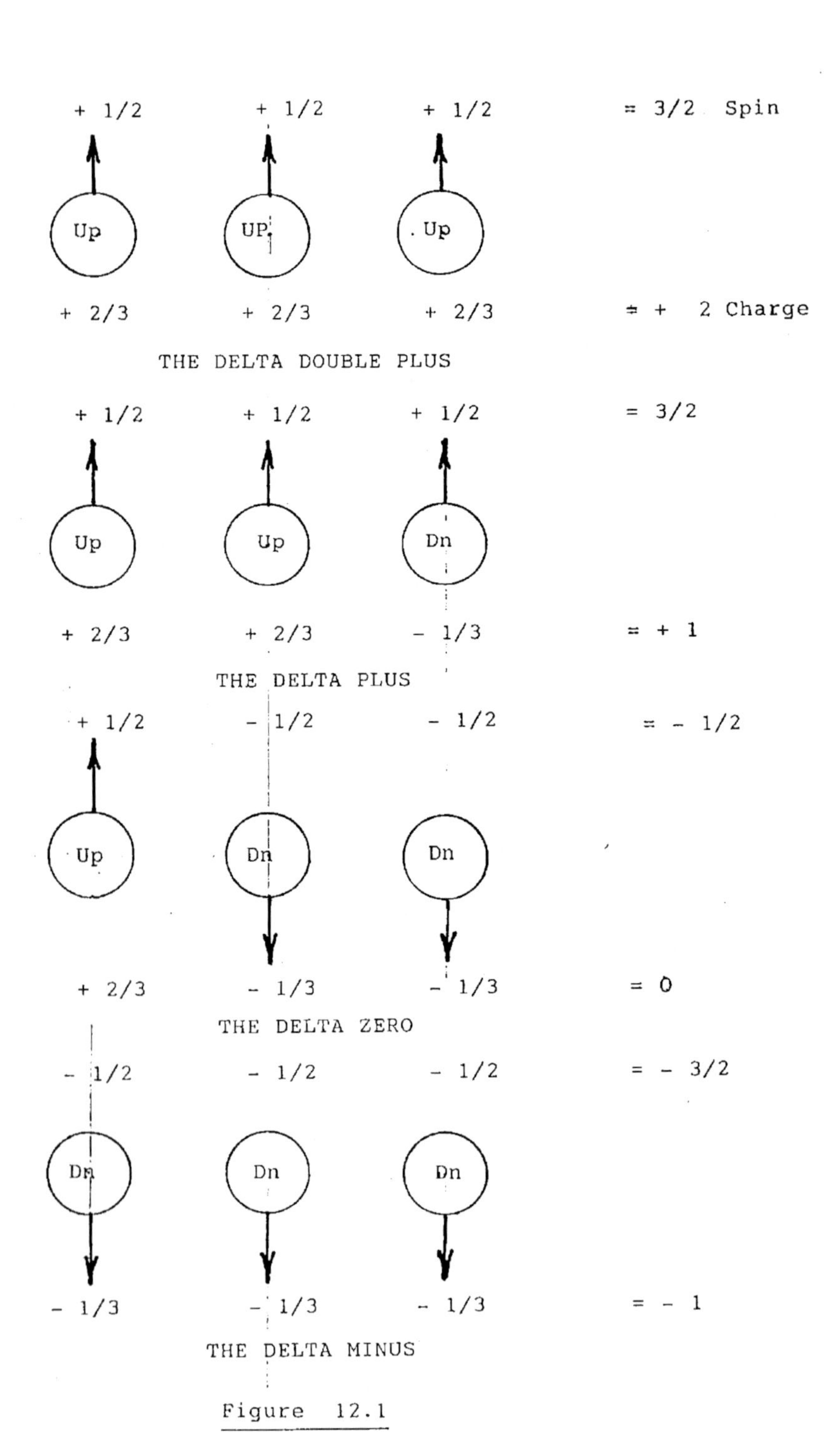
+ 1/2
+ 1/2
+ 1/2
= 3/2 Spin
Up
UP
Up
+ 2/3
+ 2/3
+ 2/3
= + 2 Charge
THE DELTA DOUBLE PLUS
+ 1/2
+ 1/2
+ 1/2
= 3/2
Up
Up
Dn
+ 2/3
+ 2/3
- 1/3
= + 1
THE DELTA PLUS
+ 1/2
- 1/2
- 1/2
= - 1/2
Up
Dn
Dn
+ 2/3
- 1/3
- 1/3
= 0
THE DELTA ZERO
- 1/2
- 1/2
- 1/2
= - 3/2
Dn
Dn
Dn
- 1/3
- 1/3
- 1/3
= - 1
THE DELTA MINUS

Figure 12.1

Behavior such as this is regarded as clear evidence of the strong force which binds the nucleons together in the heart of the atom. Once again we encounter the idea that both the strong force and the weak force are "carried" by the exchange of "virtual" particles which may be bosons. Today it is generally assumed that these two forces are in some way inter-related and are different manifestations of the same thing which existed in the earliest time of the "Big Bang" in the cosmos. How this could be still evades our full comprehension or exact definition at this point in our story, but the mathematicians are all busily working on a "Theory of Everything" or "Grand Universal Theory" (G.U.T.) just to prove it. Bosons are genuinely mysterious.

Ashkin first published the results of his research team's experiments in the *Physical Review* in 1956.

The Δ^{++} is a baryon with a "J" spin of three halves.

Today there are three additional and recognized varieties of the Delta having different electric charges: the Delta Plus (Δ^{+}) the Delta Minus (Δ^{-}) and the Delta Zero (Δ^{-}). As with all resonances their masses, at best, are only an average of laboratory observations and range from 1600 to 1632 to 1700 respectively. All four of the Deltas decay almost immediately into nucleons and pions.

By now it should have become clear to the very patient reader that the concept of hard and durable particles in the atom must be discarded. The very word "particle" has come to cover a myriad of different and peculiar attributes which can scarcely be associated with material objects in our familiar macroscopic world. The essential importance of the Delta in all this is that it became a stepping stone on the path to a better comprehension of the structure of the heart of the atom.

Section 12.3 - The K^* Meson

It may seem quite whimsical to introduce this relatively unimportant particle into our history, but it was one of the first mesons and lighter–weight resonances to be discovered in the 1960 learning decade. The asterisk superscript is a deliberate intent to avoid its confusion with the ordinary kaon which we have encountered previously.

Although the discovery that shooting negative high energy pions at hydrogen protons created resonances originated at the Lawrence Radiation Laboratory in Berkeley, California, it was the "Cosmotron" in Brookhaven, New York, which identified the K*.

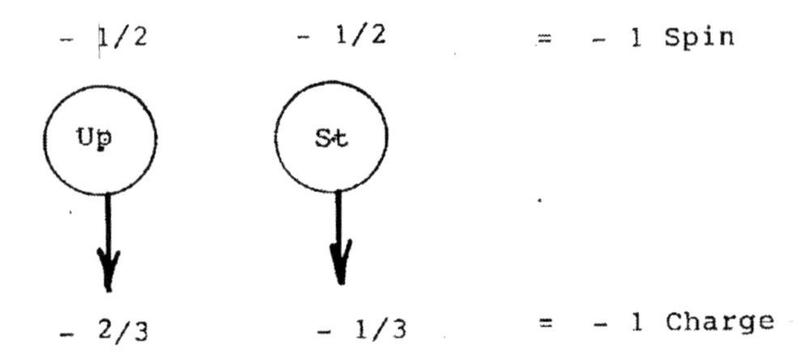

The work was under the supervision of M. Alston in 1961. Its rest mass calculated out to be 891.7 MeV or considerably heavier than the ordinary kaon, but still slightly less than the proton. The K* was peculiar indeed! Not only did it have a negative charge, but it also had a negative spin of minus one. This latter characteristic makes it eligible to be considered a vector boson. The reader may recall from Chapter 3 that this enables it to occupy the same space or location in space as other bosons at the same time so it should be no surprise there are ten other possible variations of the K^* having masses ranging from 1270 to 2045 MeV. Because of its mean lifetime of 10^{-8} seconds it is also considered a strange particle. Its usual decay mode is into an ordinary kaon and a pion, although the loss of some surplus energy by the radiation of a gamma ray photon is not impossible. Clearly something is going on here that requires a more profound insight into the formation of particles.

Section 12.4 - Bosons

Far back in Chapter 3 we discussed bosons in a rather casual manner and identified them simply as particles which had a spin, "J," of either zero or unity and its multiples. The first boson which we actually studied in detail was the light ray photon with a spin of zero. As it happens, the pion plus and its anti-particle, the pion minus, are both bosons also. The fundamental reason for this separation of all particles into either multiples of one half or unity spins is not completely understood, but it is believed to have something to do with broken symmetry. The entire subject of bosons is extremely perplexing and the uninitiated amateur is inclined to suspect that if Nature had not invented the distinction, the scientists would never have missed it. Bosons have no objection to grouping together and are exempt from Pauli's Exclusion Principle.

Unlike fermions the quantum wave function equation for bosons does not change algebraic sign when one such particle is moved in three dimensional space to the same dimension as its twin. That is, Schrodinger's wave function Ψ remains positive and symmetrical. It is quite frequently stated in reference texts that bosons are the consequence of the symmetrical wave function in quantum mechanics, but that argument is more of a circular syllogism than a good explanation. The question about bosons is also reflected in entropy and thermodynamics. Satyendra Nath Bose (1894 to 1974), in whose honor the boson was named, was attempting to explain Planck's constant by means of classical statistical analysis of gases or large numbers of molecules when he became involved in particle physics. He was the son of an accountant living under the colonial domination of the British in India. His father helped found the East India Chemical and Pharmaceutical Works and upon the advent of the great Independence movement thus assured financial security and educational opportunities for his family. After school in the province of Bengal, Satyendra attended the Presidency College in Calcutta and eventually obtained his Master of Science degree as a Mathematics Major in1915. He was a gifted student who loved poetry and was fluent in the languages of Bengali, French, and English. His first published paper was on quantum statistics in 1920. He became a professor at the University of Dacca in East Bengal in 1921. He became engrossed in proving the reality of Planck's constant for black–body radiation by means of a statistical analysis of photons of specific wave lengths or energies in a gas cloud having known volume and total number of particles. This was something that Planck was never able to do himself. Bose tried to publish his success with the mathematics in 1923 but, as an obscure Indian, he was ignored. He then appealed to Einstein and mailed him his paper. Einstein was immediately impressed and took the trouble to translate it into German and urged its publication in 1924. It happened to fit in beautifully with Einstein's own ideas on photon quanta and also their photoelectric effect on certain metals and chemicals. Bose very soon received his just recognition, and their mutual collaboration results became known worldwide as the Bose-Einstein statistics for quantum mechanics.

A consequence of their work was to demonstrate that certain specific particles such as photons or other bosons at very low temperatures and compressed volume tend to condense together in a curious low energy state with little or no increase in pressure. This behavior is not only verified by a loss in viscosity of cryogenic helium 4 isotope, which is a boson atom, but it also helps to explain the fright and genuine anxiety of some scientists when high energy particle beam accelerator machines were first being developed. They feared that the unintended condensation of super-dense matter might start a chain sequence. We already mentioned this apprehension in Chapter 10. A more positive use of bosons came much later with the discovery that boson elements helped to make superconducting wires for electric magnets and even transistors for high speed computers.

Clearly the boson has to be associated with change and instability of some kind. For that reason alone it was very early surmised to involve the transfer of the weak force. Although not entirely wrong, that conjecture was a trifle too simplistic. Yet it could not be denied that bosons were endowed with the ability to transfer mass or electric charge to other particles, as we have just learned. Moreover, unlike fermions, bosons do not necessarily conserve their spin number. Nevertheless, the vague idea of a "weak force" connection ultimately led to a new theory known as "Weak Neutral Currents" and in the 1980's decade this became the justification to find the legendary "Higgs Boson" by building an eight to eleven billion dollar underground circular "Super-collider" in the chalk strata of southern Texas. This was a sure sign that bosons had acquired crucial importance!

The idea of "carrying" forces needs some comment. It is really a reversion to the thinking of Isaac Newton and indicates the intense aversion of all scientists today to any "action at a distance without a physical cause or force" proposal. Newton himself was disturbed by this aspect of gravity in his famous dissertation and even admitted to baffled ignorance as to its real cause or mechanism in spite of his classic equations based upon its measurement. Thus students of the atom today immediately seized upon the idea that the strong and weak forces inherent in the nucleus of the stable atoms must involve the sub-microscopic movement or exchange of secondary particles. Bosons or "virtual particles" have been pictured as instantaneously bouncing back and forth between fermions like so many beach balls. This conviction had become such a compelling doctrine that as yet totally undiscovered particles have been invented in order to explain other forces, such as "gravitrons" for gravity and "magnetrons" for magnetism. Whether such a mechanistic visualization has any real justification remains to be proven in the future, but the reader is certain to encounter this approach repeatedly in the literature. Be assured, however, that the best brains among the theoreticians were not satisfied with such explanations as adequate to truthfully represent what was happening and were conferring among themselves for some better truth. Bosons, in a sense, have become the ghost operators governing the structural formation of what we know as matter from raw energy.

CHAPTER 13

SYMMETRY AND THE EIGHT-FOLD WAY

By the end of the 1960's decade the physics theoreticians were just beginning to come up with some plausible explanations for what was happening in their accelerator machines. They also fell in love with the concept of symmetry and an arcane branch of mathematics called "group theory."

Perhaps the most elegantly expressed explanation of just why the scientists became fascinated by symmetry was made clear by Chen Ying Nang in his Nobel lecture of December 11, 1957, which deserves to be quoted

"It was, however, not until the development of quantum mechanics that the use of symmetry principles began to permeate into the very language of physics. The quantum numbers which designate the states of a system are often identical with those that represent the symmetries of the system. It is indeed scarcely possible to over-emphasize the importance of the role played by the symmetry principles in quantum mechanics. To quote two examples: The general structure of the Periodic Table is essentially the direct consequence of the isotropy of Coulomb's Law. The existence of the anti-particles - namely the positron, the anti-proton, and the anti-neutron - were all anticipated as consequences of the symmetry of physical laws with respect to Lorentz Transformations. In both cases Nature seems to take advantage of the simple mathematical representations of the symmetry laws."

One of the most brilliant and imaginative of the new generation of mathematically inclined physicists was born in New York City, but later settled during his young manhood on the intellectually fertile soil of California. Murray Gell-Mann was born on September 15, 1929, to a very cultivated father who was proprietor of a language school in Manhattan. Although the child was clearly mentally gifted, his first encounter with physics during High School was not successful, even though his father urged that subject upon him as a better prospect for making a living than some of his other interests, such as archaeology and natural history. Young Murray found it boring! Notwithstanding, his grades in other courses were high enough to allow him to enter Yale University at the age of fifteen and to graduate with a Bachelor of Science in 1948. By this time he had become intrigued with physics and transferred to the Massachusetts Institute of Technology as an assistant to Victor F. Weisskopf in order to achieve his PhD in 1950 at the age when most college students are just completing their undergraduate curriculum. Thus he entered the post-World War II boom in atomic energy and particle physics at the best time, taking a position at the Institute for Nuclear Studies at the University of Chicago as an Associate Professor. He moved to the California Institute of Physics in 1955 to become a full Professor only a year later. California became his permanent home.

One of the first perplexities about mesons which attracted Gell-Mann's curiosity was what we have already designated as "strangeness" and which is demonstrated by the Kaons, Lambdas, and Sigmas. He wrote a paper on the subject in 1953 and predicted the possibility of the Xi particle, the existence of which was confirmed in 1959. The whole idea of the attribute of strangeness as an additive number, rather than a multiplier, was derived by Abraham Pais, a French emigre. Murray Gell-Mann, an American, and Kazahiko Nishijima, a Japanese, added subsequent refinements.

Strange particles are peculiar in ways other than their anomalously long lifetime periods. When one decays, the strangeness factor or number may change, although the electric charge is conserved and the mass decreases. The spin of strange mesons may change from 1/2 to 3/2 and the particle may still retain the same electric or valence charge.

During the 1960's decade there arose a great deal of speculation on the significance of mesons relative to the internal structure of the nucleons and atomic forces. This was the logical consequence of the flood of new particles and resonances being produced artificially by the accelerator machines. Several different organizational systems were suggested and enjoyed a brief vogue, only to perish as inadequate to meet all criticisms. There was a general consensus on one philosophical point. None of the mesons and their baryon offspring could be regarded as truly basic and elementary particles in the stable reality of our everyday world. Their characteristics strongly indicated some fundamental process which governed the transformation of energy into what we call matter. The ensuing ferment of free discussion between scientists would require an entire book to chronicle fairly. As is always the case in the development of human knowledge, no one man thought up all the ideas which gained ultimate consensus. Comparison of the various properties and behavior of all the newly discovered mesons and resonances, as well as anti-particles, led the scientists to the belief there was an implicit symmetry to the organization of all matter, although it was not absolutely clear why this had to be a necessity. Gradually they developed a composite theory of "CPT" symmetry involving the conservation of charge (C), conservation of Parity (P) and the applicability of time reversal (T). This latter assumption seemed so intuitively sensible and familiar in nature, but, alas, it turned out in future research that it was not always true. Notwithstanding all the later discovered complications, C P T helped to make some rational sense out of the welter of new particles and hadrons which had suddenly appeared in their scattered research laboratories.

The idea of symmetry is a very old one and is often exemplified in the beauty of a snowflake. The actual mathematical basis for the concept was first elucidated by a brilliant, but isolated and obscure, mathematician in Norway named Marius Sophus Lie, who lived between the years of 1842 and 1899. Obviously he had no knowledge of nuclear physics at that time and was primarily fascinated with all the various innovative forms and geometrical shapes in several dimensions possible as an abstract mathematical exercise. This strange connection between mathematics and the laws of our universe has inspired great wonder in many philosophers over the ages. Sophus Lie spent the greatest part of his scholarly life at the Christina, where he developed his own integration solution for partial differential equations. He also later developed a new analytical method of non-Euclidian geometry for figures in "n" number of dimensions by the translation of lines instead of points. It is to him that we owe the term "SU – 3" or special unitary group of three dimensional symmetry. These groups all held in common the feature that there may be a set of rotations of some rigid three–dimensional body which consistently preserves one fixed point.

It was Yuval Ne'eman who first perceived the mathematical relevance of symmetry studies to particle physics sixty-eight years after the death of Sophus Lie. Ne'eman was another unusually colorful and versatile figure to enter the physics era at this most important time. His career was one of the most unconventional amid a number of brilliant intellectuals who appeared on the scene as the 1960 decade began. Born in Tel Aviv, Israel, on 14 May, 1925, he obtained a Bachelor of Science degree from the Technical Institute in Haifa in 1945, and a diploma in Mechanical Engineering a year later. He then took a job in a pump factory as a design engineer in hydraulics and hydro-dynamics for two years, but became bored with that career and joined Hagana, the Jewish underground. He became a Major in the Israel Defense Forces in 1948 and advanced to the rank of Colonel in 1955, then becoming the Director of the Intelligence Division from 1955 to 1957. As part of his military career in 1952 he had been sent to the Ecole de Guerre in Paris, where he obtained a graduate degree in Physics. He ultimately became a military Attache to the Israeli Embassy in London from 1960 to 1968 and while there continued his physics studies by means of a connection with the Imperial College. While at the college in London, where he was pursuing a doctorate in theoretical physics at the age of thirty–five he had his first paper published in the European *Journal of Physics* on the application of symmetry principles to the strong force reactions between meson fragments.

By this time Ne'eman had become totally fascinated with theoretical nuclear physics and resigned his Army commission. At the age of thirty–eight he joined the teaching staff of Tel Aviv University to become the head of the Physics Department in 1963. It was during the interim period of 1961 to 1962 that his published articles on symmetry in particle physics attracted attention and he became a collaborator with Gell-Mann on the Eightfold Way theory. This led to his being invited to become a Visiting Professor at the California Institute of Technology from 1964 to 1965.

Section 13.1 - The Eight Fold Way

Murray Gell-Mann, meanwhile, had first introduced his notions on symmetry in a report given at the Rochester Conference Number 11 in Geneva in 1961. However, he subsequently had a paper published in the United States in the *Physical Review* journal in 1962 under the title of "Symmetries of Baryons and Mesons." His book on the "Eight -fold Way," which was a collection of reprints, did not actually appear until 1964 and quickly became a required textbook for all physics graduate students.

Concomitant with and emphatically an integral part of the new theory for explaining the myriad of particles was the idea of "Isospin" or Isotopic Spin, as it is sometimes called. This quantity was first proposed by Werner Heisenberg. It had become recognized that there were groups or "families "of hadrons of closely similar masses which appeared in pairs or triplets (but never more than four) and that conservation of Isotopic Spin helped to explain this odd fact. The exact reason why this should be is still not entirely understood, thus making this number somewhat of an "ad hoc"statement based upon observational statistics, rather than a theoretical deduction or intuitive insight.

Isospin is given the symbol of "I_z" and is related to and quite compatible with the ordinary spin "J" which we have been talking about all along. However, the concept of internal symmetry and combination of different spins in two dimensions (I_x and I_y) within a given meson or hadron which yield a certain combined angular momentum was introduced.

Isospin is a rather arbitrary mathematical construction which involves not only the electrical charge, but also the baryon and strangeness numbers. For a consistent image the scientists prefer to think of Isospin as the projection upon a third axis of spatial coordinates of a vector in an imaginary space of three dimensions. This vector combines three different attributes of a given particle with the electric charge weighted.

The mathematical relationship of the electro-magnetic charge (Q) to the isotopic spin (I_z) and to the baryon number (B) and the strangeness factor (S) is given by the following equation:

$$Q = I_z + \frac{B + S}{2} \quad \text{or alternatively:} \quad I_z = Q - 1/2\ (B + S)$$

B + S is often designated as "Y"or hypercharge.

The entire concept originally resulted from studies of strangeness and was introduced almost simultaneously by Kazahiko Nishima of Japan and Murray Gell-Mann of the United States. Refer to Figure 13 - 1 for an illustration of the symmetry octet for baryons, which was one of Gell-Mann's first constructions.

SU-3, or the three dimensional symmetry theory, is a highly abstract and intellectual concept which is not describing any hard, concrete object which one might touch or feel. The fundamental assumption from the start is that certain groups of mesons or baryons are somehow related to one another and can be changed or "transformed "into one another by substituting different values for a number of basic and measurable characteristics shared by this certain group of particles. It also became apparent that these related particles came in multiplets or groups of six, eight, ten, or even twelve.

As we have implied, the idea of symmetry in the world has long fascinated scientists, so it was probably inevitable that someone would eventually make the visual comparison of a three dimensional lattice similar to a natural crystal having a nuclear particle inhabiting each face of the crystal, but not the corners or apexes. In a study of perfect geological crystals it was discovered that, by means of rotating one about some selected axis, any given crystalline shape could be transformed into a new and unfamiliar two dimensional plane view. Reversing the rotation (or perhaps even continuing it) will bring it back to the its normal original appearance. These transformations themselves could be sorted into special groups. Quite naturally, this study in crystallography became known as Group Theory. Thus the mathematical analogy of conceiving of certain special particle groups as belonging to an entirely imaginary, but symmetrical, crystal becomes comprehensible. Thus the concept of symmetry in nuclear physics became popular.

Those very special attributes or characteristics of any certain particle which could be identified and quantified by numbers and then manipulated in mathematical processes are commonly called "quantum "numbers. A list is included below:

(1) Angular momentum or spin (J), which may be either numbers 1, 2, 3, or 4 or else 1/2, 3/2, 5/2 and so on depending upon whether the particle is a boson or a fermion.

(2) The baryon number (B), which may be + 1 for all particles having a mass equal to or greater than that of the proton or zero for particles with a mass less than that of the proton and - 1 for anti-baryons or anti-matter.

(3) The mass of the particle (m), which may be slightly approximate, is generally measured in millions of electron volts (MeV).

(4) The degree of strangeness (S) may vary from zero to - 3.

(5) The total magnitude of the electrical charge of the particle (Q) as measured against the electron, may be positive (+ 1 or + 2) or neutral (0) or negative (- 1 or - 2).

(6) As we already discussed previously, the Isospin relationship is determined by $I_Z = Q - 1/2\ (B + S)$.

(7) Parity (P) is either even (+ 1) or odd (- 1); but this need not concern the reader for the present and will only become important later in our story.

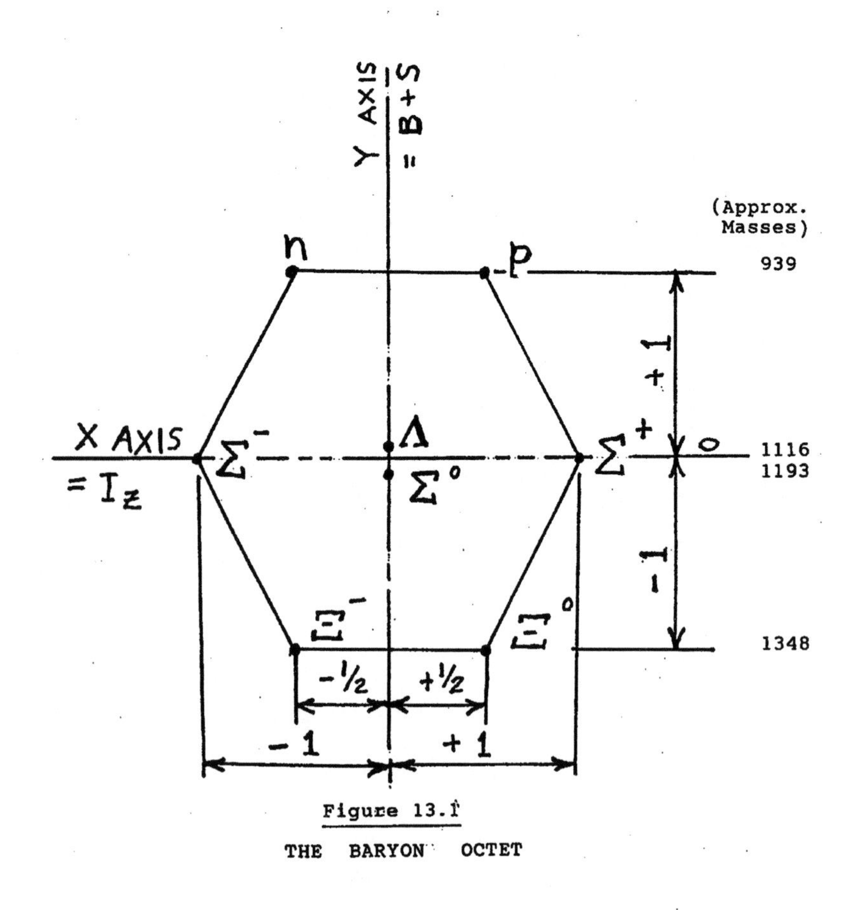

Figure 13.1

THE BARYON OCTET

The different states of related particles in any given multiplet are plotted as points on perpendicular Cartesian coordinates in a two dimensional plane on paper. The horizontal axis or abscissa is the isospin number (I_z) and the vertical axis or ordinate
(Y) is the sum of the baryon number and the strangeness number. It is sometimes called the "hyper-charge."

The major appeal of the Eight Fold Way is that it seems to work. The name was the whimsical invention of Gell-Mann. This was entirely consistent with his humorous and trenchant personality, since he was exceptionally well read in a great diversity of subjects. Not only did he have a talent in linguistics and had studied, among others, both the Sanskrit of India and the Mayan hieroglyphs of Central America, but was an enthusiast for archaeology and traveled the tropical jungles in search of wild animals. He ultimately came to pursue what became a vocational interest in natural history and evolution. The Eight Fold Way is an allusion to the eastern Indian Buddhist teachings of the possible paths to enlightenment with the hope of an ultimate culmination, after many lives, into either the status of a Bodhisattva or heavenly Nirvana. The noble Eight Fold Way or Middle Way consists of Right Views, Right Resolve or Intention, Right Speech, Right Action or Conduct, Right Livelihood, Right Effort, Right Mindfulness, and Right Concentration."Right "in this translation, of course, has the meaning of "correct" or desirable in harmony with divine laws.

Gell-Mann has always insisted that his quaint name for his organization of particles by symmetry was never meant to imply that the Hindu priests or Buddhist monks of past ages knew anything about nuclear physics and the atom. Notwithstanding, the "New Age" culture of southern California, as influenced by the infusion of eastern religions from the "Pacific Rim," has encouraged the persistence of this fiction.

Section 13.2 - The Pseudo-Scalar Mesons and the Eta Particle

A second octet proposed by Ne'eman and Gell-Mann jointly was the grouping of the strong force mesons known as pions and kaons. The known varieties of these particular mesons accounted for only seven related particles when first constructed. (See Figure 13.2.) This encouraged Gell-Mann to predict boldly the existence of an eighth meson close to the center of the Isospin coordinate at the zero station. This eventually became known as the Eta Zero or η_0 and was happily confirmed by the Berkeley Laboratories in California many months later. It is now assigned a mass of 547.45 MeV.

The actual details of the discovery of the Eta particle demonstrates that the scientists were by this time beginning to know what they were looking for and how to detect and identify it. A. Pesvner from the Johns Hopkins University and M. Block from Northwestern University went looking for the Eta Zero in photographs made of the tracks in heavy deuterium water inside a seventy–two inch bubble chamber which had been constructed by Alvarez. This three pion resonance was produced by a 1.23 GeV / c energy beam of positive pions produced by the Bevatron accelerator machine. Pesvner published his findings in a letter to the *Physical Review* journal in 1961.

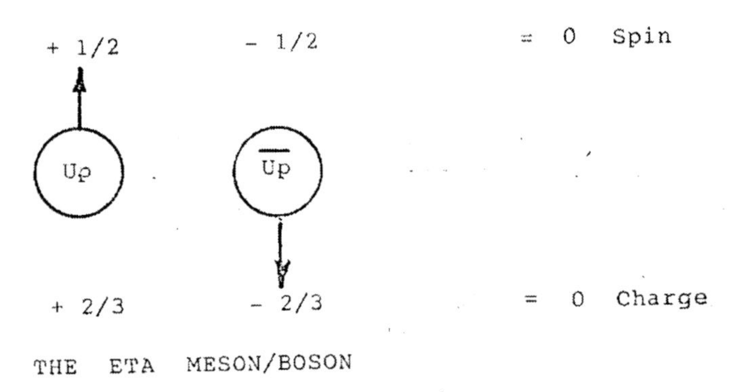

The symbol for the Eta particle is, of course, the Greek letter "η ." It possesses all three attributes necessary to qualify as a meson, a resonance, and also a boson. Moreover, it has a spin of zero and an electric charge of zero. This last combination certainly contributed to the general skepticism with which Gell-Mann's prediction was first greeted.

Section 13.3 - The Rho Resonance

The symbol for this particle is the Greek letter "ρ."

The Rho has an average mass of 769.9 MeV, which amounts to roughly three quarters that of the proton and thus makes it a meson. It has a vector spin of one, which consequently makes it a boson also—an important distinction.

The Rho meson comes in three different possible electric charges: positive, negative, and neutral.

Using symmetry rules, a student at Harvard named John J. Sakuri predicted its possible existence and Gell-Mann formulated some equations to compute its possible mass whenever discovered. A team of researchers led by A. R. Erwin verified its fleeting existence with the "Cosmotron "accelerator machine at Brookhaven, Long Island, by firing negative pions at proton targets. As we have seen, this procedure was amazingly successful in producing resonances of all kinds.

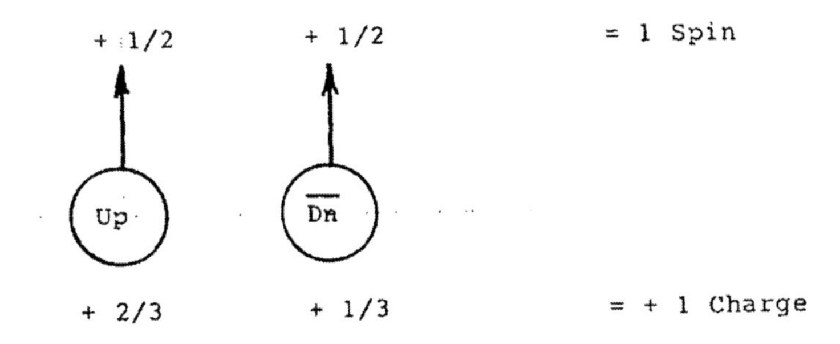

THE POSITIVE RHO MESON

The Rho most usually decays back again into two pions but occasionally splits into one pion and a gamma ray photon.

Section 13.4 - The Phi Meson

By 1962 it had occurred to the scientists to make a switch on their projectile particles in the accelerators and to discover what might happen if they bombarded hydrogen protons with kaons instead of pions. There were two separate teams, one being led by Ticho and Schlein from the University of California using the "Bevatron" machine and a seventy–two inch diameter bubble chamber and the second led by Connoly from the Brookhaven National Laboratory and Goldberg from the University of Syracuse using funding from the Atomic Energy Commission and the "Cosmotron" machine with a smaller twenty inch diameter bubble chamber. The result was a kaon conglomerate resonance which ultimately became designated by the Greek letter "Phi" or "ϕ."

It had a mass of 1019.41 MeV, which is slightly higher than that of the proton and very close to that of the Lambda particle. The electric charge was neutral or zero and the spin "J"was unity. The strangeness number, however, was calculated to be minus two.

The Phi resonance is really a fleeting amalgamation of two kaons, either one positive and the other negatively charged or else two neutral. The fact that it was given a spin of one is an oddity that puzzled the theorists for some time, since kaons individually heretofore had a spin of zero.

Not unexpectedly, the Phi particle decays back into its original components.

$$\phi \rightarrow K^{+} + K^{-}$$
$$\phi \rightarrow K^{0} + K^{0}$$

The two individual masses of the decay products of the Phi resonance total in the vicinity of 994 MeV, which is so close to the rest mass of the original mother particle as to leave almost no energy left to cause the fission at all. This circumstance, plus the unusual strangeness, was a great perplexity to George Zweig, who had obtained his PhD at the California Institute of Technology and subsequently took an assistantship job at the CERN facility in Switzerland. He independently came to the conclusion that all this could be explained if the actual building blocks of the Phi particle had *fractional* charges instead of the customary inviolate unity. These unknown and speculative sub-constituents he called "aces" and ventured his daring ideas in an internal house journal of the CERN organization in 1964. Although this was in direct competition to the unpublished notions of Gell-Mann and Serber, who were his seniors, Gell-Mann generously supported Zweig's arguments in overall concept and recommended him for the faculty at Cal-Tech. They differed at first, however, on whether to call them "aces" or "quarks"and whether they were actually physical matter or merely convenient mathematical concepts.

Section 13.5 - The Omega Minus Particle

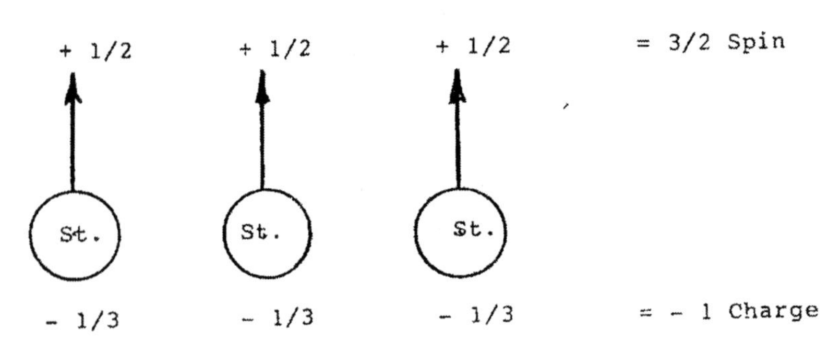

The symbol is the last letter in the Greek alphabet, whether written in the capital form of omega (Ω) or the lower case (ω). The designation was selected with the intended implication that this was the final capstone in the Eight Fold Way bridge to comprehension of nuclear forces.

In 1962 an important conference was held at Rochester, New York, in order to discuss the manifold complexities of mesons, resonances, and transitory baryons. Also under discussion was the viability of the symmetry theory to explain "multiplets "or related systems of ten or more particles instead of merely eight. The ten particle arrangement, incidentally, was given the atrocious name of "decouplet" after the Latin word "decem." Polyglot scientists of mixed nationalities are capable of abusing the English language.

During the conference discussions Murray Gell-Mann predicted from the floor the possibility of the existence of a weakly decaying particle which was not a resonance, but having a strangeness factor of minus three and an estimated mass of about 1680 MeV. Another prominent physicist, Yuval Neeman, concurred with this supposition at the same conference.

After the conference had adjourned, Nicholas Samios and Ralph Shutt at the Brookhaven National Laboratory decided to search for the Omega. Using the Alternating Gradient Synchrotron there to accelerate negative charged kaons to energies of 5 MeV, they observed the collisions in a hydrogen liquid bubble chamber. The collision of the kaon projectile with a proton always caused a shower of fragments, sometimes producing an Omega minus and a positive kaon. After two years of experiments, they were convinced that they had proof of the existence of the Omega particle and published their results in an article titled *Observations of a Hyperon with a Strangeness of Minus Three* in February of 1964.

This was a huge tribute and acknowledgment of Gell-Mann's Eight Fold Way, of course, and clinched the SU-3 symmetry theory, thus winning the inventor a Nobel prize.

The actual peak mass came out to 1672.4 MeV. The other properties were a baryon number of one, an angular momentum spin of 3/2, a negative electric charge, and a strangeness number of minus three. The mean lifetime of the Omega Minus baryon is 0.822×10^{-10} seconds, which is terribly long as things go in the world of particle fragments. The decay process was reconstructed as follows:

$$K^- + p \rightarrow \Omega^- + K^+ + K^0$$
$$\Xi^0 + \pi^-$$
$$\Lambda^+ + \pi^0 \rightarrow p + \pi^- + \text{Gamma Rays}$$

From the theorist's point of view, the mathematics demanded a decouplet of ten different particles. Refer to the 3/2 plus decouplet diagram for illustration.

The triply strange Omega minus was a hugely significant event. It captured intense international interest when the Brookhaven National Laboratory on Long Island announced that out of 80,000 photographs of tracks in their eighty inch diameter bubble chamber they were convinced they had one clear picture which identified the predicted Omega minus particle. Obviously, this convinced the scientists of the plausibility of the Eight Fold Way for a systematic arrangement of hadrons and their properties. Yet, in spite of its spectacular success, there were many questions still left unanswered. The suspicion arose that only various combinations of fractional electric charges, rather than the restriction to only the unit charge, could explain the strange particles and other difficulties in categorizing subnuclear reactions. This ultimately led to the idea of "quarks" down the line chronologically and the ultimate displacement of the "Way" as a theoretical tool and its demotion to a status only a stepping stone towards what eventually became known as the Standard Model.

This somewhat tendentious history of the gradual discovery of numerous mesons with different degrees of characteristics risks boring the average reader to distraction, but its inclusion is intended to emphasize how the ever increasing complexity of the particle studies drove the physicists to the conviction that a totally new insight into the theory was necessary. The proposal of fractional charges seemed like heresy to most of the physics fraternity.

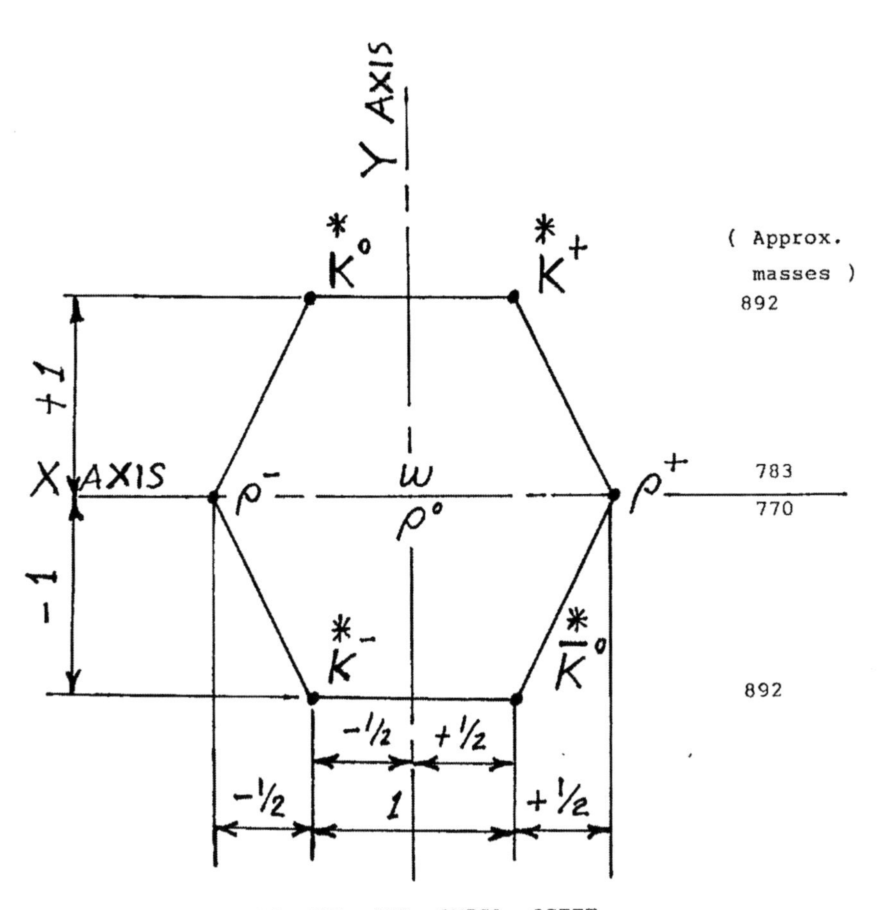

THE RHO AND OMEGA OCTET

(The asterisk * indicates a heavy Kaon resonance, as distinguished from the lighter normal.)

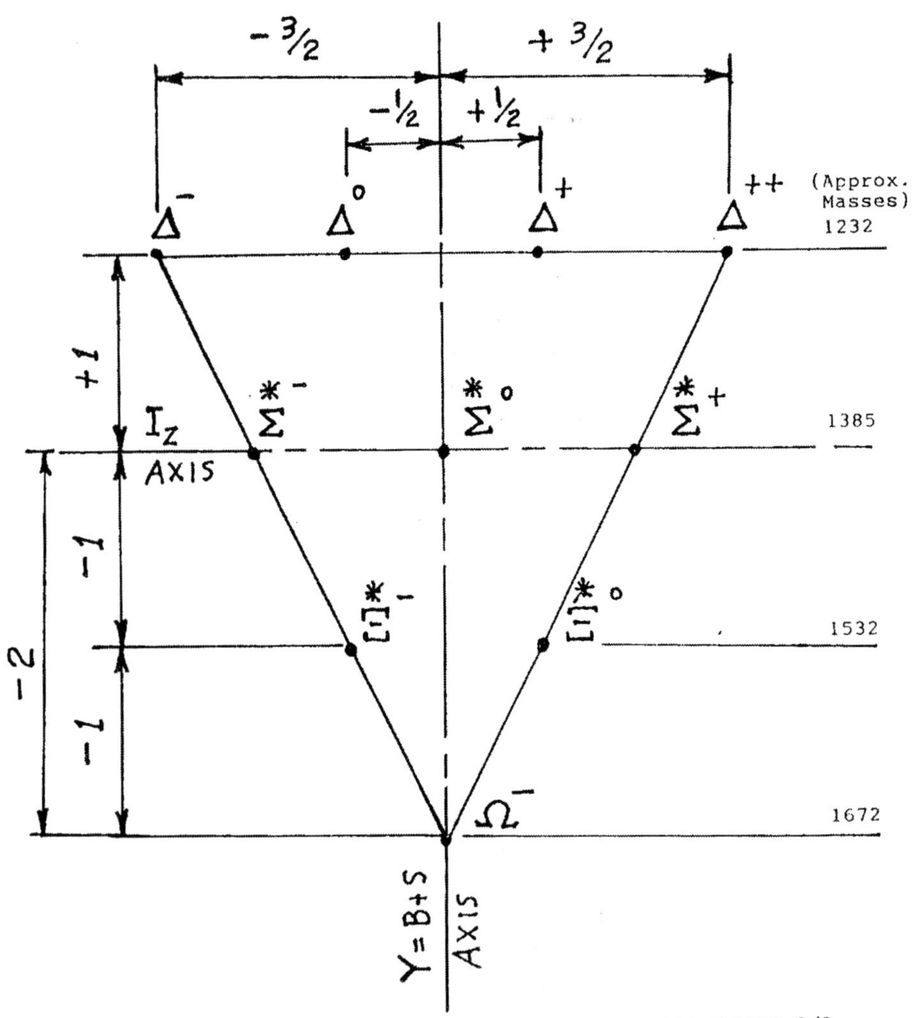

THE TEN MULTIPLET FOR OMEGA MINUS & SPIN PARITY 3/2

Figure 13.3

CHAPTER 14
SOME UNFINISHED BUSINESS

Section 14.1 - The Anti-Neutron and Anti-Matter

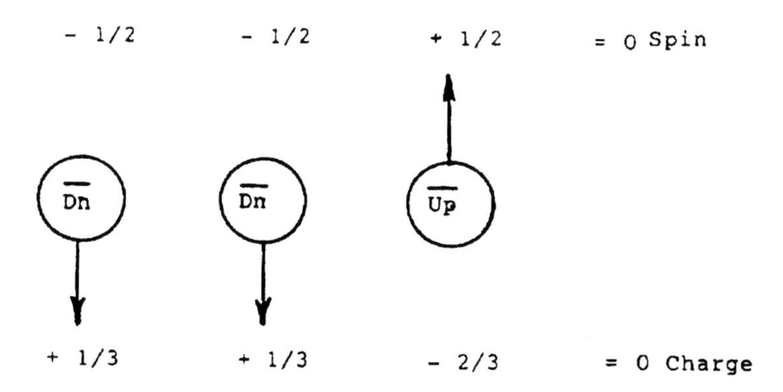

The reader should not receive the impression that all the research in particle physics of the 1950 and 1960 decades was totally preoccupied with resonances and there were not other aspects of nuclear science which deserved attention, then as well as now. Nevertheless, the specialized discipline had progressed a long way from the bewilderment of the 1930 and 1940 decades. Judging from their experience with mesons, it was now assumed that every particle was capable of having a "ghost" anti-particle of opposite electric charge and magnetic moment, although these were expected to be quickly annihilated in the hostile environment of our own world with its huge preponderance of stable and "normal" particles. Having discovered and confirmed to their satisfaction the existence of the positron and anti-proton, it quite naturally occurred to the scientists that there might be an anti-neutron.

One's first reaction to this notion might well be that an anti-particle to a zero charge or neutral particle is an absurd contradiction and impossibility, but the idea is based upon the belief that the electromagnetic charges in any particle, whether they be positive or negative, had to go somewhere and be accounted for in any decay process or collisions. In typical academic language this assumption became known as "charge exchange" or "charge conjugation." The charge of an electromagnetically neutral particle remains zero, however. The direction of spin or magnetic moment was expected to change with clockwise or right-handed spin becoming counter-clockwise or left-handed, as dictated by the prevailing notions of mirror parity.

The first experiments for investigating the reality of the anti-neutron were as early as 1955 and were the logical offshoot of the successful production of anti-protons. These negatively charged protons were the occasional and random result of colliding a dense proton beam from the Bevatron machine at Berkeley, California, with a metallic target. A team of scientists, including Goldhaber and Segre from the Lawrence Laboratory and Amaldi from Rome, "siphoned" off the anti-protons to one side magnetically. They then slowed them down by passing them into a copper block absorber and those particles passing through were directed to a thick stack of photographic emulsion plates. At rare intervals they were rewarded with a picture of a star shaped cluster representing an explosion of sub-particles and gamma ray energy. This unusual burst was clear evidence of a surplus of energy far beyond that calculated for the mass of a proton plus its added kinetic energy. Its source was construed to be the mutual annihilation of an anti- proton and a neutron. These results were eventually published in 1956 in both the United States and Italy under the name of Owen Chamberlain, who was the Director of the Berkeley Laboratory.

Stepping up the momentum of the proton beam particles to 700 MeV/c dramatically increased the effectiveness of the entire procedure and produced thirty five photographs of the anti-proton "star cluster" annihilation reaction. These results were than cautiously published in the *Physical Review* under the title of *An Example of an Anti-proton Nucleon Annihilation*. The anti-neutron was surmised.

This success inspired another physicist at Berkeley, named Bruce Cork, to seek more convincing evidence for the anti-neutron with the help of three assistants, much in the manner of Lewis Carroll's nonsense verse "The Hunting of the Snark." He elected to modify the Bevatron machine in order to make use of a newly invented "strong focusing" technique. Sometimes called a "magnetic lens," this device consisted of a four pole magnetic arrangement for both the horizontal and vertical planes which constricted the beam of accelerated particles delivered into a much narrower diameter. He also introduced the innovation of passing the fragments resulting from the collision of protons or anti-protons with a copper barrier, firstly, through a thick plate of leaded glass serving as a scintillation counter and, secondly, into a liquid toluene bubble chamber as a detector. He was rewarded with 114 clear pictures of the now famous annihilation star burst of an anti-proton with a neutron. These confirmation results were published in a letter to the *Physical Review* in 1956 with the emphatically unambiguous title of *Example of an Anti-proton - Nucleon Annihilation*. The matter had been settled.

The conventional representation of the particle transformations which take place in the reactions just described with unit charge conjugation scrupulously observed is:

$$p + p \rightarrow n + n$$

At this point the writer might introduce to the reader his own favorite reasoning as an artifice to understanding the conservation of charges, although the scientific fraternity might deride such simplistic pictures. Try to visualize the protons and neutrons in an atom as they were regarded back in the times of Neils Bohr as the basic positively charged proton and the equally fundamental zero charge neutron. Around both of these circulate clouds of negative electrons, whether real or virtual.

Pretend that: p becomes p_+ + 2 electrons

$$(p_+ + 2\ e_-) + p_+ \rightarrow (p_+ + 1\ e_-) + (p_+ + 1\ e_-)$$

Thus the end result is:

$n_o + n_o \rightarrow \gamma$ or Gamma Radiation energy burst

The important point is that an anti-neutron particle remains inherently anti-matter regardless of the lack of electric charge and will annihilate its real opposite number upon close and slow encounter. That this situation should exist and actually happen in our own present world is very strange indeed. Some nine years later a team from Columbia University were using the synchrotron in Brookhaven, Long Island, to look for the then rumored quark. They never found one, but did claim to have detected the track of a particle resulting from the collision debris of a proton beam with a copper target which was so massive as to qualify, in their opinion, as an anti-deuteron. That is, by their interpretation, it was the ephemeral mating of an anti-proton and anti-neutron together before exploding into energy. Leon Lederman, who always had an active imagination, christened it the "anti-people particle," meaning that if enough of them were concentrated together they would destroy the hydrogen atoms in the organic protoplasm of the human body. One could not always take his gruesome predictions seriously.

Nature herself does not care for the anti-neutron very much. A beam of 300 to 600 anti-protons might result in only 54 to 60 anti-neutrons at the very most. Remember that, when previously discussing the anti-proton, we learned that each single anti-proton generated in an experiment required at least 44,000 pions in order to glean it. In other words, the anti-neutron is a very rare beast which may be regarded, metaphorically speaking, as a random mutant. Nevertheless, the fact that the experimental work of Cork, Lambertson, Piccioni, and Wenzel back in 1956 made the idea of an anti-neutron plausible was a great consolation to the scientists and an encouraging indication that their theories on symmetry and the Standard Model were at least on the right track.

As with any other anti-matter particles, such as the positron or anti-proton, the anti- neutron will instantaneously annihilate any normal neutron which it encounters. If one reflects upon the anti-neutron at all, it becomes evident that it represents a profound mystery in our universe. Why does the anti-neutron or anti-anything at all need to exist? Is its opposite charge the only attribute which makes it differ from the regular neutron which composes our own familiar world? Obviously not! If one abandons his intuitive perceptions acquired since birth and adopts Dirac's habit of envisioning all particles as merely wave forms in a space-time continuum, then there is a simple solution. Two identical quantum wave trains are shifted out of phase (and thus in time) so that their peaks and troughs in amplitude exactly cancel each other out to nothing. Matter goes back to energy. Nothing could seem more simple and nothing could do more violence to our everyday experience of matter being real, durable, and intractable.

For lack of any evidence to the contrary, cosmologists complacently assumed for decades that clouds of anti-particles simply did not exists among the stars or in our own galaxy. That turns out to have been a subjective opinion based primarily upon wishful thinking. After all, there are lots of potential synchrotron accelerator machines out there in space in the form of nebula clouds or discs of gases. The advent of sensitive photoelectric sensors which can be tuned to register different radio wave lengths in the energy spectrum changed entirely what we called visible light and added a new dimension to star observations based upon perceptions gained from old-fashioned optics only. Cosmic rays originating from what is called "The Great Attractor" located somewhere near to the center of our own galaxy and hidden from us by clouds of dust and stars in the "Milky Way" yield evidence of the destruction of matter and anti-particle annihilation. Whether the engine is an enormous "Black Hole," as some believe, or not is still unknown.

Much more recently astronomers were utterly shocked to detect a vast plume of positrons or anti-electrons being ejected in one direction only from the "Milky Way" nebular disc to a distance of 3500 light years. A robot space satellite named the Compton Gamma Ray Observatory recorded the event in November of 1996. The identity of the particles in this spewed fountain of positrons was confirmed by the detection of the precisely exact energy release of 511,000 (0.511 MeV) characteristic of a positron meeting up with a normal electron and the resulting gamma rays observed by a satellite robot. This plume of positron gas was about 25,000 light years away from our solar system and measured approximately 4,000 light years across, although the density of the anti-particle cloud was actually a little more than one anti-electron per three hundred cubic meters. This would scarcely present a serious danger to any space travelers, whether from the earth or another planet in a distant galaxy but it could conceivably play havoc with their radio wave communication to their home base.

Interest in producing anti-matter artificially has gradually grown. By this is presently meant a chemical bond of an anti-proton with one or more anti-electrons protected for a measurable time from contact with normal matter for observation. Money for such an anti-proton atom has become available at the CERN facility in Geneva, Switzerland, for plans for cooling down anti-protons and anti-electrons produced in the accelerator rings there in separate traps. It is hoped to cool down the anti-protons to about 5 MeV energies and the positrons to only 0.1 Mev before bringing them together. This operation is under the direction of Dr. Gerald Gabrielse from Harvard University, the original inventor, and is not expected to actually get into operation until year 2000.

Section 14.2 - Parity and CP Violation

The capitalized letters in this Section title are a typical shorthand abbreviation among the physics fraternity for accepted theory on "Charge Conjugation and Parity." The first of these derives from their faith in the principle of symmetry and the authenticity of the quantum function equation. As Dirac's equations for antiparticles indicated, a transformation is possible for certain particles of the same mass to have a twin with exactly the same mass, but having an electric charge of the opposite sign.

This obviously applies to electrons and positrons, but may also apply to mesons, hadrons, and even baryons and quarks.

Dirac insisted that any particle of a given mass could quite possibly have an anti-particle of the same mass, but with the opposite electrical sign in the quantum equations. Its wave function was envisioned as an opposite hand mirror image of the first with the sinusoidal curves flopped upside down (or else moved sideways) so that the troughs and humps precisely cancelled each other out. In the most simplistic terms the normal and stable particle would have a parity of + 1 and the anti-particle would be designated as - 1.

Parity is a notion originally derived from the mathematics of quantum mechanics in which a specific particle is described by its wave form and the appropriate differential equations of the Fourier series commonly used in physics. The inclusion of the imaginary term -1 in the quantum equation and the necessity of assigning + or - algebraic signs for such parameters as direction of motion translation, polarity of the magnetic field, or clockwise or anti-clockwise spin of a particle undergoing a state transformation led logically to the concept of parity or mirror image.

Complete parity assumes that if the quantum wave functions were reflected in a mirror, then all known physical laws would continue to apply. The illustration very often given to this concept is that of looking at yourself in a mirror, but, in addition, you must imagine yourself like Alice in Wonderland passing through the mirror and then turning around to look out. Moreover, a common right-handed or clockwise turning screw would become left-handed or counterclockwise as well. The problem of comprehension is psychological.

Another way of thinking about parity is aligning the "Z" axis of a three dimensional coordinate system to be perpendicular to the two dimensional plane of "X" and "Y" axes composing the mirror surface. The image in the two dimensional plane of the mirror still remains symmetrical about the vertical axis.

The parity invariance rule happened to fit well with electro-magnetic wave calculations and thus was assumed to apply to many of the proliferating mesons discovered during the 1930 to 1940 decades, including anti-particles and also the pions and kaon series. The scientists for a time were lulled into believing that their convenient guess for exact parity was an immutable law of the universe. It ultimately turned out that they were wrong.

As early as 1953 a Richard Dalitz at the University of Cornell was cogitating about the relationships between particle energy, angular momentum, and the significance of parity in certain decay reactions of mesons. He noted some mathematical discrepancies, although there was no substantial evidence at that time to either substantiate or refute the notion of charge parity. Three years later two young Chinese-American physicists named Tsung Dao Lee from Columbia University in New York City and Chen Ning Yang at the Institute for Advanced Study at Princeton, New Jersey, were collaborating on the study of the so-called "strange" particles. They reached the conclusion that in the already famous Beta Decay of atoms the electrons or positrons did not invariably shoot off from the nucleus in the two opposite directions in exactly equal quantities when the weak force was prompting the action. They, themselves, fully realized that this suspicion was a form of apostate heresy; but nevertheless they published a joint paper titled *On the Question of Parity Conservation in Weak Interactions* on June 22, 1956, which challenged all current assumptions.

Their paper became the subject of heated controversy at the annual No. 6 conference at Rochester, New York, in April of 1956. Special attention was directed to two intermediate decay products of positive Kaon meson, which were the tau (τ_+) and theta (ϑ_+), both of which were unstable and rapidly decayed into pions themselves.

Quoting Richard Feynman's candid recollection of the general attitude of that earlier day

"At the meeting, when we were discussing the Tau-Theta puzzle, Oppenheimer said: "We need to hear some new and wilder ideas about this problem."

So I got up and said: "I am asking this question for Martin Block. What would be the consequences if the parity rule were wrong?" Murray Gell-Mann often teased me about this, saying that I did not have the nerve to ask the question myself. But that's not the reason. I thought that it might very well be an important idea."

"Lee, of Lee and Yang, answered something complicated, and, as usual, I didn't understand very well. At the end of the meeting Block asked me what he said. I said that I didn't know; but as far as I could tell it (the question) was still open - there was still a possibility. I didn't think that it was likely; but I thought that it was possible."

Norm Ramsey asked me if I thought that he should do an experiment looking for parity violation. I replied: "The best way to explain it to you is that I'll bet you only fifty to one that you don't find anything." He said: "That's good enough for me"; but he never made the experiment.

Murray Gell-Mann told me later that, when he was giving some talks in Russia, he used the idea of parity law violation as an example of what ridiculous and crazy ideas people were considering in order to explain the Tau-Theta puzzle.

Not surprisingly, Lee also conversed with a Chinese lady physicist named Chien-Shiung Wu in offices of the same Department at Columbia University. She came up with a proposed experiment to more precisely measure the supposed "invariance" of parity symmetry. It happens that the violation of parity can also take place with the natural radioactive decay process of atoms. Specifically, the heavy isotope of cobalt has a spin of plus five and disintegrates into the more stable metal nickel by beta decay into an electron and also a neutrino. The conventional expectation for such normal beta decay reaction was that the emitted electrons would display no preference for either left-hand or right-hand spin and thus divide evenly into a 50 - 50 distribution of numbers. The imposition of a magnetic field would simply serve to sort them into different directions. Her scheme was to chill down the Cobalt 60 by means of liquid gas cryostats to as close to absolute zero degrees as possible with the objective of reducing the jiggling dance of the cobalt atoms caused by heat in the environment and subsequently reach such a slow motion that would permit the atoms to all be aligned uniformly in their spins by the magnetic field and in the same direction. Her intent was to then measure the directions of the electron streams from the decay of the cobalt outside of the cryostat apparatus.

With some financial help from the then Atomic Energy Commission and with the cooperation of Ernest Ambler at the Bureau of Standards in Washington, D.C., which already possessed the required cryogenic facilities, Chien Wu proceeded with the first serious experiment to confirm or deny charge parity in the weak interaction. She thus became a collaborating partner in a Chinese-American triumvirate. The radio-active decay mode was well known and relatively simple:

$$^{60}\mathrm{Co} \rightarrow {}^{60}\mathrm{Ni} + e^{-} + \upsilon$$

However, the actual apparatus needed to be quite sophisticated. The cobalt specimen was enclosed in a housing of cerium magnesium nitrate with all contained inside a double walled vacuum bottle having liquid helium between its walls for cooling. The purpose of the cerium nitrate was to enhance the polarization of the cobalt nucleons in the emitter. The electrons radiated were to be detected in a 3/8 inch diameter by only 1/16th inch thick anthracene crystal located just above the cobalt specimen. The flashes of photo-electric energy from resulting photons were conducted through a glass window vertically up a four foot long lucite plastic tube to a sensitive electronic photon multiplier detector or counter. Two sodium iodide detectors were mounted alongside of the bottom of the cryostat bottle at 180 degree different horizontal axes as gamma ray scintillation counters in order to monitor carefully the polarization of the cobalt emitter and their atomic spin axis.

The author has troubled the reader with all these details of construction merely to emphasize just how painstakingly precise the experimental apparatus had to be in order to produce credible results. The ultimate net result of her repeated experiments was that charge parity, although somewhat random, was definitely violated and was not a 50 - 50 distribution up or down. The preferred direction of the emitted electrons was opposite to that of the nuclear spin of the cobalt, although the statistical deviation of the numbers was not dramatically large. Reversing the polarity of the magnetic field or directions of the poles did not significantly change the statistical results but merely the preferred direction. Parity was not observed. This was shocking news to the scientists; and for a while they blamed it all on the massless neutrino. Madame Wu published her results in a letter to *Physical Review* titled *Experimental Test of Parity Conservation in Beta Decay* in 1957.

Well before he became the Director of the Fermi National Laboratory, Leon Lederman knew and conversed with Tsung Lee regularly during their work week in New York City. At that time Lederman had just become a professor at Columbia University. He became personally fascinated with the question of parity and hit upon the idea of using mesons to test for possible violations in the decay of particles instigated by the weak force. He hoped to become a key figure in the "Chinese Puzzle" which still confounds physicists to this day. It happened that the University of Columbia had just installed a new cyclotron accelerator between 1947 and 1949 at a large estate near Irvington-on-Hudson known as *Ben Nevis*. The mansion there had once belonged to Alexander Hamilton but was given to the University by the DuPont family of Wilmington.

As a young graduate student Lederman had helped to build and assemble the machine, and it was the beginning of his career as a successful experimenter. The cyclotron first went into operation in 1950 after a dedication by General Dwight D. Eisenhower, who had just become President of the University. Using accelerated proton energies of only around 400 MeV, it succeeded in becoming an artificial source of ready-made and just newly recognized pions. The versatility of the machine had been greatly enhanced by an ingenious innovation in the system for detecting collision fragments. Instead of using the old-fashioned method of sending the particles through a stack of photographic emulsion plates Lederman and John Tinlot together persuaded the laboratory Director, Eugene Booth, to allow them to cut a hole in the stainless steel side of the cylindrical casing of the cyclotron and allow the fully accelerated protons, which had completed the spiral path to the outer circumference, to be guided through this window in a magnetically focused beam. These accelerated protons were then directed onto a graphite target. The result was a shower of pions, which in a short traverse soon decayed into muons. Normal negative muons have the peculiarity of having a relatively long life (albeit only a little over two microseconds) before themselves decay into an electron and two neutrinos, as we have already learned. The muons were collected in a cloud chamber.

After one of his frequent luncheons in a Chinese restaurant with Lee and Yang, Lederman had a sudden inspiration. It occurred to him that the decay of the pion fragments from the target into muons could be a handy means of testing for parity. He became quite excited and rushed up to *Ben Nevis* by automobile in order to tell the news to his staff there. Proton bombardment of graphite sometimes produces positive pions or the antiparticle. Lederman proposed to select these out for his experiment, anticipating that they would decay into positive muons, which are anti-particles themselves, and muon neutrinos. This pion beam was directed out of the accelerator and through a tunnel through a thick concrete shielding wall into the measurement hall of the laboratory. During their passage the scientists anticipated that at least one fifth of the pions would experience this transformation into muons and, moreover, the spin axis of most of the positive muons was expected to remain in the same direction as the parent particle as imparted to them by the magnetic field polarity. Lederman realized that the production of the anti-muon particles offered a singular opportunity to test the assumed charge parity. The muons were arrested in a second carbon block, where they usually kicked out a positron and two neutrinos. All this sequence was quite ingenious and, of course, demanded careful detection in a cloud chamber.

The final experimental results indicated that approximately sixty eight percent of the captured muons had a right-handed spin and the balance were left-handed. The expected parity was clearly violated by a significant statistical difference. The announcement at an American Physical Society convention in New York City on February 6, 1957, created major excitement and interest world wide.

It is not inconsequential that through all this period of experimentation a graduate student at Columbia named Marcel Weinrich, who was hoping to achieve a PhD degree in Physics, had to sit by patiently in some distress while his own experiment for his final thesis was put aside. His reward, however, was to have his name appear on the letter to *Physical Review* with two other seniors confirming an extraordinary revision to physics theory which projected them all into world attention and started Lederman on his career as an innovative experimenter. It had been demonstrated a second time that the weak interaction involved in the decay of mesons does not invariably obey wave phase offset or parity reflection. The quirkiness of Nature was illustrated.

For the reader with a slight turn of mind towards philosophy the lesson which might be learned from this account might be that intuitive assumptions based upon the appeal of beauty and simplicity in logic may not be always reliable. The significance of charge parity violation to the scientists was far more upsetting and serious. The attribute of "strangeness" or unusually long lifetimes for some mesons, notably the kaon, still lacked an explanation. Worse yet, there were twelve other varieties of kaon resonance having increasingly heavier masses, and the common positive or negatively charged kaon could have no less than forty–one different forms of decay. No one could call that simple by any stretch of the imagination. Clearly there was something lacking in their conventional quantum calculations. It was starkly evident to the physics theorists that something as yet unknown was going on between the strong and weak forces.

Some seven years after Lederman's achievement a team of four physicists from Princeton University in New Jersey decided to follow up on his lead with a study of kaon decay. The normal charged kaon most usually decays into three pions. These people were specifically fascinated with the neutral or zero charge kaon, which is more rare. This animal may sometimes be obtained by directing a beam of positively charged kaons into a block of copper or beryllium metal. Starting in 1963, James W. Cronin, Val L. Fitch, John H. Christenson, and R. Turlay all collaborated and received permission to use the Alternating Gradient Synchrotron at Brookhaven National Laboratory. The reaction that they wished to study was an anomaly.

$$K_1 \rightarrow \pi^+ + \pi^-$$

This raises the question immediately of how one gets two different electric charges from none at all. Much more important was the fact that the resultant pion decay products they observed ultimately and definitely disintegrated into a greater number of positrons than normal electrons. They thus had confirmed by their experiment that the weak decay of neutral kaons did not conform to absolute symmetry and charge conjugation parity. The abrogation of what the physicists thought had to be a rule was extremely perplexing to them and initiated a great deal of discussion. In the written report of the Cronin team to the *Physical Review* in July of 1964 they made the following candid statements:

"We know of no physical process which would accomplish this. ...The presence of the two pion decay mode implies that the K zero meson is not a pure eisenstate of charge parity."

In plain language, their accustomed quantum wave equations did not work. Charge Parity violation had serious implications for the mathematical theory gurus and the collapse of "invariance" remains a disconcerting perplexity to this day. The Electro-Weak theory is supposed to provide the answers.

Chapter 15
The Genesis of the Quark

Section 15.1 - Bohr Fundamentalism Of The Atom Challenged

While the accelerator laboratories were steadily adding even stranger animals to the particle zoo with disconcerting alacrity, the mathematical gurus of physics were in an intellectual ferment of puzzlement and conjecture. What were the causes of such bewildering diversity? Why should there be such a choice of electrical charges and spin? The Delta particle with a possible state having a double electric charge and the Omega Minus with a threefold strangeness number were both especially provocative.

It is probable that Murray Gell-Mann was the first physicist of major and established stature who published, however tentatively, the possibility of unknown sub-nucleon components, although it was an idea whose time had come. Private speculations among knowledgeable friends and foreign associates about such a brash and radical concept were already circulating. A theoretician formerly from Los Alamos named Robert Serber, who was then working at Columbia University, had begun playing with the triplet idea while forming the octets and decimets required by SU-3 symmetry and the Eight Fold Way.

It happened that Gell-Mann decided to represent the California Institute of Technology at a physics conference held at Columbia University, starting on Monday, March 25th, of 1963 and running all through the following week. He was scheduled to be one of the speakers both near the beginning and later at the end of these sessions. Thus it was not surprising that Serber invited him to luncheon at the Faculty Club before the second speech in the company of other physicists on the Columbia teaching staff, as related by Michael Riordan.[1] Included in this group was Tsung Dao Lee.

After alluding to symmetry and the Eight Fold Way in the conversation, Serber ventured to admit that he wondered why some combination of

1. See "The Hunting of The Quark," (Simon and Schuster, 1987)

triplets should not appear as components in the puzzle. Gell-Mann remarked immediately "That would be a funny quirk indeed!" Lee thought that it was a "terrible idea" but all agreed that for any combination of two or three quarks to add up to a whole number their electric charges had to be plus 2/3rds, plus one third, and minus one third. It was assumed that all hadrons or baryons must have an ultimate charge of unity, whether positive, negative, or zero. Fractional charges had never been observed in any known particles thus far. Gell-Mann added a further condition which needed to be met in order to make such an unusual idea feasible. Such strange point particles within the neutron, proton or meson must be permanently trapped in some kind of an imaginary "bag" inside the nucleons and not be observable in nature as a free and individual entity. In his next lecture Gell-Mann began referring to these invented sub-nucleon bits by the whimsical nonsense word "quorks" or "kworks." Nevertheless, he took the idea seriously, but it was not until much later in the year that he dared to send his suggestion to a more lenient European journal called *Physics Letters*, which published it on February 1 of 1964 under the title *A Schematic Model of Baryons and Mesons.* By strange coincidence, this also closely coincided with the announcement of the discovery of the Omega Minus particle.

Ultimately, Gell-Mann changed the name of his proposed sub-particles to "quarks" because he meant to convey by his arbitrary and bizarre word something which no one really understood. There ensued an unexpected furor of speculation and written polemics over his name for the supposed new tiny particles, much of it entirely inane and trivial. When challenged, the hard-pressed physicist produced the word in a bawdy book by James Joyce titled "Finnegan's Wake." If one refers to the last Chapter 4 of Book II of that novel, one will find a verse which starts out:

> Three quarks for Muster
> Mark
> Sure he hasn't got much of a
> bark
> And sure any he has it's all
> beside the mark.

Whether the word quark refers to a cheap cheese made in Germany from the curds of sour milk or refers poetically to the call of sea gulls or quarts of beer is not really worth debating. From the point of view of philosophers the word "quark" is simply a noumenon, which Gell-Mann well knew. A noumenon is defined as a word selected for "an object of purely rational apprehension" or, as Kant preferred, "a non-empirical concept." In other words, it is an abstract thought or concept which in no way is amenable to being seen, touched, heard, or smelled by any of our physical senses. It is rather important for us to realize that this language is precisely what Gell-Mann had in mind and that his first talks and initial published writings on the subject very carefully described the quark as a mathematical "scheme" to explain certain behaviors of the atom which we did not fully understand.

Quoting his 1964 paper introducing the notion:

> *It is fun to speculate about the way quarks would behave if they were physical particles of finite mass (instead of purely mathematical entities as they would be in the limit of infinite mass)...a search for stable quarks of charge –1/3 or +2/3 and/or stable diquarks of charge –2/3 or +1/3 or +4/3 at the highest energy accelerators would help to reassure us of the non-existence of real quarks.*

More or less simultaneously a young physics doctorate graduate, who had been a student and protege of both Gell-Mann and Feynman in California, hit upon the idea of fractional charges while puzzling over the Phi meson . His name was George Zweig and had been born in Moscow, Russia, in 1937. His family moved to the United States later, thus allowing their son to obtain a Bachelor of Science degree from the University of Michigan in 1959. Working in the experimental staff at CERN in Switzerland in 1963, he finally completed his dissertation for a PhD and published his ideas on fractionally charged sub-nucleon particles, which he called "aces," in 1964. To Zweig his aces were real and substantial particles, and his justification for believing this was that they helped to explain the Phi meson. Thus he helped to clinch the idea of quarks, aces, or partons in the minds of physicists, although he was ignored in the subsequent dispensation of Nobel prizes.

Gell-Mann, however, continued to hedge cautiously on this decision. In a speech at a physics conference in Berkeley, California, in September of 1966, the actual quotation of his statement was"whether or not real quarks exist," the "q" and the "$\bar{q}$" (anti-quark) we have been talking about are mathematical in particular, I would guess that they are mathematical entities that arise when we construct representations of current algebra...." This casual remark eventually caused a storm of debate and protest. Many of the younger physicists, remembering Rutherford and his scattering of alpha particles by the tiny nucleus inside the atom, were predisposed to accept the physical reality of the quark notion. In the 1966 Proceedings of the International Conference on High Energy Physics Gell-Mann firmly defended his opinion and defined his word "mathematical" as meaning "the limit of an infinite confining potential or a limit to an infinite mass and binding energy." He was deeply hurt and offended by insinuations that he was equivocating on this issue just in case the entire idea should prove false and by what he described as "terrible insults as the result of being right." In view of the fact that at this historical stage the existence of the quark had not yet been conclusively demonstrated by laboratory experiments, nor had the physics community at large yet reached a majority endorsement, his caution seems entirely reasonable. The writer's sympathies are quite emphatically with Gell-Mann. Although he gradually modified his opinion over a few years in the face of more evidence, he may yet be proven technically right in the distant future. To quote an opinion of Werner Heisenberg printed in a collection of essays published by the Princeton University Press in 1983: "today in the physics of elementary particles, good physics is unconsciously being spoiled by bad philosophy."

The whole idea of quarks has stimulated some wild speculations. Two computer experts, namely Edward Fredkin of the Massachusetts Institute of Technology and Digital Equipment fame and Stephen Wolfram from the Princeton Institute for Advanced Study, went so far as to draw an analogy of the new quarks to bytes of coded information in a computer giving instructions to a passive pool of energy in the universe to activate the formation of physical particles. Following a similar train of thought, a PhD student at Princeton named Jack Beckenstein wrote his thesis in 1972 on "Black Hole" singularities. It is considered by cosmologists that particles such as atoms, nucleons, and mesons lose all their identity and disappear once they pass the event horizon and fall into the gravity trap. Beckenstein's assumption was that this apparent phenomenon is a form of entropy and is analogous to scrambling bits of coded information in a communication field into meaningless chaos. We shall leave all that to science fiction.

Sheldon Glashow of Harvard, who was an enthusiastic and vocal believer in quarks, nevertheless felt it necessary to add a mild disclaimer in his 1988 book *Interactions.* "Quarks cannot be isolated nor can the gluons which bind them. Therefore quarks must be studied *in situ* and the evidence for their reality, while compelling, is necessarily indirect."

Another person who stalwartly endured the scoffing of his confederates during the 1960 decade and who supported the quark idea was Richard Dalitz at Oxford University in England. As a theorist, he became convinced and endorsed the quark model publicly in 1965. So also did an Italian at the University of Genoa named Giacomo Murpurgo, who in the same year wrote a paper describing quarks as real and material, but very massive entities. On the strength of his own convictions Murpurgo and some assistants built an apparatus by which they could suspend small granules of graphite in a magnetic field where they could bathe them in ultra-violet light at will. The objective was to drive off electrons from the carbon atoms and measure whether any fractional charges might remain. The experiment failed.

A factor which contributed greatly to the tolerance of some nuclear scientists to the idea of a sub-nucleon particle was the common knowledge of experiments conducted by Robert Hofstader at Stanford University in California as far back as 1950. He was greatly admired for using a third generation linear accelerator machine known as Mark III, which had a modest length of only 300 feet. The laboratory staff, under Hofstader's encouragement and supervision, had been firing electrons at energies of only 1 GeV at all kinds of different atoms. This bombardment of protons with electrons produced scattering angles similar to Rutherford's old experiments with alpha particles on gold foil and confirmed that the nucleus was a dense and hard object of extremely small size. Studies of the ricocheting electron patterns off a hydrogen proton led to the conclusion that, though point-like, it actually had a dimensional size of approximately 10^{-13} centimeters in diameter. This apparently satisfied a great many scientists that it was large enough to argue reasonably in favor of still smaller components. Hofstader received a Nobel prize for this measurement in 1961, some four years after he and R. W. McAllister published their results.

Another personality in this history who very significantly enhanced the pursuit of the quark theory was a brainy young theoretician at Stanford University named James Bjorken. He was intimately associated with all the SLAC accelerator operating staff from its beginning experiments and also had access to the raw data sheets. In his early graduate studies days he had attended the Neils Bohr Institute in Copenhagen. Starting as early as 1965 he ran intricate and preliminary calculations on the possible results of collisions of high velocity electrons upon hydrogen ions based upon the "current algebra" and "sum rule" techniques, as they were known. The correct interpretation of the scattering results from the detector was crucial to the team's success. As with any atom, metallic or not, a beam of electrons can have their energy first dissipated in the cloud of electrons which surround the nucleus (called W_1) and then, secondly, by the sphere of a proton itself (called W_2). Selecting a target of liquid hydrogen instead of a metal was one obvious way of simplifying the problem, but that left a remaining question to be answered. Was the positive electric charge inside the proton diffused like ice in a snowball or raisins in Rutherford's "plum pudding" analogy or as a hard point-like object? Bjorken opted for the three point or quark possibility and worked out the possible effects of that assumption. His paperwork so convinced him of the latter eventuality that in September of 1967 at the International Symposium on Electron and Photon Interactions in Rochester, New York, he introduced the subject of deep inelastic scattering by electrons and tentatively brought up the subject of quarks. His lecture was not well received. One of the unconvinced skeptics from CERN even made the unkind remark that "he was making such a fool of himself."

In fact, he was a genius unrecognized who had a vision of the future, but he failed to make a good case for his reasons. He predicted that one clue to the truth of the quarks proposition would be the detection of an unusually high number of electrons (and occasionally pions) bouncing off at large deflected angles from these point-like objects within the proton as it was being bombarded. He also talked a lot about something called "scaling" and presented a graph of expected recoils versus beam energy which might have a broad base instead of the sharp peaks usually found with resonances. On that last particular he was quite mistaken, but relatively few scientists in the audience really understood him at that time.

A crude mental picture of what he was attempting to describe is to imagine a target of two or three small hard beads strung together on a web of fine silk threads at random spacing with someone shooting at it with shotgun pellets. The caliber of the shotgun or whether it had one or two barrels, within limits, did not make a huge difference in the probability of hitting one of the beads, although the velocity of the pellets might. There would be some difference, yes, but not great. His "F" graphs were supposed to illustrate the point. All his predictions were close to the eventual truth, although well before any laboratory proof. Many subsequent publications on this subject deferred to his great intelligence and acknowledged his tremendous contribution to the total effort but he never received a Nobel prize.

Quite naturally, a new and younger group of experimenters at the Stanford Linear Accelerator Center perceived an opportunity in 1967 and decided to instigate a search for evidence of the quark particles in collaboration with physicists from the Massachusetts Institute of Technology in Boston. Although construction had been started back in 1962 with the aid of Federal government funds, Stanford University had only just completed its radical new linear accelerator underground with a length of two miles in a straight line. The acceptance of government funds was important, since it established a precedent for encouraging the policy of cooperation with faculty members from other universities. This inter-collegiate idea was not kindly received by some Stanford administrators or professors. Wolfgang K. H. Panofsky, familiarly known as "Pief," then became the new overall Director. The "Monster," as the entirely new accelerator machine was soon called, came into actual operation in 1967 and quickly reached its intended design top energy of 20 GeV. It was a remarkable achievement for the United States.

The technique planned for finding quarks was simply to fire electrons at high kinetic energies between 8 to 15 GeV strength into target protons provided by liquid hydrogen, and to look for inelastic or " hardball " scattering by means of two huge spectrometers. If one remembers the duality of both photons and fast electrons, that they can behave as either particles or waves, then it becomes clear that the significance of the higher beam energies is their shorter apparent wave lengths, making for a better microscope. This was not simple to accomplish nor was interpreting the data easy either. The results were not as expected, and it began to look as though the proton was mostly empty space if it were not for a few resonance peaks followed by a large bump on their graph and a suddenly increased number of events at the higher and so-called "deep inelastic" region. Late in August of 1968 a very cautious report of the data collected and a description of some of the most puzzling events was presented to the High Energy Physics symposium in Vienna by Richard Taylor and Jerome Friedman from California. A graph of functions which were calculated from the laboratory scattering results and plotted by Henry Kendall, who came from MIT, was an important feature of their presentation. They did not officially announce the discovery of quarks at all, but at the close of the conference in September Wolfgang Panofsky surprised his own staff by letting the cat out of the bag, figuratively speaking. After explaining their laboratory data, he made the following statements: "The qualitatively striking fact is that these (scattering) cross sections are very large and decrease much more slowly with momentum transfer than elastic scattering cross sections...Therefore, theoretical speculations are focused on the possibility that these data might give evidence of point-like charged structures within the proton." His speech did not create any great stir as anticipated. Apparently, the audience was not listening!

Two months later it chanced that Richard Phillips Feynman came in order to visit his sister in California and check in with the SLAC staff activities for a second time since June. Feynman (1918 to 1988) was born in a then village called Far Rockaway near the southern shoreline beach of Long Island. His father, Melville Feynman, had been brought to the United States at the age of five by his immigrating parents in 1885. His mother was the daughter of a prosperous milliner. As a young teenager, Richard was fascinated with electricity and radio and tinkered with spare parts and old radios with zest, as well as having a proclivity towards science which his father encouraged. The son excelled at mathematics. He also had a sister, Joan, to whom he was devoted. After High School, he applied to both Columbia University and MIT and was accepted by that last college as an undergraduate in 1936 despite the existing quotas on Jewish applications then prevalent. By the time of his graduation in 1939 young Feynman had decided on a career in physics, which was *not* noted as a lucrative field of endeavor then. Having been told by his admiring professors that it would be better to seek his graduate degrees at a different university than MIT, he applied to Princeton University and was accepted. As a graduate student he was recruited into World War II efforts on the Manhattan Project but still received his Doctorate degree in 1942. It was then natural for him to go to Los Alamos and work on the development of an atom bomb until 1945, at which time he accepted a teaching position as professor of theoretical physics at Cornell University in Ithaca, New York. In 1950 he was enticed to move to the California Institute of Technology in Pasadena, where he remained for the duration of his life. Feynman had a vivid imagination and a knack for illustrating physics principles with everyday articles or materials behavior. Although obviously highly educated, he had a personal inclination to remain somewhat of an autodidact in the sense that he preferred to work out his own original explanations for the many laws of physics. These were often highly ingenious and dramatic. Although his sex life is not our concern, his behavior with most women was highly amoral and irresponsible. This defect won him a number of enemies. Nevertheless, he was generally regarded as an eccentric genius.

He was well informed on the experiments which had been going on since 1967 at the Stanford Linear Accelerator Center using a high energy electron beam. His active imagination had already led him to the notion of hadrons being composed of sub-particles which he dubbed "partons." He had been attempting calculations on the results of projectiles such as electrons or protons on other protons. Thus it was only natural that when he was shown the measured results which came from the SLAC drift chamber detector and also the "F" graphs prepared by Kendall, he became wildly excited. He regarded the data as an absolute verification of the fact that there must be real Dirac type point-like sub-nucleon constituents in all hadrons and baryons and that they shared between them the total momentum of the larger particle. Although superficially similar to Gell-Mann's quarks in concept, partons were not identical. Feynman's most valuable contribution to the debate was probably to introduce the idea that the graphs of Freidman and Bjorken could represent the probable feasibility of assigning fractions of the total momentum of the proton, as well as the electrical charge, to the sub-components. In his characteristically picturesque way in his explanations to his fellow physicists, Feynman would represent the proton as a pancake flattened by the relativistic speeds of the electron projectile and his fist as the electron rushing towards it for collision. In other words, he was imagining himself as a passenger on the electron perceiving the approaching proton with both time and dimension shortened by Einstein's Relativity relationship to velocity.

In order to be strictly correct and not to mislead the reader, the writer must admit to a crude simplification in his previous representation of scattering. From the theoretical purist point of view, electrons do not strictly "bounce" off quarks or protons. The electron may emit a photon of virtual energy which may be eaten or absorbed by the quark with the result that the electron moves off in a different vector path with reduced energy of motion to become measured, hopefully, by a detector.

Regardless of the theoretical and mathematical niceties, Feynman's endorsement of the notion of particles within nucleons and fast electrons ricocheting off hard point charges inside the proton accelerated the general interest and acceptance of many physicists in this radical new development. It would be another six or seven years before the various arguments between partons and quarks adherents would be settled to everyone's satisfaction. There were, however, certain mutually agreed upon requisites for their attributes or properties.

(1) Both in order to build electric charges to unity, whether plus or minus, in both hadrons or baryons there had to be fractional electric charges measured in one thirds.

(2) The spins of each quark had to be 1/2 in order to make them fermions and thus satisfy the Pauli Exclusion Principle, for otherwise they could not co-exist together in such close proximity.

(3) The "Up" or "Down" nomenclature refer to the direction or orientation of any individual particle spin axis when subjected to a magnetic field.

The final result, for starters, for the basic definitions for the surmised three essential and basic quarks which would be required to construct either a proton or neutron or meson thus fell out from the table below:

Name	Symbol	Electric Charge	Spin J	Strange Number	Baryon Number	Mass in MeV
UP	u	+ 2/3	1/2	0	1/3	330
DOWN	d	-1/3	1/2	0	1/3	333
STRANGE	s	-1/3	1/2	-1	1/3	486

The added " strange " quark was first proposed by Murray Gell-Mann.

All this theoretical work was duly reported in conference lectures in both the United States and Europe. On February 3rd of 1969 James Bjorken gave a lecture at the American Physical Society in New York City. The published follow-up was a written article for the *Physical Review* in April of 1969 titled *Inelastic Electron-Proton Scattering and the Structure of the Nucleon* by J. D. Bjorken and E. A. Paschos of Stanford University. However, it was not until the Electron-Photon Symposium held in September of 1969 at the seaport town of Liverpool in England that the remarks of the three California scientists (Bjorken, Taylor, and Friedman) finally attracted the attention they deserved from the general physics community.

Using the three new building blocks, it became possible to represent the proton as consisting of an Up quark (+ 2/3rds charge) plus another Up quark (+ 2/3rds charge) plus a Down quark (- 1/3rd charge) for a total electric charge of positive one.Similarly, one could describe the neutron as comprising one Down quark (- 1/3rd charge) plus another Down quark (- 1/3rd charge) plus one Up quark (+ 2/3rds charge) with a total electric charge of zero.

Two officially written reports on *High Energy Inelastic Scattering* representing the separate California and Boston teams were published in short sequence in 1969. The first was on August 19th and gave credits to Bloom, Coward, De Staebler, Drees, Miller, Mo, and Taylor for the SLAC team and also included separately were Breidenbach, Freidman, Hartman, and Kendall for the MIT team. The second letter to *Physical Review* was received on August 22nd and listed only Breidenbach, Freidman, and Kendall with no change in the California line up. So much for the lone genius myth in modern science.

Richard Taylor, Jerome Freidman, and Henry Kendall all received a joint Nobel prize for their work in 1990. The gauntlet had been thrown down to the world bravely, and it was now time for the judgement of the scientific community to agree or disagree on these new sub-nucleon particles. The process would take time to achieve unanimity.

CHAPTER 16

CHARM AND CONFIRMATION OF THE QUARK

Section 16 - 1 Gargamelle

About this time two physicists on the staff of the Stanford Linear Accelerator, Sidney Drell and Tun Mo Yan, began wondering whether there might be some possible methods for measuring or evaluating the distribution of momentum that Feynman was talking about and the magnitude of the electric charges among his hypothetical "partons." Their best choice was to use neutrinos as the test projectiles, since they were electrically neutral, tiny, and almost massless and thus could be expected to penetrate the nucleus more easily. Not coincidentally, the Center for European Nuclear Research (CERN) in Switzerland was desperately anxious to get involved in the quark/parton question itself. Not only that, but CERN had just invested huge sums of money and engineering expertise in building a neutrino study detector and was interested in attracting some of the California "Post-Doc" people who were looking for jobs. The big troubles were that one had to devise a method of collecting a neutrino beam in the first place and the entire scheme for detection of them and the collision results was technically difficult.

The construction of a huge bubble chamber measuring fifteen feet long and holding 3,170 gallons of liquid freon marked the serious interest of the European scientific community in neutrino and anti-neutrino research. Costing about ten million United States dollars, this gigantic detector machine with its surrounding magnets and computer logged photo-electric cells was nick-named *Gargamelle* after the mother of the fictional giant *Gargantua* in a French novel. Its very large size was intended to increase the probability of a neutrino hitting something and producing a detectable shower of charged fragments or new particles.

Although the CERN administrators had approved the use of the new machine for the neutrino effort as early as 1971, partially under the ardent urgings of Donald Perkins and other associates from Oxford, England, the task proved difficult indeed with numerous changes. One started with a standard proton to proton collision in order to derive a stream of pions, which naturally had to decay into muons eventually. The muons then needed to be filtered out by a passage through common earth or iron blocks in order to leave a beam of undeflected muon type neutrinos. This beam of neutrinos was aimed to enter a tank of liquid freon. Those very few which happened to collide with a proton or neutron usually caused the creation of a new charged muon whose track could be detected by photo-electric sensors.

All this experimental operation resulted in verifiable collisions and rebound at a maximum rate of one per minute using an intense neutrino beam of relatively low kinetic energy and an immense target. Something like 174,000 recorded events had to be sifted through in order to cull out roughly 200 satisfactory pictures. Nevertheless, the physicists there concluded that the neutrino projectiles were indeed striking hard, point-like objects inside the protons, and that these internal particles represented roughly one half of the momentum energy of the proton. This was comforting to the SLAC people in California; but they were "scarcely overwhelmed" in the words of Michael Riordan, who was a graduate assistant at the time. The report of CERN's findings were duly submitted to *Physical Letters* in Europe in August of 1973 with a rather dull statistical analysis and acknowledgment of "scaling" effects in the scattering which substantially agreed with the early California conjectures.

Section 16 - 2 The Advent of the Pi Meson or J / Ψ Particle

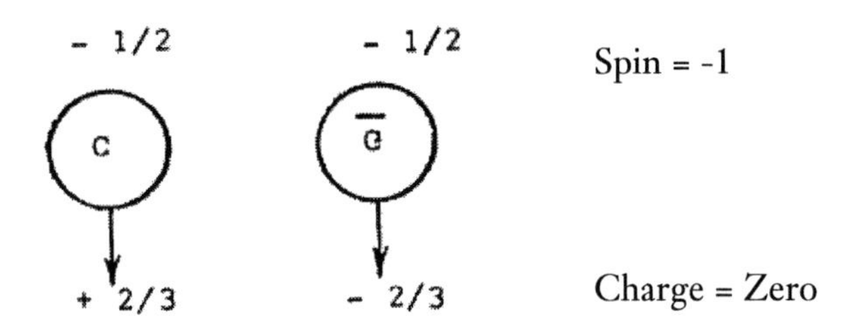

This entirely new particle of pivotal importance was discovered almost simultaneously by two different and famous laboratories at opposite edges of the North American continent. Sam C. C. Ting of the Brookhaven National Laboratory in Bethpage on Long Island first learned he had a serious contender for the first discoverer when he met Burton Richter of the Stanford Linear Accelerator Center at a physics conference in California on November 11th of 1974. We will get to Ting's story subsequently. Richter there formally announced their identification of a strange resonance which they called the Psi particle. Ting immediately rushed home and upon his return claimed his own discovery of the same particle back in October, justifying the delay in his announcement because of their group's extreme caution in establishing careful confirmation of their observational results. Ting had selected the letter "J" for its symbol, allegedly because that letter was close to representing his own name in Chinese script. His statement of the cause for delay was eventually corroborated as genuine by Victor Weisskopf and others. This extraordinary coincidence of discovery created intense personal frictions and resentment at first, but cooler heads quickly decided to name the new resonance "J / Psi" as a compromise and eventually a gentlemanly reconciliation ensued.

The J / ψ is a very peculiar meson resonance which exhibited several anomalies and ultimately led the physicists to conclude that three quarks were not adequate to explain everything, as they first believed. It is electrically neutral. It has a relatively huge mass of approximately 3097 MeV or three and one third times that of the proton. Its basic (J) spin is minus one, theoretically making it a boson. Because its spin was not one half, it could not be considered a baryon and hence zero. Its strangeness number was totally out of sight to a degree which put it into a class by itself. Moreover, it was not really predicted by Gell Mann's Eightfold Way, making its discovery essentially a total surprise.

Burton Richter at Stanford University had been agitating for a circular storage ring to be added to their famous Linear Accelerator since before 1971. His purpose was to allow the collision of electrons with positrons or anti-electrons. Inasmuch as these leptons had opposite electric charges, it seemed feasible to circulate them magnetically in the same ring in opposite directions, providing the two beams were kept separated and well apart, and thus accelerate them to higher velocities. This new addition to the older existing linear accelerator was built on a parking lot area and became known as the Stanford Positron Electron Asymmetric Ring or SPEAR collider. It became operational in 1971 for the bargain cost of $5,300,000.

The positrons are obtained by deflecting some of the electrons in the beam being accelerated in the linear tube and colliding them with a copper block. The shower of positrons thus created is collected and focused to be circulated in the new ring tunnel. This toroid beam pipe is known as the storage ring because it can be slowly filled with bunches of fast moving electrons or positrons circulating continuously around the circumference of the ring in opposite directions for a period of two or three hours. The entire scheme was extraordinarily ingenious and not at all simple to achieve because of the need for a lot of skillful juggling of the polarity of the magnets by switching the electric current directions around with micro-second timing. Accompanying all this new machinery was a Mark I particle detector built around the spot where the two opposing beams were planned to impact. Everything had to be orchestrated like a symphony in order to produce evidence of the scattered results of the collision of objects which no one has ever seen for themselves! Their existence had to be construed from fog chamber tracks and momentum calculations.

It should be recognized that in all these laboratory team efforts which involve graduate students, post-docs and technicians working around high voltage electrical apparatus and detectors, the latter sometimes contain inflammable or potentially explosive liquids, there is a certain element of danger. We scarcely need to mention the risk of ever present radio waves, gamma rays, and stray neutrons. Safety is a prime consideration today and good shielding is assumed, but a few accidents have occurred nevertheless. For obvious reasons these incidents in experimental work are seldom mentioned in official public announcements. A liquid hydrogen bubble chamber once leaked and the gas was ignited by an electric spark, killing one man and injuring eight others in Cambridge, Massachusetts, in July of 1965. At SLAC itself in Stanford the staff were testing spectrometer magnets in an underground chamber one day and the operator twirled the control dial for the electrical voltage too casually. A loosely bolted cable clamp connection triggered a short circuit voltage overload of thousands of amperes which vaporized the bolt with a loud explosion which could be heard through four feet of concrete roofing. The unprepared technicians staggered out of the smoke filled chamber stunned, but very fortunately no one was seriously injured.

The announced purpose of all this intricate apparatus was to study the impact of electrons upon anti-electrons at moderately low kinetic energies of 2.4 to 5 GeV and thus gain some knowledge of what and how any material particles might be formed from the pure energy of annihilation. The result was a shopping list of all kinds of hadrons and their resonances. These observations kept the scientific staff happy for two years, but they were also aware that the Germans in Hamburg were working the same territory with their DESY machine. By October of 1974 rumors had begun circulating among all the graduate students of some very strange and inconsistent results obtained in the 3 GeV energy band. Although the Stanford University crowd was still working with electron versus positron collisions in the 4.8 GeV energy range, Goldhaber and Schwittzers, who were familiar with the German experiments, urged them to crank back down the linear accelerator to the 3.0 GeV range in order to take another look in spite of objections from other members of the staff. Richter, as Director, convened a conference and finally approved the change on November 4th, 1974. Only five days later the experimental team began to find an unexpected burst of hadrons in the narrow 3.1 to 3.14 GeV energy range. The detector counting rate doubled again that night as the physicists continued to work late. By ten o'clock the next

Sunday morning they had definitely identified a stream of hadron fragments at 3.105 GeV. This was an extremely narrow resonance with the number of events amounting to about ten times what had been considered normal in past laboratory experiments at other energies. The crew was ecstatic with excitement and self-congratulation; and their leaders rushed immediately to fire off a brief report of their observations to the *Physical Review* journal, which was received on November 13th of 1974.

This was the beginning of what was subsequently dubbed as the famous "November Revolution," which started the acceptance of the quark concept and shocked the physics theorists worldwide. Compared with the customary resonances of past experience, this new discovery was a colossal pinnacle with an unbelievably narrow width on the energy to mass graph. The reader may recall that the narrower this base of the graphical pyramid becomes in energy spread, the longer becomes the mean lifetime of the particle and the more stable it is. The great mystery, however, was the question of what exactly was prohibiting the expected more rapid decay. Burton Richter, after a long debate, decided to christen this new resonance the "Psi," although his first choice was the "Iota" meaning "small" in Greek.

In the meantime on the east coast of the United States an entirely different team of scientists from the Massachusetts Institute of Technology in Boston had received an allotment of time on the Alternating Gradient Synchrotron (AGS) at Brookhaven National Laboratory in Long Island for their own distinctive project. Doctor Samuel Chao Ching Ting was the supervising Director and brain. He had come to the United States in 1956 from Taiwan, the reactionary island off the coast of China, as a young student aspiring to get a scientific education at the University of Michigan. This same university had become famous during World War II for its correspondence courses for the United States troops. He completed his undergraduate and graduate degrees in record time and had already gone on to laboratory work in both Germany and Switzerland when his reputation for brilliance came to the attention of Victor Weisskopf, who recruited him for MIT in 1967.

Ting wished to use the 30 GeV energy accelerator machine at Brookhaven for a pet project of his to shoot only 7 GeV protons at a target of beryllium metal in a very meticulous search for what he called "neutral vector mesons." It would be hard to imagine an experiment more different from the California "SPEAR" concept. The Rho, Omega, and Phi particles are typical of these; and Ting hoped to identfy such similar reactions by the detection of electron and positron pairs resulting from such collisions of protons with other proton or neutrons. It was considered a difficult and ambitious experiment. In order to protect the personnel from dangerous levels of radiation while sitting in a trailer provided for logging the collision data registered in their detectors, it became necessary to surround the detector and target with hundreds of tons of solid concrete blocks and another five tons of borax soap.

By April of 1974 the equipment was in satisfactory operation and the laboratory staff had identified the already known "Phi" meson at a mass of 1020 MeV by its characteristic peak. At that point the experiments were shut down until mid July. Subsequent searches through the 3.5 to 5.5 GeV energy range were disappointing; so Ting decided to back down for a look at the 2.5 to 4.0 GeV band instead. Not until September 10th of that year did his team begin to suspect a strange "bump" in the 3.5 GeV energy level. Their allotted time on the accelerator had run out, so they were compelled to argue with the managers of the Brookhaven facility in order to get another time allotment in October. Once this was promptly granted, then Ting's trusted deputies, Min Chen Y. Lee and Ulrich Becker (who was a former German colleague from Hamburg on the DESY machine) went back to work again. They confirmed the existence of the elusive resonance spike at a 3.1 GeV beam energy on the morning of 13th of October. By chance it happened that during the following week dozens of physics dignitaries converged on Boston from all over the world in order to celebrate a retirement party for Victor Weisskopf of Harvard. Most serious work stopped, but the MIT crew of Ting's were all told to keep their mouths shut about the new discovery.

Ting, who was the epitome of the proverbial perfectionist, first required more time in order to have the data checked and rechecked for authenticity. However, he had another mental block. There was nothing in the established theory that could possibly explain such a peculiar resonance. He was ultimately persuaded by his associates to send a written report to the *Physical Review Letters* titled *Experimental Observation of a Heavy Particle "J."* Why "J"? It has been alleged that this letter of our own alphabet had a phonetic similarity with his own name in Chinese calligraphy, but that story may be apocryphal.

The "J / ψ," as we shall call it henceforth, most of the time eventually decays into a veritable cornucopia of both stable and unstable hadrons in a bewildering array of eighty–seven or more different possibilities. Its mean lifetime is estimated to be around 10^{-20} seconds. Only in approximately twelve percent of the time does it decay into either an electron and positron pair, a reaction which is indicative of the weak force, or else a muon and anti-muon pair. The long lifetime was similar to the Lambda and Sigma particles but many times greater.

The scientists were now confronted with an enigma having the following peculiarities:

- Its lifetime or strangeness number for the new resonance was off the board from anything known previously.

- It had an isotopic spin of minus one, which made it a boson and disqualified it from being a baryon or being acted upon by the strong force.
- With a baryon number of zero any combination of two quarks which were assumed necessary for all hadrons thus required the two quarks to be one normal quark and one anti-quark. Baryons have three quarks.
- With an electric charge of zero, these two assumed quarks, which have fractional charges remember, must cancel each other out entirely. That is:

  $$Q = \text{Electric charge} = +\,2/3 + (-\,2/3) = \text{Zero}$$

- Its very large particle mass implied that the two component masses had to have individual masses well in excess of those calculated for those of the protons or neutrons. That fact was the clincher for considering the J/ψ as a very unique animal.

By this time in our history the general idea of quarks rattling around inside an invisible "bag" to form mesons, hadrons and baryons had gained credence with most theoreticians. It also helped those who were groping to find a consistent mathematical explanation for the apparent dichotomy between the weak and strong forces in the atom. All sorts of propositions, such as "gauge theory," "intermediate weak neutral vector bosons," and "weak neutral currents" were being conjured up to give a compatible systematic explanation for both the electric charges and the two internal forces in the atom. In spite of the best brains in Harvard and other world universities working on the theories cooperatively, the solution to this puzzle had thus far eluded them. Thus it happened that the J/ψ particle seemed to be a propitious key to the overall formation of what ultimately became known as "The Standard Model" for all nuclear physics. The theorists were in a delirium of delight and excitement at its arrival.

Back in Southern California., at the urging of a Harvard doctoral graduate named Terrence Goldman, the physicists were proceeding upon the notion that the two infinitesimally tiny quarks were orbiting about each other. This was a concept which led to the possibility that other similar resonances being possible with greater masses. As the acknowledged leader of the Stanford Linear Accelerator, Martin Breidenbach decided to crank up the electron beam to the 3.6 to 3.7 GeV energy range and see whether they might discover another sharp resonance peak. He applied for permission to hook up the SLAC main frame computer to the particle detector readouts for a short time, but all arrangements were not made until just before midnight of November 21st. The laboratory crew kept working continuously day and night for ten days and were eventually rewarded with a second narrow based and sharply pointed " spike " at an energy level of about 3.7 GeV, just as predicted. This new resonance was instantly christened the Psi Prime (ψ '); and this time the Stanford people had no competition. Its actual mass/energy content was calculated more accurately to be 3686 GeV. The fact that the Psi Prime decays into the basic and original Jay/Psi with positively and negatively charged pions flying off in curved tracks branching off into opposite directions to give a cloud chamber track similar to the Greek letter "ψ" and thus accounts for one of its names.

By the end of another two years other experimental laboratories in the world had reported finding four other resonances with mass variations of ascending masses, as shown in the table below. Eventually the two original particles were redesignated ψ_{S-1} and ψ_{S-2}. This was a nostalgic reversion to the memory of the old days of Balmer lines and spectroscopy for the lowest energy levels of electrons in the hydrogen atom. The entire group of six soon became known by the term " charmonium " invented by the Spanish physicist Alvaro De Rujula.

CC "CHARM" MESONS

SYMBOL	MASS IN MeV
Ψ	4415
Ψ	4160
Ψ	4040
Ψ	3770
Ψ (2 - S)	3686
Ψ (1 - S)	3097

Section 16- 3 The Confirmation of the Charm Quark

The rather ridiculous term "charm" was invented as far back as 1963 by Sheldon Glashow, who had been a former pupil of John Schwinger at Harvard University. Glashow was one of a significant number of respected theoretical physicists the world over who were struggling with the formation of a coherent and plausible explanation in mathematical terms for the weak, strong, and electro-magnetic forces within the nucleus of the atom.

The history of their efforts really deserves an entire separate book by written by one of the qualified participants. Glashow made an attempt in his biographical book titled *Interactions* (written with the assistance of professional writer Ben Bova) to describe the development of their theory for the public in non-mathematical terms with only partial success. He claims the intent of his new word "charm" was to indicate some magical, and as yet poorly understood, influence or property which might cause the observed behaviors of particles found in accelerator laboratories which were puzzling all the theorists. These included reactions which produced a pair consisting of a normal electron with an anti-electron (or positron) simultaneously and also, similarly, decays which produced a pair consisting of a muon and anti-muon together. There was also the riddle that electron beam reactions produced distinctive electron neutrinos whereas muon reactions produced only muon neutrinos. Both of these two neutrinos are not interchangeable, so far as we know, and their masses are considered to be different, regardless of the fact that they are so small as to be virtually immeasurable. Charm was also expected to explain a great number of decay reactions which were considered to be theoretically possible, but which never happen. The most glaring anomaly and mystery of all these questions was how the strange particles with abnormally long lives managed to tumble in stages from a strangeness number of three down to only one or less in successive decays.

For all these reasons mentioned and many more intricately complex justifications, the idea of quarks seemed to present an immediate solution. The advent of the J/ψ particle was a kind of culmination of thought which resulted in its being hailed immediately as evidence for a fourth type of quark to be known hereafter as "charm." It filled a void and became a companion for the singular strange quark. mall wonder that the year of 1974 became known among the scientists as the "November Revolution," although this should not be construed as indicating that physicists are secret Communists. They just like jokes. The charm quark must therefore be added to our previous list given in Chapter 15 as having the following properties:

NAME	Symbol	Electric Charge	Spin J	Charm Number	Baryon Number	Mass in MeV / c^2
Charm	C	+ 2/3	1/2	1	0	1650

The reader should be aware that the masses given for individual quarks in this book are approximate only and that they may change somewhat in practice. The writer hopes this will not add to his or her bewilderment. The J/Psi particle is considered to be a meson composed of a normal charm quark and an anti-charm quark in close juxtaposition. The fact that they do not instantly annihilate each other is a major miracle in itself. One physicist's answer is that extra residual energy causes them to orbit around each other, vaguely like twin stars in a pulsar. Ultimately probability, or even perhaps gravity, catches up with them and they come too close, releasing all their energy in a shower of decay fragments. It must be admitted that this explanation has an inherent weakness.

CHAPTER 17
MORE PIECES TO THE PUZZLE

Section 17.1 - The "D" Particles

Given the widespread acceptance of the reality of the charm quark, the next perfectly logical question was "Why cannot a charm quark team up with either an Up or Down quark and thus form another kind of hadron?" The answer turned out to be that they can and do. Several accelerator installations in the world, including SLAC in California, DESY in Hamburg, Germany, CERN in Geneva and the "Cosmotron" in Brookhaven, Long Island, all set out to find such mesons in their detectors. The confirmation took another two years with four principal varieties of link-ups identified by 1976. These became classified as the "D" meson mixtures. All have slightly different masses and decidedly different electric charges.

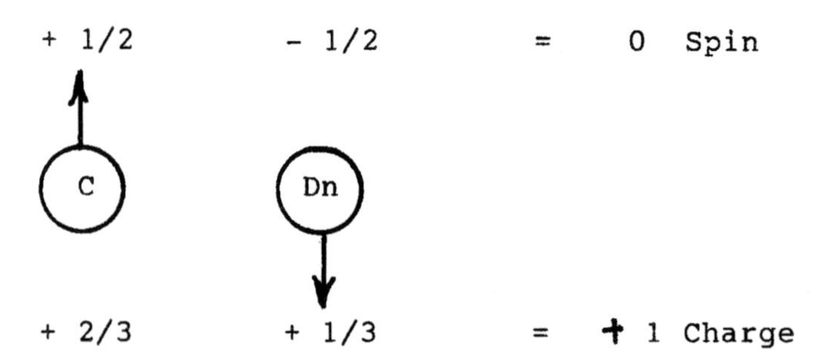

The D Plus Meson

Its mass is 1869.4 or roughly twice that of the proton.
Its mean lifetime is 1.057 X 10^{12} seconds.

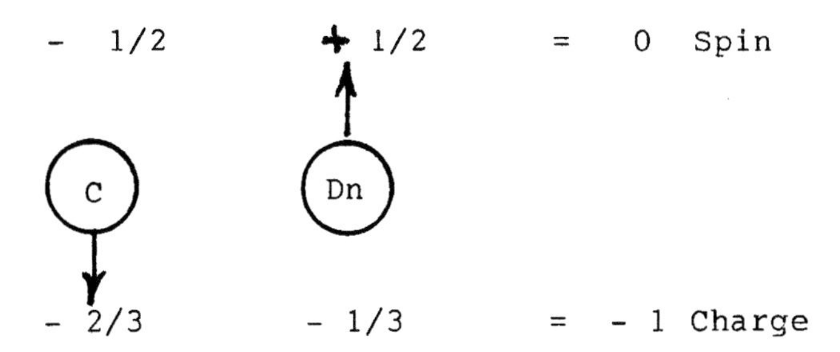

The D Minus Meson

The mass and the mean lifetime of the D Minus meson is identical with that of the D Plus shown previously.

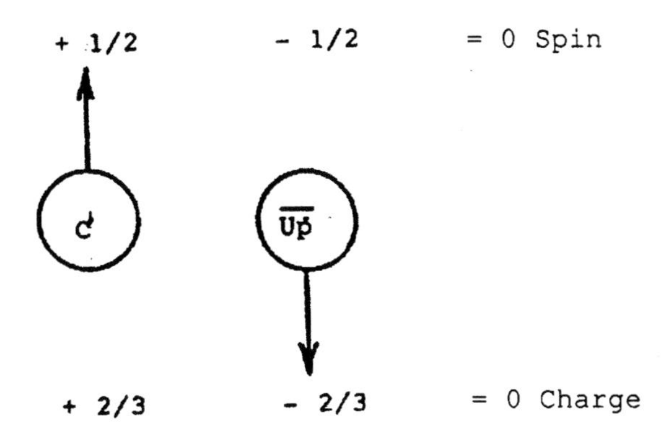

The Normal Charm Quark D^o Meson

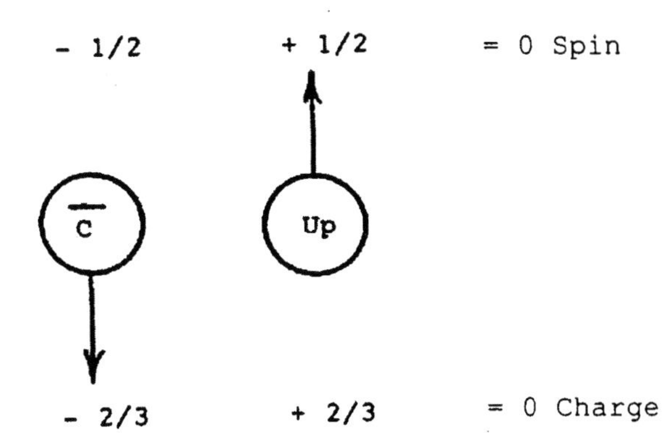

The Anti-charm Quark D^o Meson

The mass of each of the two neutral D Zero varieties is 1864.6 MeV. Their mean lifetimes are 0. 415 x 10^{12} second.

The possible decay products of any and all of the described charm particles by the Weak Interaction are so varied and multitudinous as to make any listing useless for the purpose of this book. It is only necessary for the reader to realize that the charm quark does, in fact, possess the magical capacity to change like a chameleon into other quarks or combination of quarks to a bewildering degree.

The study of charm quark hadrons continued for quite a few years in attempt to find other combinations. The pairing of a normal charm quark with one anti-strange quark was recorded in 1983 and confirmed in 1986. First called the "F" meson, it is now listed as the D_S Positive Meson. Since the identification of such weird and unusual combinations depends very heavily on a complete identification of their decay products, whether kaons or muons or pions, it places a severe burden on the reliability of the detector apparatus and the diligence of the laboratory personnel and computer readers. The mass of the "F" meson was placed at 1968.5 MeV but its mean lifetime was not significantly longer than its cousins. Eventually the physicists came to the conclusion that they had learned enough about the charmed meson to desist from any further expensive searches. The lesson was clearly that quarks have the ability to coalesce with others in almost any combination. Two quarks, however, is the minimum. A free and independent quark is evidently an impossibility.

Section 17.2 - The Tau Lepton

The symbol for the tauon, as it is called, is the Greek letter "τ" or tau, meaning the "third."

The discovery and confirmation of the J / ψ particle having been accomplished, that left the 8 GeV energy SPEAR accelerator machine in California available to Martin Perl. At the age of forty–seven, he was one of the senior and most thoroughly experienced members of the laboratory staff. He was already waiting for the opportunity to pursue his own project. Perl had become curious about occasional and confusing signals in the Mark II detector which seemed to indicate a negatively charged particle having a mass of at least twice that of the proton or four hundred times that of the electron and seventeen times the mass of the well known muon. It was not the Jay Psi particle but it took him until 1975 to become quite certain. A reaction at relatively low collision energies which he felt could make the Tau lepton identifiable is its decay into a negative muon and a positron with an accompanying neutrino, although an alternate decay process which can also happen is the negative tau producing an electron and two neutrinos.

It was Perl's intuition and perseverance which really brought about the discovery of the tauon. The muddle of different resonances connected with the earlier laboratory observations connected with the J / ψ charm particle described in Chapter 16 contributed to a general skepticism about a third family of leptons. The strange and charm quarks were conceded to be the second of two known families; but few people were anxious to introduce a third complication, although it was logically possible.

Milton L. Perl had been a student and protege of Isadore I. Rabi at Columbia University and thus was well aware of that scientist's interest and amazement at the discovery of the muon. Perl was born in Brooklyn, New York, in 1927. After High School and before going to college he became involved with the Coast Guard and later in World War II years saw service in the Army. He eventually returned for studies at the Brooklyn Polytechnic Institute to achieve a Bachelor of Science in Chemical Engineering in 1948. Needing money, he immediately went to work for the General Electric Company in Schenectady, New York. A physics teacher at Union College, where Perl was taking random graduate courses, advised him that his real interests seemed to lie in the area of physics, so he decided to apply for admission to Columbia University. He completed his thesis there on atomic beam measurements of nuclear quadrupole moments and obtained his Doctorate in Science and Physics in 1950. Having by this time become fascinated with the so-called high energy physics field for his future, Perl accepted a faculty position at the University of Michigan and there participated in pion scattering experiments using the Berkeley "Bevatron" machine. He there made the acquaintance of Samuel Ting, whose reputation he found intimidating. The chance to work with the brand new two mile long linear accelerator lured him to Stanford University in 1964.

His team persisted in their study of the results of collisions between electrons and protons at beam energies in the range of 3.6 to 4 GeV. Basically, their expectations were built upon the discovery of the positron and the known decay of the muon. They theorized they might get the production of a negative taon and a positive anti-tauon pair which would instantly annihilate each other with the first decaying into a normal negative electron and two neutrinos and the second yielding a positron and two neutrinos. Missing from these fragments were any other hadrons or miscellaneous photons or gamma rays, thus eliminating any other dissipation of energy in the conservation equation. Another probable alternative reaction would yield a normal muon and a positive muon plus neutrinos. Their search was successful and out of 34,850 test runs the team selected 86 signatures which they were convinced represented the tau lepton. They identified the unknown particle as having a mass somewhere between 1600 and 2000 Mev, which happened to be a remarkably good prediction. The presumed tau neutrino in the decay products, which represented the missing energy, had a tiny theoretical mass of only 65 electron volts.

The first written announcement of this discovery was sent by Martin Perl in August of 1975 to the *Physical Review Letters* with a scrupulous inclusion of thirty four of his associate workers under the title *Evidence for Anomalous Lepton Production in Positron-Electron Annihilation.*

Once the news was out and circulated, the Germans with their DESY accelerator machine in Hamburg, which also collides positrons with electrons, jumped onto the band-wagon. R. Brandelik of the Heidelberg University submitted their confirming evidence in 1978. Milton Perl, along with co-winner Frederick Reines, ultimately was rewarded with a Nobel prize some twenty years later in 1995 for the studies in the three leptons and their associated neutrinos. The existence of the Tan neutrino was confirmed at the Fermi National Laboratory in July of the year 2000.

The Tauon has a spin of one half and an electrical charge of minus one, thus duplicating the other two leptons. It has a mass of 1777.1 MeV/c^2 and a mean lifetime of 295.6 x 10^{15} seconds, which was just long enough to detect.

The possible decay modes for the Tau lepton are extremely diverse, depending upon the total energy released in the collision of the two opposing beams. This great variety instigated a great deal of discussion in the ongoing theoretical work on electro-weak reactions.

For the normal decay one would expect for leptons the Tau particle may decide to split into an electron and two neutrinos or else a muon and two neutrinos. The latter particle is, of course, unstable itself. However, just to confuse the scientists, it can also decay into a pion hadron plus a neutrino.

It is even possible for the quark and anti-quark pair found in the pion to appear with the accompanying neutrino having the Tau "flavor." In other words, with a given energy range a variety of weird and unexpected results were found. Perl's own discovery group registered the production of a normal negative muon in company with a positron and undetected neutrinos or, conversely, an anti-muon or positive muon paired with a normal electron plus neutrinos.They concluded in their written report that this behavior could not be explained by "presently known particles" or "well understood reactions." In 1974 this was a perfectly correct statement.

If the existence of the muon and muon neutrino once seemed superfluous and unnecessary for the construction of matter in our everyday world of matter, then the new Tauon added a second and even higher degree of irrelevancy. Why these particles exist, even for the briefest fraction of a second, and what they imply for the universe is still a mystery for the scientists. All they really know now is that the three leptons obviously reflect and would be compatible with three families of quarks. Some people consider them as evidence to justify a leap to the assumption that other worlds comprised of matter different from our own may be possible. A more obvious and less fanciful conjecture was that there must be a third family of quarks having even greater masses than the two then known that remained to be discovered. And so it was.

As with all the other particles, the transitory existence of an anti-tauon was taken for granted. Moreover, the neutrino which customarily accompanies any reaction which involves that anti-particle and accounts for a tiny fraction of the energy carries a special "flavor" attribute peculiar only to the tauon. If that does not seem peculiar to the reader, as it does to the writer, then he must have become indifferent to such elegant nuances. The name "flavor" is another ridiculous term invented by the physicists. It is supposed to designate unique characteristics singular to the tau lepton alone and serve as its "tag." The neutrino's mass may be the only difference from any other neutrino but we are not absolutely sure of that.

As explained in Chapter 9, the credible detection and identification of a neutrino of any particular flavor from amid a variety of spurious signals from extraneous sources is a painstaking task of precision requiring a dedicated patience on the part of the observer.

Neutrinos can skip through the bulk of the earth's diameter and never once impact on an atomic nucleus for any kind of a reaction, since they are absolutely immune to electric charge or magnetic influence and also essentially undeflected by gravitational forces which might exist in the atom. As a consequence the scientists were first obliged to resort to cosmic rays or radiation beams from the heart of a nuclear power reactor for a reliable source. They also were required to use huge containers of liquid, such as water, freon, carbon tetrachloride or other exotic materials in hopes of catching that one brief and tiny flash of light which signals an impact reaction. Transfer of momentum is the only tool available for flavor identification unless the reaction should happen to yield an electron or muon. Considering all the difficulties encountered in the search for neutrinos, whether electron or muon or tauon type, one can only marvel at the stubborn persistence of some physicists like Frederick Reines.

The mass of the tauon neutrino is thought to be less than 31 MeV. An alternative artificial creation of a neutrino in the laboratory having a particle accelerator machine always involves the simultaneous creation of its companion lepton. This means that a very high beam energy from the accelerator is required in order to achieve a detectable reaction in spite of the neutrino's trivial mass. As one moves up the scale from electron through muon to the tauon neutrino the task becomes progressively more difficult. On top of all the previous problems there is the constant risk of false or spurious signals in the detector due to natural background radiation in the environment. The usual defense against this possibility is to locate the detectors deep underground in mines or caverns or tunnels. An indication of just how seriously this need for protection against outside random causes is taken by the researchers was evidenced by the Italians. They carved a large room out of solid rock off the side of the Gran Sasso vehicular highway tunnel 4600 feet under the slopes of the Appenine Mountains. It came to their attention that archaeologists had discovered the remains of a wooden cargo ship which sank in a storm in the first half of the first century Before Christ off an island near Sardinia. It turned out that the sailing ship was carrying 1500 cast ingots of lead destined for Rome, each ingot weighing 73 pounds. They were retrieved by salvage entrepreneurs. The metal lead is only very slightly radioactive when mined; but laboratory tests on these samples revealed that, because of the half-life of lead, their age was such that the ingot lead had become virtually inert and pure. The Italians bought all of them for use as shielding around their neutrino detector.

Section 17.3 - The Quark Numbers Game

By this time the alert reader may well be forgiven for believing that he can enjoy playing with the alphabet soup or anagram game with the four quarks we know about almost as well as the scientists. Recall that we now have, up to this point in accepted physics anyhow, the following choices in our repertoire:

Name of Quark	Fractional Charge	Very Rough Mass Fraction of the Hadron Total
UP	+ 2 / 3	1 / 3
DOWN	- 1 / 3	1 / 3
STRANGE	- 1 / 3	1 / 2
CHARM	- 2 / 3	1

By definition, all mesons must have a baryon number of zero and an integral spin of either zero, one or two. This means a limit of only two quarks; and with two quarks the easiest way to build a meson with a zero or neutral electric charge is to pair a quark with an anti-quark.

$$\text{Electric charge} = +\ 1/3 + (-\ 1/3) = 0$$
$$\text{Spin} = +\ 1/2 + (-\ 1/2) = 0$$

Supposing, however, we wished to create a meson with a strangeness of zero but an electric charge of one. The positive electric charge requirement eliminates the possibility of an identical quark and anti-quark pair. If the reader turns back to Chapter 13 for a moment, he will discover that these qualifications actually fit one of the three possible Rho particles. One Up quark and one anti-Down quark will do the job:

Electric Charge = + 2 / 3 + (+ 1 / 3) = 1

It is not the purpose of the writer to drag the patient reader through more of these combinations and permutations at this point; but it becomes apparent why the physicists entered into a frenzy of predictions for as yet undiscovered particles with dreams of a Nobel prize dancing in their heads.

Section 17.4 - The Upsilon Meson And The Bottom Quark

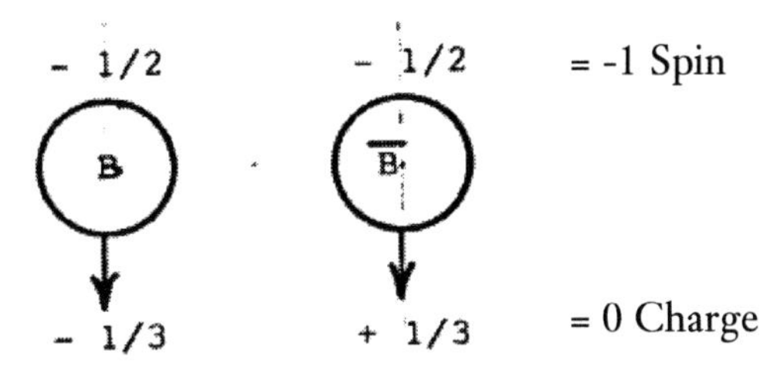

The symbol is the Greek letter Upsilon (ϒ).

Its discovery was not entirely an accident, since many theorists had been speculating upon the muon plus anti-muon type of collision reaction and its possible resonance by-products. Moreover, in 1975 the famous experimenter, Leon Lederman, along with Stephen Herb, had temporarily moved from Columbia University over to work with the Fermilab in the outskirts of Chicago. He had been greatly impressed by the work done by Stanford University in isolating the newly found J / ψ particle with precision and he was not a little jealous of their success. He had come very close to finding that resonance "shoulder," as he called it, when using the Brookhaven AGS machine in Long Island and later in 1972 when working at the CERN Intersecting Storage Rings facility in Switzerland. Not only that, but very early in his career back in 1957 he had experience with tracking muons with a low energy Cyclotron at Irvington-on-Hudson when attempting to verify parity violation. Thus his chagrin at being only a spectator to the discovery of the J / ψ was perfectly natural. Thus it happened that, with the approval of the original Director of the Fermilab, Robert Wilson, Lederman decided in 1972 to upgrade the National Laboratory accelerator for a second time since its start-up in 1972 in order to have a beam of protons with a peak collision energy on a metal target of 400 GeV. The target remained uranium; but the absorber stage, which was used in previous experiments to slow down and eliminate unwanted hadrons, was changed to the metal beryllium instead of iron. The reason for this revision was that beryllium has less diffusion effect on the beam trajectories of the muons in their passage to the detector. The technique for detecting and measuring the expected muon pairs remained essentially the same, although the detection spectrometer apparatus was vastly improved by an electrical engineer named William Sippach. The new detector consisted of two entirely separate arms, measuring approximately six feet high by six feet wide by one hundred feet long, and arranged in a Vee pattern with their apex point terminating inside the steel and beryllium target box. Each detector arm array consisted of a drift chamber and gas-filled Cerenkov spark counter followed by a series of eleven electrostatically charged wire screens (called Proportional Wire Chambers) and seven scintillation counters called Hodoscopes. The intent of all this was to catch and identify the simultaneous passage of lepton pairs, whether electron-positron or muon with anti-muon. This was considered the evidence of the original decay of a heavy meson resonance particle.

Data began accumulating in May of 1976. The experimental operating team usually comprised a minimum of sixteen people selected from a personnel pool of sixteen experts comprised of six individuals from Columbia University, six representing the permanent staff of the Fermi National Accelerator Laboratory, and four representing the New York State University at Stony Brook, New York. This clearly demonstrates the collaboration typically required for science today.

Their original objective was to find another possible peak resonance at a higher energy level similar to the J / ψ. At first they thought that they had found one in the data, but the evidence disappeared as unsubstantial in closer analysis. However, in the summer of 1976 they noticed a promising "hump" in their by-product graphs of scattering versus mass in the vicinity of 9.5 GeV (9500 MeV). The staff was so encouraged that John Yoh placed a bottle of champagne in an onsite refrigerator in anticipation of a celebration.

Because of a slight electrical malfunction in the machine the actual confirmation of this resonance peak had to wait until May of 1977. A small electrical device over-heated and caught fire, spreading the fire to nearby cable insulation. That was a relatively minor accident; but the real damage was done when the brave emergency fireman summoned for help sprayed water over everything, thus chemically creating hydrogen chloride fumes which combined with the water to make hydrochloric acid. A Dutch salvage expert recommended by the CERN people had to be aroused from his home in Spain and flown to California in order to clean up the mess in a great rush. Ten days later they were back in operation again and satisfied themselves that their resonance discovery was genuine. Their tersely worded report titled

Observation of a Di-muon Resonance was mailed off to the *Physical Review Letters* by the very end of June, 1977.

The physicists at the Deutches Elekronen Synchrotron at Hamburg, using an electron-positron accelerator ring, quickly jumped onto the bandwagon and duplicated the accomplishment in May of 1978. They were using a new Pluto detection counter. They named the resonance the "B" meson after the word "beauty" on the conviction that it represented the first of two new family quarks which the theorists were predicting and which had been whimsically christened in advance "Truth" and "Beauty." Someone must have been enthusiastic about his English poetry. Two years later the Cornell electron-storage ring at Ithica, New York, also succeeded in repeating the experiment. Leon Lederman shared his long sought Nobel prize with two other men in October of 1988.

The Upsilon resonance, as it was subsequently re-named by popular majority, has a mass of 9.46 GeV in its first or "ground" state, which is ten times that of the proton. It has a spin of minus one and no electrical charge whatsoever. These facts came as a surprise to the scientists. There was no way that they could add two or three known quarks at that time together and come up with such an unexpectedly large mass. Moreover, the mean lifetime of the Upsilon was remarkably long. Since the recently discovered Tauon had already pointed the way to a third family of leptons, a consensus was quickly reached that the Upsilon had to be a pair of entirely new quarks. In other words, it was conceded that the Upsilon had to be a very brief coalescence of a new Bottom quark and an anti-bottom quark similar to any other meson. The Top quark to go with it remained entirely speculative, but the physicists were confident that it must exist and be discovered in the future.

The existence of four other companion varieties of the Upsilon having slightly higher and ascending masses in subsequent years clinched the acceptance of the bottom quark and also indicated strongly that the imaginary top quark had to have a spectacularly higher mass, quite beyond the energy capacities of the existing collider machines then available.
It was decided that the newly found Bottom quark had the following properties, remembering that its share of the momentum mass is never ascertained as a precise number and varies slightly with the particular resonance:

NAME	SYMBOL	ELECTRIC CHARGE	SPIN	BEAUTY	BARYON NUMBER	MASS IN MeV
Bottom	B	- 1 / 3	- 1 / 2	- 1	+ 1 / 3	4500

Ultimately, the discovery of the anticipated Top quark would give us a total of six different kinds of quarks.

Section 17.5 - Implications and Complications

Although the discovery of the sub-elemental quark particles, if point fractional electric charges which cause collision scattering could really be called that, appeared to present a wonderful solution to the perplexity of a myriad of mesons, hadrons and baryons in the particle zoo it was also somewhat of a mixed blessing to the physicists. Their existence raised a host of new questions, as well as some apprehensions among the theorists. Whatever happened to the much vaunted "beautiful simplicity" to the laws of Nature to which most classical scientists adhered as the ideal? Was there a possibility of another family of quarks? Worse yet, did all this presage a chain of some infinitesimal and unimaginably tiny structures below the quarks like the skins of an onion reaching toward an infinite smallness? This last wild notion was expressed by people like Gerard Hooft of the Institute of Theoretical Physics in Utrecht, the Netherlands, and Buchmuller at the Max Planck Institute for Physics in Munich, Germany. It survives today in the much publicized String Theory or "Theory of Everything." Curiously, this speculation also harks back to Schrodinger's Wave Function equation, which yields only infinities without the renormalization technique of inserting known physical parameters. The sneaking suspicion that quarks may not be the ultimately small building blocks of nature has not entirely gone away yet, and a few imaginative disciples even coined a new word for such sub-sub-particles - Preons.

One consequence of the fourth quark was that physicists then began to wonder about what they called "families" of elementary fermion particles, and this led them into the quandary of just how many families were possible anyway.

This notion of "families" is rather vague with respect to any rigorous definition and is more of a human intellectual effort to group some obvious similarities for the purposes of memory and comprehension of complexity. Actually one will not often find this term in the index to serious physics textbooks. The principal difference between the so-called "families" is one of mass.

Consider the facts with which the reader is now familiar.

CLASSIFICATION OF FERMION	FIRST FAMILY	SECOND FAMILY	THIRD FAMILY
Leptons	Electrons	Muons	Tauons
Neutrinos	Electron Neutrino	Muon Neutrino	Tauon Neutrino
Quarks	Up and Down	Strange and Charm	Top and Bottom

There will be more about that later. Some persistent metaphysicians began to wonder if quarks were the end of the splitting down of the proton or neutron and whether it could have many layers like an onion. Thus far, that notion is not taken very seriously and evidence that such a disturbing possibility is real has not yet appeared.

Insofar as we refrain from Super-String theories and metaphysics, there is no need to offend the reader's psychology and commonsense perceptions of his physical environment with such speculative fantasies. Fortunately, things have settled down in the field to a general acceptance that quarks are real and that further sub-divisions are most improbable in high energy physics within practical demonstration. Thus far the scheme of quarks works so well logically that physicists are very loath to discard it. The astonishing success with the quark concept in explaining all the varieties of discovered particles and even in predicting ones to find in the laboratory won enthusiastic support by the majority. The ultimate endorsement was the inclusion of the quark code (u, d, s, c, and b) in the *Review of Particle Properties*, as edited by the Particle Data Group in Los Alamos, New Mexico.

One can indeed be intellectually comfortable with the idea of quarks being some kind of intermediate building blocks in the process of raw energy "freezing" into matter, somewhat like moisture in the air turning into ice crystals and organizing themselves into various patterns of snow flakes when chilled to low temperatures. Yet many questions which lack answers stubbornly persist. Some of these are listed below:

1. Are the fractional electric charges (2 / 3 and 1 / 3 plus or minus) actually real or are they only a semblance of some total unity which we deduce as a sharing process, but do not fully understand?
2. Why has no one in any laboratory ever observed or detected a single lone quark existing all by itself independently outside of a conglomerate of two or three?
3. Can quarks rotate or move about each other in their special confinement or spatial "bag" limitation?
4. What strong mechanism or congealing force constrains two or three quarks within their invisible envelope of tiny space which comprises a meson or nucleon?
5. Why do certain combinations of quarks which seemed to physicists perfectly logical and possible, as conceived by the theorists, seem not to exist and have defied any detection?
6. Why does the strange quark (and others) have the peculiar ability to transform itself into a different variety of quark, such as the "Down," for instance?
7. What makes quarks form or come into being in the first place and how does all this relate to the strong and weak forces and bosons in general?
8. Is there a serious possibility that there may be more than three families of quarks? In fact, why does Nature find it necessary at all to

present us with the possibility of three families when it seems to us that only the first would suffice?

9. Why is this conjectural Top quark so difficult to verify anyway? Is it merely that its estimated mass is large enough to tax the energy capacities of most existing accelerators?

In the next chapter we shall attempt to cope with these questions without delving into the necessarily complicated and abstract mathematics which govern the explanations of the scientists.

Chapter 18
Gluons, Chromo-Dynamics and Electro-weak Currents

Section 18.1 - Quark Properties and Transformations

The world of the quark turned out to be not quite as simple as some people first thought, although this development may not have come as a complete surprise to a few of its inventors. To use a metaphorical illustration, the situation of the physicists then might be compared to a small band of determined explorers who climbed to the peaks of a mountain range in front of them only to be confronted with a view over an intervening valley to yet another range of mountains before them. Quarks are very strange entities which do not conform to all the rules already collected in the book prior to their discovery. Beginning as early as 1971 theoreticians such as Murray Gell-Mann and Harold Fritzsch were working to try to explain their peculiar characteristics.

It was also fortuitous that the ingenious invention of the storage ring addition to the Stanford Linear Accelerator, which allowed the contra-circulation of two different and oppositely charged ion beams, occurred. This single idea started a revolution in the entire engineering of accelerator machines, which the Germans were very quick to realize. They instantly raised some money and built the Proton Electron Tandem Ring Accelerator (PETRA). These new machines actually made the demonstration of quarks possible. There were many reasons. The first and most obvious one is that the collision of two opposing beams at high velocity almost doubles the momentum energy at their impact as compared to one beam hitting a stationary target. Think of two Volkswagen automobiles hitting each other head-on instead of one Volkswagen hitting another parked vehicle sideways from the side. This analogy helps to understand a second advantage of the secondary acceleration ring. The momentum of a single high velocity projectile hitting a stationary target tends to produce a bunched bundle of parts and fragments all moving downstream together; whereas two projectiles coming from opposite directions not only produce more debris, but the pieces and fragments are apt to fly off from a single impact center in all directions, thus making their detection and identification far simpler than for a cluster.

Finally, there is a third and much less obvious reason for the advantages of the storage ring and one which we have not emphasized in previous chapters. Since quarks are produced from raw energy alone, it is theoretically possible to create them from an apparent vacuum by colliding photons or (easier and much more practical) electrons with positrons. Here we are dealing only with leptons. The amazing thing is that this actually happens in the laboratory. Thus when an electron and a positron collide head-on, the center of their masses remains essentially stationary and the energies distributed between all the resultant shower of particles, as revealed by their tracks and momenta, must sum up to the original total collision energy. That makes the calculations very much simpler. There is a cost penalty to be paid for in exchange for this wonderful tool, however. The accelerated leptons in their motion emit synchrotron energy, which increases as the fourth exponential power of the velocity and is inversely proportional to the radius of the accelerator circulating ring. This constant energy loss must be compensated for by radio frequency power boosters spaced at regular intervals and also higher strength magnets for bending their tracks into a circle. All this involves huge electric power costs and places a premium on ever larger diameters of the storage rings. Nonetheless, quark research would probably never been possible without these new machines.

Quixotically, physicists among themselves frequently refer to "the family problem" when discussing quarks or leptons. A stranger might receive the impression of a group of social workers debating a particularly difficult adoption case. In reality, they are pondering the reason why Nature seems to have created three different sets of quarks and accompanying leptons and neutrinos with increasing masses and decreasing life spans. Moreover, why did she stop at three? From the point of view of us humans one family alone would have been enough to populate the cosmos with atoms and molecules and thus ourselves. The question of whether there could be more than three families of quarks possible is not yet settled unequivocally. The answer seems to be dependent on whether the question is predicated upon the billion degree temperatures assumed in the "Big Bang" creation of the cosmos or in the relatively cold background temperatures of the moons or planets. All research and observations at the date of this writing appear to result in a negative answer to the speculation that there might be a fourth, or even fifth, family of fundamental particles still beyond our discovery. For a while the cosmologists speculated that a fourth family might solve their problem of the mysterious "dark matter" around nebulae, but this idea is presently unpopular. From the point of view of the astro-physicist the number of quark or lepton families is vitally important and linked with the quantity of helium found in the early universe. More than three families makes their calculations come out with too large an amount of He. The masses of the quarks in any conjectural family beyond three, as projected from the known scale, would be unbelievably heavy and no credible evidence of them yet appears in our solar system. Nevertheless, the physicists still keep the European names of "Truth" and "Beauty" in reserve in their glossary kit bag for a possible fourth pair of quarks, just in case. The suspicion that they are theoretically possible and may once have existed persists.

Even accepting the restriction to three families, if one includes the possibility of an anti-quark with an opposite electrical charge for each quark identity listed, the result is an expected total number of twelve different quarks. This does not sound like a very simple explanation for the fundamental building blocks of nature to some people.

As we already know, the quarks inside a baryon or meson share between them the total momentum of the nucleon or particle in some distribution determined by their so-called "flavor," whether Up, Down, Strange, or whatever else. Physicists, in the interest of absolute accuracy, are in the habit of identifying quarks by their moment distribution factor instead of mass, since these momentum numbers change whenever a quark is struck by an electron projectile in the accelerator. It is somewhat of an annoyance to the scientists that the individual rest mass cannot be sorted out and predicted exactly by calculations. Actual observations in the accelerator laboratories are required in order to determine the precise contribution of each quark. Notwithstanding the persistent echoes of debate about the real nature of quarks as used in the lexicon of physics, they have become "real" in the minds of the younger generation of physicists as we approach the twenty first century.

It is conceded that, while inside their confining envelope, quarks can move around or orbit and change their positions with respect to one another within limitations imposed by symmetry and the forces which bind them together. This idea was not really original, since Neils Bohr and John Wheeler together in Denmark around 1934 worked on the mathematics of the nucleons inside the atom behaving as particles in a drop of water. This fact becomes important for a number of reasons. It is surmised that three quarks within a baryon can orbit around each other in various configurations which have different energy levels. As John Polkinghorne described it several years back, "in the point-like limit, quarks in the color gauge theory behave as if they were freely rattling around and not tightly (or rigidly) bound to each other." Metaphorically speaking, they behave as if they were marbles confined in a bag! Just how and why quarks were confined was never exactly explained to everyone's satisfaction.

Quarks, being considered fermions, are also assumed to have a nominal spin or magnetic moment of one half. This assumption was very convenient for the theoreticians and, at first, seemed superficially true, but difficulties soon arose. Quarks are not as simple as first supposed. In the period between 1987 and 1971 numerous experiments in what were called "high energy deep inelastic scattering" studies were conducted both at the Stanford University in California and the CERN facility in Switzerland. They used electrons or muons as the ion projectiles to bombard a proton target. They determined that the spins of the three quarks within their envelope that form a proton or neutron contribute only between twenty and thirty percent (and possibly less) to the total spin of the resulting nucleon. This came as a big surprise and became known temporarily as "the spin crisis." Obviously, the binding forces of the three quarks within their "bag" needed to be better understood. The conjecture was that the missing remainder most probably came from the orbital angular momentum of the quarks themselves plus the effect of spin one for the boson forces which glued them all together. The concept was that the three quarks must all revolve as a group. The proper mass of a proton can be greater than the combined masses of the quarks; but less than the relativistic masses of the constituent orbiting quarks in high speed motion.

Another degree of freedom which quarks possess is the ability to change their "flavor" or identity. We understand the word "flavor" as an unsatisfactory nomenclature for that property of a quark to share a certain range of momentum or mass and to still retain a certain fixed fractional electric charge of whatever polarity, plus or minus. A prime example of this versatility of the quark is the Beta decay process for a neutron becoming a proton with the emission of an electron and neutrino, as discussed back in Chapter 6. Obviously any reduction in mass must require an emission of energy.

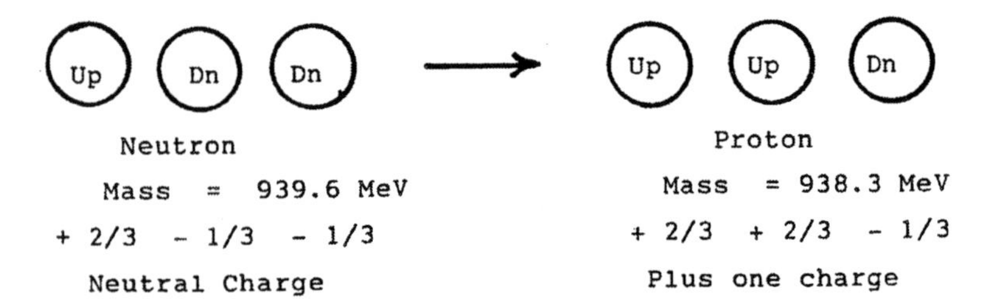

Another vivid example is the doubly charged positive Delta (+ +) decaying into the singly charged Delta Plus (+).

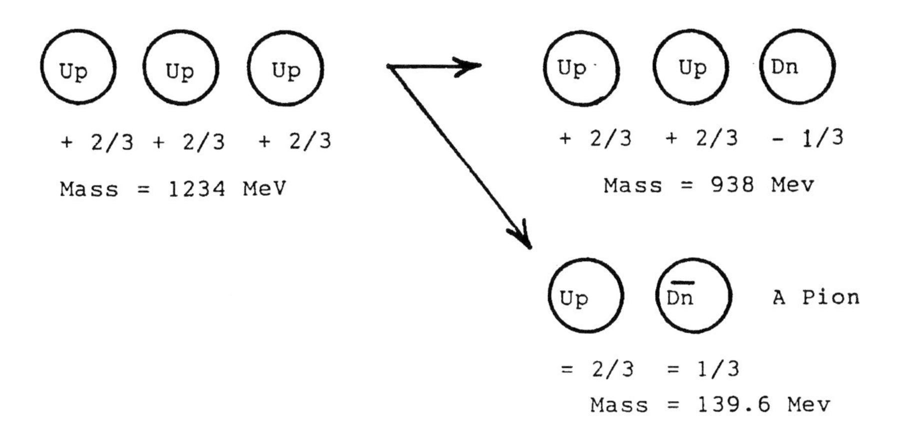

As shown above without the directions of spin for each quark delineated the reader could easily confuse the Delta Plus with a proton. The difference is the spin direction of the down quark.

The change in the electrical charge of a quark is usually accompanied by a reversal or "flip" of isotopic spin direction. Thus the two different attributes must be considered together. Of course, the masses of the quarks also may change, making the problem more complicated.

In all the illustrations given above the conservation of electrical charge and mass energy is maintained by the emission of a pion meson.

In comparison, the decay of the strange quarks, on the other hand, proceeds very slowly. This peculiarity is interpreted as an indication that their downward cascade is governed by the weak force. As one would quite sensibly expect, as in any degeneration of a resonance particle, a change in the flavor of a quark proceeds downward to a lesser mass or momentum factor. Consider, for instance, the decay of the positively charged Sigma Plus.

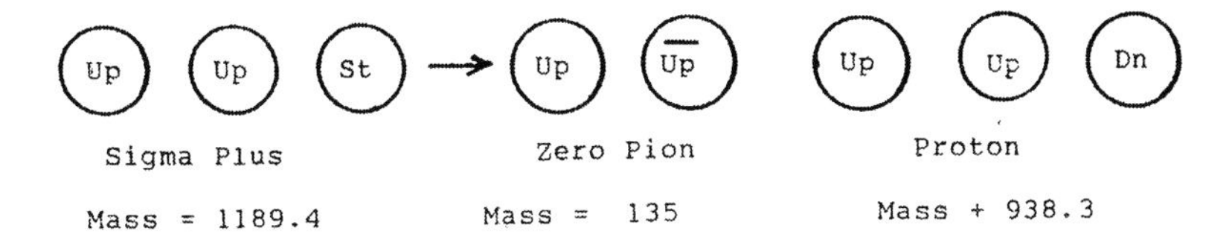

The reader may notice that the final end result is often the most reliably stable neutron or proton particle. Since the pion is taken as evidence of the strong force, we should therefore recognize that both the strong force and the weak force may operate together on quarks. The writer added masses to the diagrams above in order to allow the readers to play with the conservation of energy sums, as do the professional physicists. The game is quite possible, but seldom exact in practice. Momentum energy imparted by an accelerator can leak off in many ways, including gamma photons or neutrinos, as well as that found in the velocity of moving fragments. This is one reason why the physicists tend to talk somewhat ambiguously about "momentum sharing" of quarks and why even today a few professionals prefer to use Richard Feynman's word "parton" in order to describe the process. A further complication, as we pointed out, is that a quark may sometimes flip one hundred and eighty degrees on its spin axis orientation, thus changing its total magnetic moment.

The fact that the collision of high velocity electrons with positrons in accelerator rings could produce bound quark pairs or mesons was really a momentous revelation to the scientists. They already knew about the Rho positively charged meson, which we mentioned in Chapter 13. It consists of the combination of an Up quark with an anti-Down quark with both spins in the same direction and was created by smashing high velocity pions into a proton target. The decay of the Rho into two pions gave them a valuable clue about resonances and also to the various combination possibilities for mesons, including having their spins opposite to one another. The chief requirement for inventing mesons was to make certain that the 1/3rd and 2/3rd electrical charges were additive to unity. Thus the appearance of the J / ψ in our Chapter 16 was in the nature of an epiphany for the theorists. The pieces of the mosaic were beginning to fall into place and traces of the grand picture were emerging. In that case, moreover, the similarity of a couple consisting of the Charm and anti-Charm quark with the already encountered electron plus positron pair, as well as the muon plus anti-muon annihilation pair, was too obvious to ignore. The conjecture was that there had to be a close relationship between leptons and quarks with a quantum electrodynamic commonalty. The three families of leptons and neutrinos with quarks could no longer be considered a strange accident. Refer to Table 18 - 1 in the text.

The actual precise masses of the three possible neutrinos (electron, muon and tauon) are still very debatable, although the astro-physicists seem to agree that none of them should exceed 65 electron volts. This assumes that the electron neutrino has any mass at all and they all remain fixed.

A CHART OF THE FUNDAMENTAL PARTICLE FAMILIES

CATEGORY	LEPTONS	QUARKS	QUARKS	NEUTRINOS
Electric Charge	- 1	- 1/3	+ 2/3	0
First Basic Family	Ordinary Electron	Down Quark	Up Quark	The Electron Neutrino
Second Family	Muon	Strange Quark	Charm Quark	The Muon Neutrino
Third Family	Tauon	Bottom Quark	Top Quark	The Tauon Neutrino

TABLE 18 - 1

The fact that the different families of the fundamental particles came in steps of increasing masses led to the inference that there must be some scaling law of increasing magnitudes which, if not perfectly linear, had to behave within mathematical limitations for such a structured behavior. The word "scaling" was convenient for expressing this concept. We have already encountered the use of this word with regard to the laboratory observations of the richochet fragments obtained by bombarding protons with high velocity electrons, muons, or positrons. Here, however, it has a slightly different connotation.

The incremental increases in quark masses between families appears mathematically to be distinctly more exponential than linear. Nevertheless, there are countless examples of such a relationship in the natural history world. Studies have been made, for instance, on the amount of air admitted into the internal contents of chicken or bird eggs due to the porosity of the egg shell. A certain amount of air is required to keep the embryo alive; and such studies have been made on the eggs of sparrows up to those of ostriches having, obviously, increasing masses. Similar calculations have been made on the metabolic rate or amount of heat produced by their food consumption necessary to sustain a healthy, normal body temperature for mammals ranging from mice to humans to elephants. Possibly one of the most popularly recognized scaling rules of a significant practical nature is the marine architecture formula to determine the maximum speed of a displacement vessel through the water. For any water borne boat of appreciable draft and weight, whether a rowing shell or an ocean liner and regardless of the means of propulsion, the formula indicates that the top speed achievable is principally proportional to the waterline length.

The mathematical trick to finding such a formula depends upon identifying the various crucial variable factors which can be measured and arranging them so that the various units all cancel one another out to result in a dimensionless number. The equation usually defines the slope of the curve for a quantity being studied under various conditions with the curve approaching a straight line to some degree. The fascination for this statistical game for physicists began with the application of such scaling techniques to quantum numbers in order to achieve renormalization of the quantum wave function equation and avoid the infinity answers which it frequently yields. In the case of the three quark family masses which we have been discussing here, scaling is applied in a most simplistic way. One can make graphs of the known observed masses for the different quark flavors and then make a projection of the curve in order to arrive at a guesstimate of the masses for a conjectured Top quark or a supposed fourth family that might conceivably exist. The writer, however, does not wish to exaggerate the importance of scaling in the reader's mind but it is used as a particle prediction device by the physicists.

Section 18.2 - Quarks and Color

A more intractable perplexity arose with the conventional quark pictures for various particles. Quite a few of the three quark pictures surmised for baryon resonances clearly violated Paul's Exclusion Principle, inasmuch as their isotopic spins all lined up parallel with one another with their rotation axes all oriented in the same direction. In other words, the total spin of the resonance or hadron became three or even five halves! A most conspicuous case was the Delta ++ particle with a double positive electric charge of 2 and having three Up quarks, as pictured in Chapter 12. The Omega baryon, which consists of three Strange quarks, as displayed in Chapter 13, is an exactly similar case. The total spin was considered to be three halves, which is normal and permitted for fermions; but for this to happen with quantum identities in such extremely close contiguity contradicts a long established rule. Quarks are, of course, still free to change their polarity or spin axis orientation and they may do this quite readily and quickly. However, that still does not solve the dilemma for the theorists as to how three quarks in such close conjunction can all share the same spin direction for a short time without separating. The entire quark theory came under re-examination.

The ultimate conclusion was that, in addition to flavor, each individual quark had to possess some other special and hidden attribute or else were inexplicably behaving as bosons. While on a sabbatical leave from California in 1971 Murray Gell-Mann visited CERN in Switzerland to study another puzzle as to why the neutral pion decayed nine times faster than its sisters and more rapidly than expected. This conspired to lead him back to the current turmoil in the quark theory and also to an association with a brilliant young man named Harald Fritzch. The latter had only recently managed to escape from Russian–occupied East Germany by ostensibly going for a vacation in Bulgaria. Once there he took a kayak and paddled alone across the Black Sea to Turkey. Since he had been a qualified physicist under the Communist regime, he was given sanctuary. Subsequently he wrote a book in German about his adventures with the title *Flight to Leipzig*. A translation into English does not seem to be available.

At any rate, these two collaborators soon came to the conclusion that any quarks, whether Up, Down, or Strange, were capable of having three different attributes (dubbed "color" for no particular reason except Gell-Mann's penchant for whimsy). His first choice for these three colors was the nationalistic and patriotic red, white, and blue; but the international scientific community eventually settled on red, green, and blue. Aside from politics, another reason for this selection is that they are integral components which taken together form white light. For the uninitiated, the thought of university science professors seriously proposing colors for sub-microscopic entities confined within the proton or neutron and which can never be seen suggests that they may have been watching too much "virtual reality" on television. Nevertheless, a large body of mathematical calculations support the contention of an extra and unobservable attribute of quarks which has no counterpart in our own familiar macroscopic world. Names, therefore, become purely arbitrary.

In order for most of us to comprehend why such an invention became necessary to the physicists, this writer would beg to suggest an analogy which involves both symmetry and the probabilities so essential to quantum mechanics. Consider the situation of two people each, in turn, tossing a copper penny in the air and letting it fall on the flat surface of a table. Let us further suppose that these two individuals doing the casting are very young people in Middle School classes. Each penny, as is customary in currency, has a different embossed figure or picture on each of its two sides which in general parlance in the United States are called "heads" and "tails." After they have practiced tossing the pennies, their teacher asks his students how many different combinations of heads or tails are possible. They could very well answer three, as shown in Figure 18.2 on the following page.

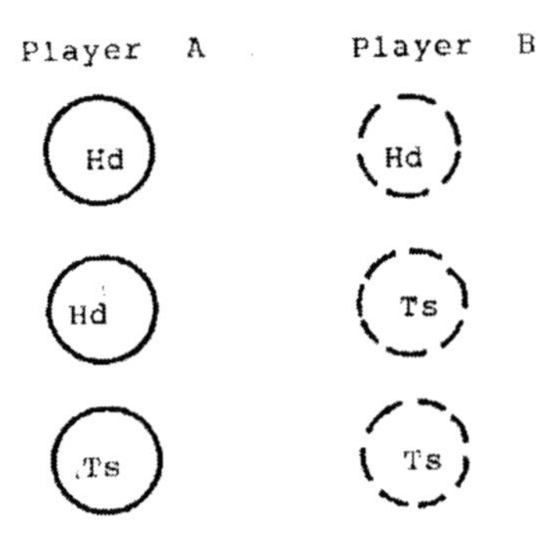

Figure 18. 2

The same two children, when they eventually reach the subject of combinations and permutations in the advanced algebra class at High School level, will be told that the maximum number of toss arrangements with two pennies is supposed to be two squared or four! How would you, as the teacher, explain why they were wrong in the first example? Painting one coin red and the other green would help tremendously.

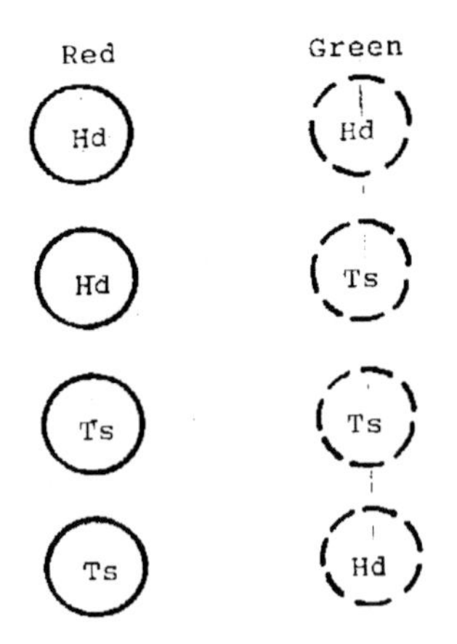

Figure 18. 3

The answer, of course, is that, as long as each coin maintains its separate identity, then the number of flips or turns to the obverse side of the coin made by each coin must be equal for both players in order to preserve absolute symmetry. Color makes this evident for comprehension. Outrageous as this example may seem, it emphasizes the importance of symmetry to physicists in their understanding of the behavior of nuclear particles and even of quantum mechanics.

Section 18.3 - Bosons and Gluons

There still remains one very perplexing conundrum about quarks which we have not really addressed. What force is it that binds two together in a meson or holds three quarks inside the confined space of a baryon? How does it happen that no single quark can exist independently outside of an envelope which is no larger than 10^{-13} centimeters in diameter? Nor has one ever been observed.

It is axiomatic in today's elaborate theoretical structure for particle physics that the forces of either attraction or repulsion and electrical charge or even, in some cases, mass or momentum, which are exchanged between fermions of spin one half are carried by or moderated by separate and different spin one entities known as bosons. The simplest known elementary boson is the photon, which moderates electrical charges and originates electromagnetic wave phenomena associated especially with light and electrons. Back in Chapter 3 the reader was told quite didactically that all bosons, whether nuclear or atomic, had a spin which was some multiple of one. That is still true; but the concept of a boson is strange and by no means simple. In Volume III of his Lecture Series dealing with Quantum Mechanics Richard Feynman made the following statements:

"Why is it that particles with half integral spins are Fermi particles whose amplitudes add with a minus sign, whereas particles with a an integral spin are Bose particles whose amplitudes add with a positive sign? We apologize for the fact that we cannot give you an elementary explanation. An explanation has been worked out by Pauli from complicated arguments of quantum field theory and relativity. He has shown that the two must necessarily go together; but we have not found a way to reproduce his arguments ... This probably means that we do not have a complete understanding of the fundamental principle involved."

Feynman was referring to the fact that, in order to make the Schrodinger quantum wave function equation produce any sensible results, it was necessary to give fermions a negative spin of - 1/2. One reason that Feynman got away with such disarmingly candid comments like the previous quotation was that if he did not understand it, then it was very probable that no one else did either. Professor E. C. D. Sudarshan at the Center for Theoretical Physics for the University of Texas at Austin, Texas, has spent a great deal of time and thought on this problem. He says: "We do not know why (such) anti-commuting particles are required to exist. We are glad that they do, the point being that none of us would be here... to worry about it if they didn't." From the writer's unqualified judgement, the explanation of Ian Duck and Sudarsan in a joint article depended upon the interpretation of Dirac's " Holes in a Sea of Negative Energy " as merely the recognition of the existence of our " normal " particles and anti-particles, application of the Principle of Least Action, and whether time could be considered as going either forward or backward. It is doubtful whether that contribution is helpful to the reader, however.

Today, in the age of quarks, bosons are regarded as the consequence of Non-Abelian Gauge Theory and Gauge Invariance. The Spin-Statistics Theorem states that identical integral spin particles satisfy the Bose-Einstein statistics which permit any number of such particles in each quantum state. Professor David B. Cline of the University of California in Los Angeles puts the entire question succinctly and dogmatically. "The Standard Model in Physics is based on the assumption that ordinary matter is composed of two types of particles — quarks and leptons — and that the forces between them are transmitted by a third category of particles called bosons."

The forces which bind quarks together were christened "gluons" for obvious associative reasons. Gluons are massless vector bosons with a spin of one and their interaction with quarks can change the color of a quark. Thus gluons are to quarks what bosons are to protons and electrons, but with one huge difference. The attractive force becomes very weak and approaches zero when the quarks are close together, but becomes larger and approaches infinity whenever the two quarks are separated by a distance. That is totally contradictory to our Newtonian world, of course; and opposite to the behavior of gravity or electro-magnetic attraction. Eventually the stretched 'elastic' will snap, but only when enough energy has been put into the system **to create two 'new' quarks ($E = mc^2$ again!), one on each side of the break.**

The process is reminiscent of trying to separate a north magnetic pole from a south magnetic pole by sawing a bar magnet in half. Every time you break the two poles apart, you find you are left with two new bar magnets, each with a north pole and a south pole, instead of two separated poles. This most unique behavior is casually dismissed by physicists as an asymptotic curve, as in descriptive geometry. It is this behavior of the gluon which not only confines the quarks, whether two or three, into their very small envelope of space, but also makes it impossible to extract any single quark from its confinement without the intrusion of large amounts of energy which completely changes the matter composition of both target and fragments.

Physicists talk of quarks having different colors. All baryons are supposed to contain three quarks in the colors of red, green and blue respectively. Making an analogous comparison to the color spectrum in light rays, this conglomeration of three primary colors is regarded as becoming "colorless," as with "white" light. This so-called "color" attribute for quarks and gluons has nothing whatsoever to do with the real world of our perceptions of different frequencies of light rays. It is an arbitrary name for a hidden variable which cannot be isolated outside of a meson or hadron. Nor can a lone quark be isolated, for that matter. At this point in history we have six recognized quarks having different flavors plus six anti-quarks. Now we are saying that a quark may have one of three possible different colors. Obviously this makes for a profusion of possible combinations. It becomes starkly apparent to even the novice student that there has to be a great number of different gluons to fill the bill. The quark versus boson picture has suddenly become much more complex. We can simplify things considerably by remembering that there cannot be more than three quarks in any baryon or hadron. That reduces the required number of gluons down to 3 x 2 x 3 or eighteen. If we stop regarding the anti-quarks as peculiar and that the color attribute of gluons makes no distinction between the normal and anti-quark in a hadron, then the number of required colored gluons is further reduced to only nine. These possible color connections between quarks are displayed in the subsequent chart.

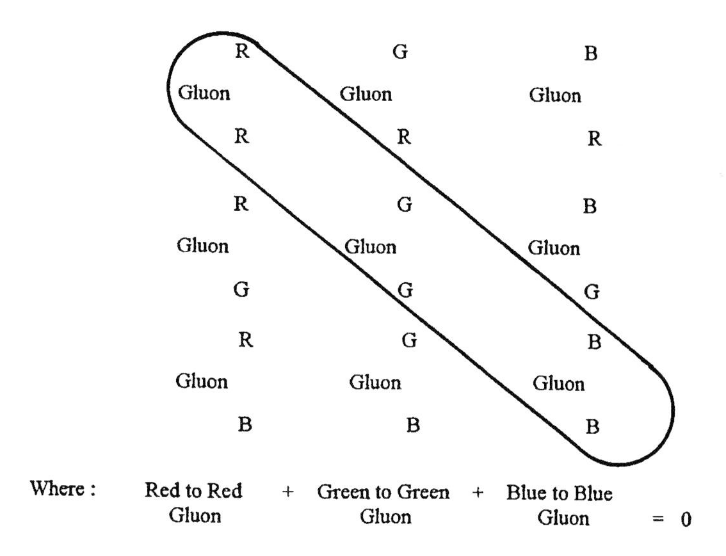

The three primary complementary colors indicated inside the diagonal oval merge together in what the physicists call an SU-3 singlet which they visualize as white light containing all possible colors. Such a combination can produce no change in a quark color, so it follows that at least one of the color pairs shown within the diagonal group must be eliminated. The scientists

appeal for physicists is that renormalization of the fundamental parameters, such as masses and coupling constants, is possible in order to obtain reasonable calculation results for a range of low to moderate energies. Nevertheless, perturbation corrections must be introduced, since a really rigorous field theory which incorporates gravity and relativity is presently beyond their ability. Even so, the theorists can describe the structure of matter down to a size of 10^{-16} centimeters. It is interesting to the writer that K. Grotz and H. V. Klapdor of the German Max Planck Institute between them wrote an entire book on Weak Interaction Theory in 1990. They included the following innocuous sentence in their Chapter 4: "It is very possible that the gauge principle may be replaced one day for a more fundamental (and complete?) principle." This implies a secret philosophical worry that the explanation may still be inadequate in some way to provide a concrete visual picture satisfying our own need for objective reality.

Section 18.5 - Quantum Chromodynamics

Having devised the system of colors and gluons in order to explain how the quarks were bound together, the scientists were then confronted with the task of explaining how such a system actually worked and to produce the particles observed in the laboratory. After all, that is the basis of the revered "scientific method."

To quote W. S. C. Williams for a concise definition found in his excellent textbook *Nuclear and Particle Physics*:

"The fundamental principle of quantum chromodynamics is that the fields describing the quarks are required to have a form of gauge invariance. The consequence is that there must exist, in this case, eight massless spin one bosons which interact with the quarks and are responsible for the forces between them. These are the gluons.... Thus gauge invariance is now considered to be an essential element of all quantized field theories."

One should start with some basic assumptions.

1. There are eight different "color" possibilities for individual gluons.
2. Gluons all behave as massless virtual energy quanta or bosons with spin one somewhat similar to the electro-magnetic photon.
3. The different gluons may have different binding magnitudes and color charges.
4. Quarks can only be bound together in symmetry singlets for SU-3 to satisfy gauge group theory. This is the "white light" analogy.
5. A gluon coupling may change the color of a single quark with which it is coupled in order to allow it to violate the usual Pauli Exclusion Principle.

6. Gluons are electrically neutral. They carry only binding tensile energy and can impart momentum characteristics to a single quark.
7. Gluons with a color charge can and do react with one another.
8. The gluon binding charge between two quarks increases as the distance between them increases. This means that it requires a great amount of energy to separate two quarks by a distance such as one centimeter. Conversely, the magnitude of the binding force between quarks at sub-nucleon dimensions of 10^{-13} centimeter diminishes as the quarks come closer together.
9. At the extremely small distances within a nucleus the quarks gain what is called asymptotic freedom and can move around as almost independent of one another.
10. Breaking the chromodynamic force between quarks in the laboratory is extremely difficult. It was first thought that dislodging quarks from their invisible space envelope was impossible. Ultimately the physicists modified their first opinion to concede that a quark may be ejected from its confinement shell but the energy required to accomplish this is instantaneously converted into another quark, thus making a pair and creating a new meson or hadron or even, in rare cases, an electron and positron pair. Alternatively, the escaped quark may instantly join and become part of a nearby particle, thereby prompting a rapid decay disintegration with a shower of fragments. Obviously, this type of reaction confuses the detector pictures and the physicist responsible for interpreting what happened. However, the conversion of the stretched gluon attached to an impacted quark into pure energy causes a very distinctive "jet" picture of some particle which serves as a clue.

The entire color gluon scheme is actually a very clever analogy to facilitate the solutions for SU-3 symmetry relationships for the binding forces between quarks and thus avoid the abstract nature of group theory.

The real mystery to the result of knocking a quark out of a nucleus by means of high velocity projectiles is where did the increase in boson energy originate? Some texts liken the increase in the binding force to elastic rubber bands which constrict closer together in the lines of force instead of arcing through a wider spectrum like the lines of iron filings on a piece of paper resting on top of a bar magnet. This is a convenient mental picture, but is strictly inaccurate and begs the question of the actual source of the increase in boson energy. Some scientists speculate that there are swarms of virtual particles which wink in and out of the vacuum of space and provide this energy. Others prefer to believe that the process may arise from severe bending or curvature of space in a short time frame. Whatever the reasons, these mysterious bosons are an integral and important part of the creation of matter from energy and moderate the process. To most of us who are unfamiliar with such abstract speculations we tend to regard the instant materialization of "virtual" bosons from empty space as voodoo magic.

The actual theory of gluons was first worked out early in 1973 by Hugh David Politzer of Harvard University with David J. Gross and Frank Wilczek of Princeton University. Justification for the gluon theories in the laboratory had to wait until 1976 when the newly upgraded Super Proton Synchrotron ring at the CERN facility in Switzerland went into operation. The assistance of a long and elaborate CDHS detector designed by engineer Jack Steinberger was invaluable. Credence in the entire gluon idea was strongly encouraged by their careful studies of the behavior of the neutral or zero pion responsible for the strong force. The theory helped to explain why this two quark meson had a mean lifetime which was nine times shorter than the 7.5×10^{-16} second time interval formerly calculated by the theoreticians. It had also been noticed that in the electron versus positron beam collisions at the SLAC facility in California the ratio of the quantitative production of hadrons divided by the counted production of quark plus anti-quark paired mesons could readily be explained by assuming the existence of three gluons of different colors.

The first real experimental confirmation of the existence of gluons most probably came in 1979 from the Deutsches Elektronen Synchrotron or DESY machine in Hamburg, Germany under the direction of Herwig Schopper. Unfortunately, the credit for this achievement was partially obscured by competitive publicity and claims by Americans from both Berkeley, California, and Brookhaven, New York. Nobel Prize fever is quite contagious.

To quote the assessment of Andrew Pickering of the University of Illinois, "By this stage, though, the quark-parton model was in danger of becoming more elaborate than the data it was intended to explain!"

Eventually, however, the colored gluon scheme for quantum chromodynamics became the accepted doctrine in spite of some qualifications from the experts. Richard Feynman, for instance, is quoted by his biographer, James Gleick, as saying: "I don't get any physics out it! .... The QCD theory with six flavors of quarks, with three colors, each represented by a Dirac spinor of four components, and with eight four-vector gluons, is a quantum theory of amplitudes for configurations each of which is 104 numbers at each point of space and time. To visualize all this qualitatively is too difficult." [1] The reader may say "Amen."

A more measured, but equally candid, judgment was offered by Professor W. S. C. Williams of the University of Oxford and author of a respected textbook on Particle Physics:[2]

---

1. James Gleick, *Genius.* (New York, Pantheon Press, 1992.)
2. W.S.C. Williams, *Nuclear and Particle Physics.* (Oxford, U.K., 1991)

"It is believed to be the correct theory of strong interactions of quarks and gluons. However, in spite of that confidence, the theory has many difficulties of calculations which made it impossible to obtain sound quantitative predictions, except in a few cases."

Translated, that means it is all we've got and it seems to work tolerably well when we have the laboratory facts to guide us.

Section 18.6 - Beta Decay, The Weak Force and Neutral Currents

Beta Decay is the process created by the so-called weak force which causes a neutron in the core of an atom to spontaneously decay into a proton plus an electron plus an electron anti-neutrino. We should, however, note that a similar decay can take place in atoms in other modes, especially with isotopes of helium for instance. The reaction takes place extremely slowly, and first became evident in laboratory experiments with naturally radioactive substances as far back as 1896. Enrico Fermi attempted to define a theory to explain it in 1933. He fully realized the need for a connection between electro-dynamics and relationships between electric charges, as well as the pressing need for some rational explanation for the origins of the weak and strong forces within the atom. A small part of any satisfactory explanation had to involve the relative orientations of the spin axes of the fragmented particles. For a time, however, his ideas were dismissed as being incomplete and prone to infinities in the calculations.

The puzzle of the weak force continued to attract the attention of the heavy thinkers in the field. As early as 1968 a former Pakistani resident and child prodigy now living in Trieste named Abdus Salam started publishing his ideas in physics journals, sometimes in collaboration with John Ward in England. Salam held the eminent position of Director of the International Center for Theoretical Physics in Trieste, Italy, and, moreover, appointed to with a Chair in theoretical physics at the Imperial College in London. He later collaborated with Stephen Weinberg, now of Texas, and Sheldon Glashow of Harvard on the origin and behavior of the Charm quark and also a proposed electro-weak theory of neutral currents. The eventual outcome of this work was his sharing a Nobel prize with them in 1979. A very important and basic assumption for explaining both the Weak Force and Strong Force was the existence of massive vector bosons with a spin of unity. These were designated W^+, W^-, and the Z zero. All these were in addition to the massless photon and gluons. The role of bosons in forming matter from energy had acquired a major status.

Compared to the strong force and electro-magnetic attraction or repulsion, the weak force is indeed very feeble. The only thing weaker inside the atom is gravity. Nevertheless, we know from the long list of naturally radioactive elements and equally long list of natural and artificially created isotopes that this force is capable of loosening the bonds which hold an aggregation of neutrons and protons in the nucleus of an atom. This fact, in itself, was enough to suggest to the scientists that there could be a relationship between the weak and strong forces, however obscure to them. The weak neutral current makes itself evident in a variety of ways in addition to the familiar Beta decay of the neutron which we have already discussed in other chapters. One of these which the physicists focused upon was leptonic reactions, such as the decay of the muon or tauon. Similarly, the relationship between the three leptons and their neutrinos came under examination. The study of neutrinos is a specialty in itself on which vast sums of money are still being spent for experimental test facilities. In this effort both nuclear physicists and astronomers quite agree, which enhances their political clout for funding. The arguments for the amount and reasons for the differences between the electron neutrino and the muon and tauon neutrinos still rage, aggravated by the fact that the predicted radiation quantities from the sun, as detected on the earth surface, are significantly wrong. Those detected in existing neutrino detectors thus far do not in any satisfactory way compare with the amounts predicted by current fusion energy theory and the scientists are eagerly awaiting the operation of a new heavy water reactor in Canada for corroboration.

The importance of all this to astronomers is intense; but the decay processes of the muon and tau leptons bring the question down to earth, so to speak. Why should the reaction shown below occur?

$$\mu^- \rightarrow e^- + \nu_\mu + \nu_e$$

Today this reaction is regarded as an example of " weak neutral currents " and a weak force decay mode involving the "W" boson. The Feynman diagram shown below gives a picture which is compatible with the standard Gauge Theory doctrine.

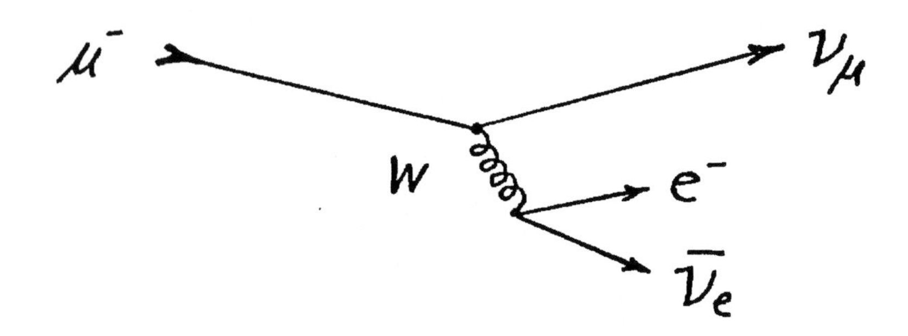

A second category of weak force decay of prime importance must be added. This is the violation of the strong force holding particles together, whether hadrons or mesons. At first, this statement seems to be a contradiction in terms but the rapid and widespread acceptance of the quark concept and the discovery of the various two quark mesons led quite logically to the consideration of this possibility.

Mesons, the reader may remember, are a combination of one quark with another companion anti-quark which is not necessarily of the same flavor. Their decay is obvious. In order to explain the decay of heavier hadrons into smaller and less massive fragments, however, one must confront the probability that some of the internal quarks must exchange charge and energy to become a new quark of entirely different flavor. This implies a possible change in the electric charge of a quark, which introduces a new dimension into the picture. Quarks are bound together by gluons, which are another variety of strong force. Decay of the strong force is more unusual and happens rapidly; whereas decay by the weak force is much more common and proceeds slowly.

Returning to the subject of Beta Decay once again, the most commonplace example is the decay of a neutron into a proton. This is now regarded as a change of one of the three quarks into a different flavor, specifically a Down quark into an Up quark.

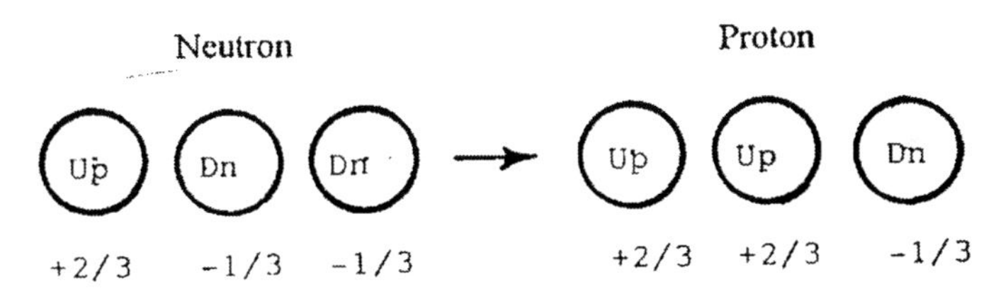

In this process the following internal transformation in the neutron is necessary:

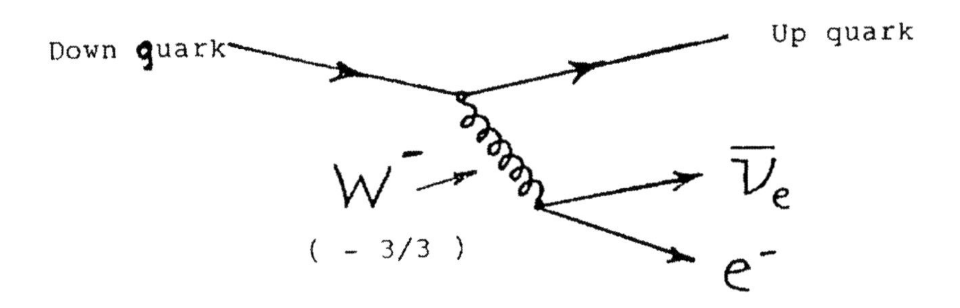

Notice that in all these Feynman diagrams of the transformations of particles there are always various types of neutrinos flying around. That is fortuitous for the scientists. It happens that all "normal" or relatively stable neutrinos have their direction of spin pointing against or in opposition to their direction of motion or negative. The anti-neutrinos, on the other hand, have their spin direction pointing along the direction of motion, which is called positive. The consequence is that a study of the neutrinos in any reaction gives the theorists clues and the laboratory technicians a demonstration of what is probably going on in a given target reaction. This explains the physicist's own interest in neutrinos which coincides with the astronomer's. However, neutrinos are extremely difficult to detect, having negligible mass and no electrical charge. One can only identify their presence when a neutrino happens to impact directly upon a particle in the environment with a high energy to impart.

The alert reader may also have observed that the author has not launched into an explanation of the electro-magnetic charge changing the nature and origin of the "weak currents." His reasons are simple. It took some of the smartest professional physicists more than twelve years to develop a satisfactory mathematical theory which is a technological thicket of bewildering intellectual subtleties into which no sensible person should venture without a doctorate degree in the subject. The subject was replete with "renormalizations," "perturbations" involving both strange and charm quarks, and intricate calculations which must be made on large main-frame computers. Sheldon Glashow admitted that back in 1961 he had to break the symmetry of his Weak Neutral Current or Z_0 model by "putting the masses in by hand." Weak interactions, such as the decay of the neutron into a proton and electron and neutrino, involve some type of redistribution of electric charges; but what and how remained unclear. This is attributed to an intermediate vector boson which was designated the "W" boson and may be either positive or negative. On the other hand, degenerative influences which disrupt the strong force, but do not involve any electric charge change, are attributed to a Z neutral or Z zero boson. None of these three bosons can be simply explained, especially with regard to their origin or energy source.

The small fraternity of physics theorists recognized these difficulties, especially with regard to the Weak Interactions or "W" bosons.

It needs to be acknowledged that two Dutch physicists, Gerard 't Hooft and Martinus J. G. Veltman, labored together in collaboration and relative obscurity to clean up the mathematics of the Electro-Weak Theory and, in particular, to eliminate the need for renormalization tricks in order to avoid infinities and account for the W^+ and W^- bosons. Veltman, who was born in Waalwijk, Netherlands, in 1931, and had obtained his PhD at the University of Utrecht in the Netherlands, in 1957 worked at the University of Michigan and 't Hooft was part of the faculty at the University of Utrecht. They published their results in 1972.

As happens so often in the course of human events, their achievement did not achieve public fame until 1999, when they were both jointly awarded a Nobel prize, twenty years after the theory introduction by the Salam-Weinberg-Glashow team was recognized by the same organization in Sweden.

To some people the "W" and "Z" bosons simply represent a redistribution of electrical charges, and thus forces, between quarks or several nucleons within the core of an atom. Others claim that they come into being from space filled with a cloud of "virtual" particles winking into and out of existence and appear temporarily whenever there is the slightest sign of instability in the structure of a nucleon assembly. Gauge theory supports their existence and they may transfer momentum, spin, and electric charge between the sub-particles of the atom. Depending upon the energy required to exert a given messenger force, these " virtual field quanta " particles can be very massive indeed - many times that of the proton. By the year of 1991 the calculated masses were as follows:

$$\text{For the } W = 84.41 \quad GeV / c^2$$
$$\text{For the } Z^0 = 91 \quad GeV / c^2$$

The decay of the mesons can also be due to the weak force interaction. As one instance, the Kaon in particular afforded the researchers a veritable playground on which to try out their theories. The decay process proceeds downward in mass, of course; but the process often involves a change in the flavor of one quark. The positive kaon alone has several possible modes involving the production of a pion having potentially different electric charges. One example is shown below.

$$K^+ \rightarrow \pi^0 + e^+ + \nu$$

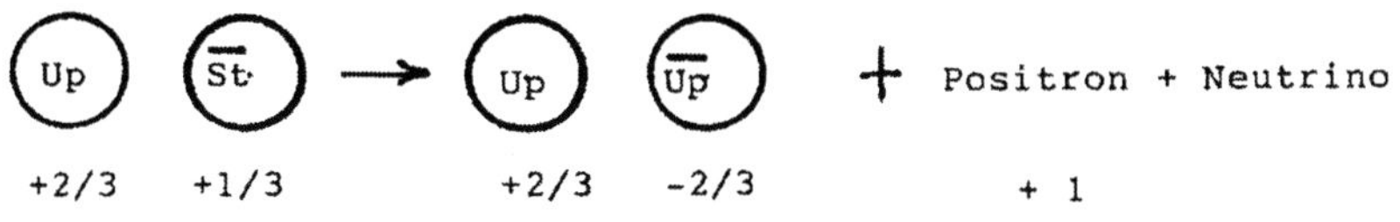

(Up) ($\overline{\text{St}}$) → (Up) ($\overline{\text{Up}}$) + Positron + Neutrino

+2/3 +1/3 +2/3 -2/3 + 1

This reaction can again be pictured schematically by Feynman's shorthand method.

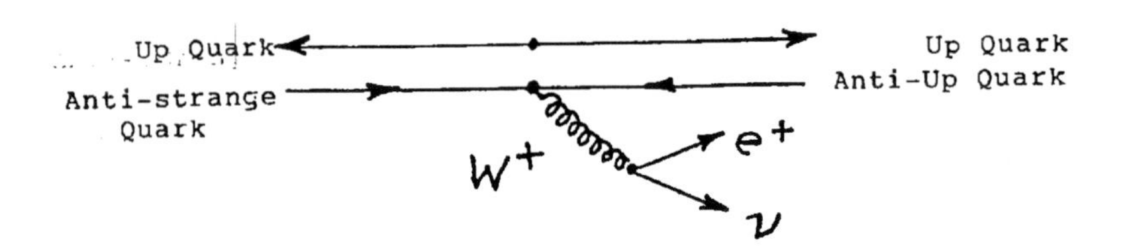

Actually, the more frequent method of decay for the positive Kaon is into a positive muon or anti-muon, but the principle is still the same as our arbitrary selection.

An appropriate example for the operation of the Z zero boson might be the very common annihilation of a normal electron with a positron or its anti–particle into two different neutrinos. This can happen naturally in the stars. The principle is the same.

This brings us to the end of this chapter, since the persevering reader has already assimilated more knowledge of the subject than most people who do not intend to earn their living in physics. By the end of the 1980's decade the science profession became quite encouraged and pleased with their comprehension of nuclear and particle physics and with their ability to predict the probable outcomes of reactions in the laboratory, or, at least, to identify the remaining subjects on which to expend research funds for better comprehension. If the bemused, but patient, reader should happen to feel somewhat lonely in his perplexed skepticism about Quantum Chromo-Dynamics, then he might derive some solace from the following statement by two skeptical professionals: "The use of QCD in treating the hadronic world has become an overwhelming trend in particle physics. Perhaps it is the first time in the history of physics that a theory which is neither precisely defined nor proved to have the right to exist as a consistent theory has become so popular."

Chapter 19
Laboratory Confirmations of Boson Theory

Section 19.1 - The Standard Model

By this time in our story the physicists had begun to talk of the "Standard Model" as an explanation for particles as though they had assembled all the pieces of the jigsaw puzzle into a rational and coherent picture. There was a bit of conceit in this assurance, since it gave outsiders and college students the impression that they had solved all the really difficult problems. This was not actually true. Even their master, Murray Gell-Mann conceded this point in his contemplative and philosophical book[1] *The Quark and the Jaguar*. To quote him:

> "...There are a number of reasons why it (the Standard Model) cannot be the ultimate theory of the elementary particles."

In the opinion of astronomer / editor of "Science" magazine, David Lindley,

> "The present theoretical structure is both an impressive achievement for the variety of particles and phenomena which it includes and also an unholy mess full of loose ends and replete with odd and ingenious mechanisms."

Notwithstanding such honest criticisms, the Standard Model has proven astonishingly successful in describing how particle physics works, based upon quarks as the ultimate elementary units. Moreover, it has become possible to make reasonable predictions for the laboratory experimenters to confirm, as well as making fairly close estimates of the anticipated masses for new or as yet undiscovered particles. The true or more precise mass values still need to

1. Murray Gell-Mann, *The Quark and the Jaguar*. (New York, W. H. Freeman, 1994).

be verified by actual experiments. However, the number and complexity of the assumed fundamental quarks and their flavors and colors still troubles the theorists and challenges their philosophical credo that the basic laws of physics should be essentially simple. The problem becomes worse when attempts are made to include Einstein's theories of Relativity into the calculations.

There are other embarrassing gaps in our knowledge which clearly indicate that the Standard Model is not yet in any terminal stage. The origins of virtual particles and gauge bosons, which are assumed to inhabit the vacuum of empty space, are not really clear or unequivocally explained. Another piece of unfinished business is the postulated Higgs Boson, named after its inventor. Thus far the energy level required for a synchrotron accelerator needed to confirm the actual existence of this particle is well beyond the capacity of our present machines.

Established and widely respected as the Standard Model may have become among physicists today, this writer was quite startled to hear a lecturing professor from Columbia University, who had served his apprenticeship at the CERN facility in Europe, say that we "should get rid of that dead albatross!" Of course, he was alluding to the "Rime of the Ancient Mariner" by Samuel Taylor Coleridge. Having heard how wonderfully well the Standard Model worked for years, the author was unprepared for such iconoclastic attitudes. One may surmise that the younger members of his academic group have become impatient with some of the theory's quirkier aspects and would prefer more license to search for other explanations or newer mathematics.

As we all know by now, the result of this general dissatisfaction with the status quo of particle physics has led to the proliferation of very abstruse mathematical constructs for the "theories of everything." These include Super-Symmetry with a world of ten dimensions and the "Strings" theory. This relatively recent trend has also encouraged several popular science books such as *The End of Physics* or *The End of Science*.. The authors of these books do not actually believe the titles selected by their publishers; but are really deploring the current situation in their field of high energy physics. They are uneasy about too much reliance on very complicated and intellectually demanding mathematical hypotheses without the ability to test these ideas with the old-fashioned and pragmatic methods of physical demonstration in a laboratory. To that extent their complaints are genuine.

Section 19.2 - Jets

With the advent of large high energy collider machines suddenly being financed by government funds throughout Europe and the North American continent under the stimulation of the Cold War with Russia and optimistic dreams of cheap nuclear power the opportunities for particle physics research suddenly escalated. The consequence of significantly higher kinetic energies in the accelerators resulted in the appearance of new and strange phenomena. A list of dates when these huge accelerator machines first became operational tells the story of a steady progress towards higher kinetic beam energies.

- In 1959 the first Proton Synchrotron Accelerator ring with an energy rating of 28 GeV was completed at CERN in Switzerland. It went into actual operation in November of 1960 and performed beautifully.
- In January of 1968 the Russian proton synchrotron accelerator at Serpukhov went into operation at 76 GeV energy levels.
- In May of 1972 the United States national "Fermilab" in Illinois reached a 200 Gev energy reading under the direction of Robert R. Wilson.
- In August of 1972 the Meson Physics Facility at Los Alamos in New Mexico were jubilant that their own small proton accelerator reached 800 MeV.
- In February of 1974 the CERN facility in southern Europe starting testing their new storage ring addition to the older original synchrotron for what they christened the Super Proton Synchrotron or S. P. S., which was expected to have a design power of 400 GeV sometime in the near future.
- In September of 1976 the "K.E.K" laboratory outside Tokyo in Japan was under construction for the modest expectation of 12 GeV energies.
- In August of 1977 the University of California started the actual construction of a new Positron Electron Project, called " PEP" for short, which was fed high velocity particles from the older linear accelerator.
- By 1983 the addition of two new accelerator and storage rings to the older "Tevatron" synchrotron at Fermilab raised the collision energy level of protons with anti-protons to 512 GeV by means of two separate and opposing beams of 270 GeV each.

One of the first and most peculiar of the new phenomena observed with these large machines was what became called " jets." These were beams of puzzling particles emanating in two opposite directions from a common collision point.

As we explained previously, impact on a stationary target is not always the best method for detecting diverse concentrated tracks of fragments because the momentum of the incoming high velocity projectile tends to push everything before it in one direction. It had already been deduced from the quark and gluon theories that directly hitting a quark inside of a nucleon such as a proton should separate it from its two companion quarks. This should produce one or more hadrons in a downstream direction, not to mention photons or electrons; but their tracks were all pushed in one general direction. In Chapter 18 we said it was axiomatic that there can be no such thing as a solitary quark existing outside of an atomic nucleus and retaining its original identity. The gluon forces within the nucleon restrain all quarks. As their separation is effected by whatever outside means, the potential energy between them also increases with the distance of separation. This storage of energy is often pictured as being similar to that of a stretched steel spring. The continued separation of a quark from its sisters causes the storage of enough energy to enable the creation of a new and additional quark or particles when the bond breaks. We now have a new anti-quark of some flavor appearing which is free to unite or pair up with any other quarks in the vicinity and does so immediately. The result of this fragmentation can be a shower of hadrons or mesons forming a jet emanating from the target vertex.

The addition of secondary, or even tertiary, storage and acceleration rings to the end point of an older synchrotron was a stroke of genius which enabled the physicists to accomplish two significant improvements. First, they were able with the help of strong focusing magnets to collide two pencil beams of particles having opposite electrical charges coming from opposite directions at a selected impact site. This meant their individual momenta more or less canceled each other out and the collision site remained essentially stationary inside the detector chamber. Second, the high velocity energies of the two opposing beams combined together to double the total energy involved in the reaction. Thus the shower of resulting fragments from the collision seemed like a veritable explosion from one obvious center. The team from Stanford University and University of California decided to collide electrons with positrons. The CERN facility in Switzerland, on the other hand, decided to collide protons with anti-protons. Thus the stakes in the game were immensely increased.

An alert young woman named Gail Hanson perceived this jet possibility and adopted it as her personal goal of investigation. She had acquired a PhD degree from the Massachusetts Institute of Technology and started her experimental work with the ill-fated and unreliable Cambridge Electronic Accelerator in 1968.Frustrated by its demise, she then moved west to California to join Richter's team there. Although hampered by the then relatively low energy capacity of the SLAC linear machine, she did succeed in producing evidence for jets in 1975. The fact that the collision of high velocity and very light weight leptons can produce more massive and fundamental particles deserves our amazement. The large annihilation energy released by the collision of a normal electron with a positron may result in a photon or, more probably, a boson. The latter may quickly decay into two quarks of opposite charge speeding away in two different directions.

In the meantime, an English physics theorist named John Ellis, who was working at CERN with the Italians, was also arguing that the breaking of a gluon binding force and consequent release of energy would not only produce two jets but might also even create a third. This latter possibility was based on his assumption that it would be needed to account for any uncharged or neutral particles. In 1977 he finally managed to persuade the management of the DESY facility in Germany to allot time for his research program on their electron versus positron collider machine. The planning for this effort was assigned to a pretty young woman from Hong Kong named Sau Lan Wu, who first entered the physics field at Vassar College in Poughkeepsie, New York State. Her first experimental runs were only partially successful; but after the beam energy of the German machine had been raised to 27 GeV in 1979 things changed. Wu produced computer data and photographs made by their new two–armed Solenoidal Spectrometer detector which showed clear pictures of three–pronged jet events. Bjorn Wiik, the Director of the Hamburg facility, became convinced. The era of what became known as "hard scattering" had begun.

The announcement of this discovery precipitated a bad international scene involving Samuel Ting. It could be that he was still smarting from his previous loss of precedence in the first revelation of the J / ψ particle; but he challenged the Germans on their claim to credit for confirming the existence of jets. He managed to place a telephone call to Walter Sullivan, who was then the highly regarded Science editor of the New York Times. Sullivan was somewhat annoyed, since he was on vacation at a secluded lake in New Hampshire. He told Ting to talk to his assistant in New York City. Not content with that, Ting dispatched a graduate student from Cambridge, Massachusetts, by automobile with explicit instructions to deliver the professor's written press release to Sullivan in person. Unfortunately for the hapless messenger, this meant rowing across the lake in a rainstorm in a small boat to an isolated cabin, and returning home sodden. When eventually published by the newspaper, the Europeans were not favorably impressed by the American style of competitive physics. However, the international excitement over jets was genuine inasmuch as they were regarded as a visible vindication of the quark-gluon theories and the widespread belief in quarks.

Section 19.3 - Assumptions Made in Boson Theory

At best, the entire idea of bosons is not only completely unfamiliar to the non-physicist but is apt to try the imagination of the ordinary citizen and tax his credulity to the breaking point. When we assign them mass and the mysterious ability to change the mass or electric charge of mesons or hadrons, then it seems like summoning up magic. Where do bosons come from and why? The strange behavior of helium heavy isotopes as bosons when chilled to a liquid state, which we mentioned previously in earlier chapters, is easier to understand than the bosons which are considered to participate in Weak Neutral Currents and quantum electrodynamics. The entire concept of the "W" and "Z" bosons springs from three very basic ideas which are extraordinary in themselves. Firstly, as Albert Einstein courageously predicted, raw energy of any nature can become what we know as matter having mass. This concept is really only the reverse of an easier process which mankind only recently discovered how to accomplish in a crude and inefficient way – namely to create raw radiant heat energy from certain radioactive isotopes of atomic elements. As is characteristic for our misguided species, they first applied the discovery to make a super-weapon for killing people.

The second fundamental assumption was that a particle of matter can be created for an unimaginably short time interval even when no extra energy is artificially supplied from the outside. These are called "virtual" particles and the process is allowed by the "Uncertainty Principle" of quantum mechanics. This stated principle, as we emphasized previously in this book, should not be thought of as some unavoidable tiny error made in the scientist's observational measurements. It cannot be overcome by practical improvements in the laboratory. Rather, it is a function of Planck's Constant and a very strange limitation of Nature to an extremely small interval of time below which anything unpredictable can happen. It is a period of time where it is impossible to calculate the appearance or existence of a particle of matter as "real" or substantial in our familiar world. The Uncertainty Principle of quantum mechanics demands that the mass energy (Δ E) of any predictable or real particle times its incremental time of duration (Δ t) must be equal to or greater than Planck's Constant (h). The mathematical equation to express this condition is shown below:

$$\Delta E \text{ times } \Delta t \geq h$$

Given the assumption that everything happens at or near to the speed of light (c) and that Δ E is equal to m c^2 by Einstein's famous formula and also given that a distance or range of action (R) must equal the velocity (c) times some time interval (Δt), it becomes a simple algebraic operation to demonstrate that the range of action of a given boson reduces to the following values:

$$R = c \text{ times } \Delta t = c\,h / mc^2 = h / mc$$

The more energy which is required to create a transitory boson means that the time interval during which it is able to exist becomes briefer. This means that a virtual particle having a heavier mass may wink in and out of our world existence in an unbelievably short time whereas a lighter mass virtual particle or boson may hang around for a longer time, as is demonstrated by the photon. Photons are the ancient and most multitudinous examples of such a creation. Time in all these examples given is shorter than anything which we biological organisms can even picture. Paradoxically, it is also a function of the previous equation that the shorter the distance that a boson force must act the greater the mass must become by an inverse ratio. For the electro-weak interaction to happen between quarks inside of a neutron, for instance, the force to be exerted by the virtual particle shrinks down to about 10^{-17} meters. Consequently the mass of the "W" boson must become greater than that of the familiar pion and is estimated to be roughly 100 times that of the normal proton. Actually, the early predictions for boson masses needed to be adjusted slightly upwards in order to allow for some interaction between bosons themselves. One can perceive the energy demands which the physicists and engineers must confront in order to design ever more powerful accelerator machines.

Every charged particle, whether electron or not, can be pictured as continuously emitting and then reabsorbing some type of virtual particle; hence the attractive strong force between closely packed protons in the nucleus of an atom is now pictured as the exchange of a virtual pion. This is the third important assumption about bosons. They can exert a force upon other existing particles of matter. We return to the older crude analogy of two basketball players passing the ball back and forth between them. Depending upon the nature of the electric charges, the force exerted by a boson mediator may be attractive or repulsive. It was an easy step for physicists to imagine a charged particle generating a cloud of virtual particles in space around itself, thus allowing forces of different magnitudes. The answer to the question of "where does the boson's energy come from?" seems to boil down to two possibilities. The energy is either contributed by some applied external source or else it is stolen from existing particles within the atom having mass. A third and new possibility is the conjectured Higgs boson.

Section 19.4 The Neutral Z Zero Boson

Its symbol is Z^0 with the superscript of zero indicating the lack of any electrical charge. Strictly speaking, the use of the word particle for the Z^0 is inappropriate. It is categorized as an Intermediate Vector Boson and has a spin of one, which means 1h/2π.

The Z^0 is one of the heaviest elementary particles confidently known at the present time, having a rest mass of 91.16 GeV /c^2. That approaches the weight of an atom of silver. Its first hypothetical suggestion must probably be credited to Sheldon Glashow at Harvard University. He had intuitively adopted the belief that the strong, weak, and electromagnetic forces in the atom were all originally derived from a common ancestor at the very beginning of the creation of our universe. Proceeding upon that assumption, he predicted a boson such as the Z^0 particle as an additional carrier for the Weak Force in 1960 long before its general acceptance. The Strong Force carrier, as previously postulated by Hideki Yukawa in Japan, remained the pion (π) with a mass varying from 135 to 139 MeV, depending upon its electrical charge or lack thereof. Since the reader was first introduced to the photon as the only known boson and was told that it had a zero mass, the discrepancy must have occurred to him. This simple inconsistency was the origin of a great deal of intellectual ferment among the theorists who originated the Electro-Weak ideas. Their solution was a super-massive particle called the Higgs boson, named after its inventor. It is, as yet, still undiscovered and unconfirmed, but it is supposed to be the mechanism by which W^+, W^-, and Z^0 all obtained their masses. The Gauge Invariance and Yang-Mills theories made it all possible.

As part of the Weak Neutral Currents theories the Z zero is considered to be the carrier of the Weak Force between leptons, quarks, or even other particles in the atom whenever there is no change in electrical charges involved. It thus also becomes the source of natural radioactive decay of the heavy atoms. In such rare cases when a Z^0 boson is created alone and unabsorbed its lifetime is extraordinarily brief (something like 3×10^{-25}), which is hard for us humans to imagine. Being unusually massive, it has the latent ability to decay simultaneously into a number of paired fragments or possible combinations. These may include a very high velocity neutrino and also an anti-neutrino, or an electron and a positron, or even a normal muon and an anti-muon as well. In many accelerator particle collisions it may manifest itself as an Up quark and its anti-quark of opposite electrical charge or a Down quark and its anti-partner. All these options obviously make for the possibility of considerable confusion when interpreting the data from detectors arrayed around the collision point.

As the simplest possible means of identifying the production of a Z zero boson in the laboratory, its decay into a positively charged muon and a negatively charged muon attracted attention. Its incredibly short lifetime and the fact that its neutral charge leaves no cloud chamber or scintillation detector track makes its identification extremely difficult. Its actual existence can only be surmised from the instantaneous creation of its decay products. The laboratory detectors must be extremely sensitive and placed in a dense three–dimensional array around the collision point. Moreover, reams of computer print-outs holding the data for hundreds of experimental runs must be carefully scanned and analyzed by experienced graduate school technicians. It is hardly surprising that all this complexity places the validity of some results at risk. The final accepted proof of a laboratory demonstration for its existence must be subjected to the group judgment of scientific peers.

Section 19.5 Strategic Planning for the Verification of the W and Z Bosons

Now that the leading brains in the physics community had predicated the existence of the W and Z bosons as active and important players in the ongoing drama of the conversion of energy into matter, tradition confronted them with the obligation to prove in the laboratory that these particles were real and not merely phantasmagoria. This was no simple task; but the incentive for the application of money, ingenuity, and effort was there. Proving out the Electro-Weak and Neutral Currents theories suddenly became the road to a possible Nobel prize in the 1980's. It also instigated an appraisal of the available particle acceleration machines and the kinetic energy requirements. Although national chauvinism remained a factor, the consideration of which of the existing international machines was the most suitable for this specific task became dominant.

It was clear from the very beginning that the high energy collision of electrons with positrons was theoretically the simplest and most promising avenue for obtaining Z^0 bosons. This made a good case for the Lawrence Berkeley Laboratory in California and also the LEP synchrotron at the CERN facility in Switzerland. The planned

P E T RA machine at Hamburg, Germany, might have been a candidate also, but lack of money and construction delays prohibited its best operation. Moreover, its management was really more interested in smashing electrons into protons in order to study the behavior of quarks and possibly find the Top quark. The necessary high kinetic energies inhibited the achievement of their goal.

At casual first thought one might have concluded the proton versus anti-proton or proton versus proton beams then possible at both the enhanced Tevatron accelerator at the Fermi National Laboratory in Batavia, Illinois, or the Super Proton Accelerator at CERN in Europe might have been good possibilities. This would seem especially feasible for the W^+ or W^- bosons. There were at least three major difficulties to these options. Firstly, protons do not stay conveniently in bunched pencil beams very long but tend to scatter apart and wiggle around and resist concentration. This means more controlling magnets around the periphery of the ring in order to keep the beam compressed. Secondly, colliding two protons together at high velocities is the equivalent of firing two opposing shotguns at each other in order to discover what happens when the lead shot (six quarks) hit each other. It may possibly produce a W or Z boson once in a while, but this event would be masked by a shower of entirely new fragments which would require an extremely sophisticated detection system. Thirdly, the two possible W bosons both obviously carry an electrical charge, whether negative or positive, and the law of conservation of electrical charge prevents the production of only one W boson alone by itself. They usually are produced in pairs with opposite charges. This effectively doubles the required beam energy in the accelerator machine.

In any event, the managers of the largest experimental laboratories for particle physics in the world began scheming in 1976 for money and the design modifications which might make the W and Z bosons idea feasible. At this point an energetic and volatile Italian with a brilliant mind and a prodigious memory for all aspects of nuclear research took up the challenge. This was Carlo Rubia, who was born on March 31, 1934, in the town of Gorizia in northern Italy. His father was a telephone company worker and his mother was a school teacher. By the time the child was only eleven years old he was scavenging telephone and radio parts from abandoned World War II equipment. In 1945 his father's job required the family to move to Pisa, which gave them the opportunity to enroll their young genius in the prestigious local Normal School "Superiore" affiliated with the University of Pisa. Quite naturally, Carlo Rubia continued his education there and eventually earned a PhD degree in 1958. He took graduate courses at the University of Columbia in New York City and then returned to Rome in order to continue post-doctoral studies there. He soon quickly accepted an appointment to the European CERN facility at Geneva. His combination of a thorough theoretical knowledge of physics with a great talent for practical electronics made his move to a research laboratory almost imperative. The discovery of the W and Z bosons, then merely an nascent idea, became his obsession. In 1969 Rubia moved to the United States in order to join the team at the Fermi National Laboratory, but subsequent international competition and jealousies then prevailing impelled him to return to Switzerland when his American entrance visa expired in 1974. At this point in his career he joined with David Cline, Peter McIntyre, and Simon Van De Meer as a group of compatible thinkers who devised an entirely novel approach to generating W bosons by 1976. Peter McIntyre from Harvard had proposed back in 1972 that, if it were possible to manufacture enough anti-protons and then circulate them in a counter–rotating direction so that they could collide with a beam of ordinary protons, then the combination of all the quarks having opposite electrical charges might result in single W bosons of either plus, negative, or zero charges. Not only that, the total momentum energy of the collision would be almost ten times greater than what had been managed previously. There were significant possible hitches to this idea, however. One was the practical difficulty of bunching the moving protons or anti-protons into narrowly focused pencil beams. Another was the task of producing the anti-protons in enough quantity to make a sustainable beam of adequate density. The managers of the CERN facility were nevertheless impressed and boldly voted to gamble the funds for a major alteration of the Geneva synchrotron. Work started in late 1981.

Although it had already been demonstrated that anti-protons could be reliably manufactured and chilled down to a relatively quiescent state where they could be stored in a cryogenic thermos bottle, the contribution of the Dutch born Van der Meer was to devise a method of mass production of anti-protons that was the key to success for the planned experiment. He had been with the CERN organization since 1970 and had engineered the world's first Intersecting Storage Rings. This made it possible to have two accelerating rings of protons circulating in opposite directions and then direct them to a collision of proton versus proton at some intersection crossing point. The energy yield for this collision was consequently doubled from 270 to 540 GeV - a huge gain. His second great achievement, however, was to introduce a method of "stochastic cooling," which he had conceived and proposed as far back as 1968. This was a fancy name for a technique for concentrating the protons and anti-protons in a narrow pencil beam inside a helium cooled pipe and controlling their tendency to have a random sideways movement towards the walls of the tube due to heat energy. He accomplished this by very fine tuned or "stochastically timed" bursts of electrical power applied to toroid magnets spaced at regular intervals around the circumference of the accelerator ring. A third demonstration of Van der Meer's engineering genius was a semi-circular ring of very powerful magnets around a tube in which he trapped and stored temporarily the newly–made and low 8 GeV energy anti-protons. By this means he created a controllable gate or shutter by which he could introduce the anti-protons into the giant accelerator ring in short bursts in a direction of motion contrary to that of the normal protons.

The Directorship of this amazing team of technical skills was assumed by Carlos Rubia, whose ebullient enthusiasm for the task was matched by a mercurial temper and impatient irascibility with subordinates whom he thought slow or uncooperative. Not a few graduate students who worked for him briefly for laboratory experience found him too demanding and arrogant and quit the CERN facility assignment. Notwithstanding all this, his brilliance in the fields of particle physics and electronics was universally conceded and his personal energy was proverbial. In addition to supervising the complete alteration of the existing proton accelerator and its operations he was, at one time, regularly flying back and forth across the Atlantic Ocean in order to deliver teaching lectures at Harvard University.

In the meantime the people at the Stanford Linear Accelerator Center in southern California were making plans to upgrade their own machine and, in particular, to design a huge and very expensive detector called the Mark II for the specific purpose of finding the Z boson.

Section 19.6 - Discovery of the W Boson

Although it was scheduled to take only two years to complete, the task of converting the CERN proton accelerator to a proton versus anti-proton collider was technically daunting and truly monumental. It required a large assembled team of experts and encountered during the work, quite understandably, several crises and electro-mechanical difficulties before ever reaching a satisfactory performance under test. It was Rubia's job, eagerly assumed, to orchestrate the entire effort.

A crucial key to the success of the entire mission was the particle detector. At the CERN major primary accelerator ring, which was housed in an underground circular tunnel having a circumference of 27 kilometers or 16.75 miles, there were three large underground chambers spaced around the ring which were designated for detectors of different design. Two of these, which were designated UA-1 and UA-2, were assigned exclusively for the new proton versus anti-proton experiment. The UA-2 was a smaller and simpler version given to Pierre Darriolot of France for his direction. The UA-1, however, was the exclusive brainchild of Carlos Rubia and became an essential part of the new Super Proton Synchrotron. The entire detector assembly stood ten meters high and weighed at least 2000 tons. It had to be erected and the pieces assembled inside a cavernous subterranean room having a ceiling height of sixty feet. The entire unit was huge, dwarfing the human figures of the many technicians swarming around, inside or on top of it. The actual box enclosure which housed the proton beam collision point was mounted on wheeled tracks in order to permit it to be slid sideways away from the beam tunnel for maintenance or alterations.

The purpose of all this expensive construction, of course, was to register and identify any W or Z bosons that might possibly appear. Actually accomplishing this task involved some very complicated physics. The collision of high velocity protons with a beam of opposing anti-proton particles was expected to produce a veritable explosion of hadrons such as pions and kaons plus a shower of leptons going off in all directions. It was known that, in addition to all the above, the possible release of quarks and the formation of new combinations of these basic particles could conceivably create a W^- boson. Its detection might be accomplished by finding its immediate decay into a pair of oppositely charged leptons, whether an electron and positron or, alternatively, a muon and an anti-muon with accompanying neutrinos. Given that the mass of the W^- boson is estimated to be in the range of 85 GeV, this would mean that even muons would come off in this reaction at very high velocities and it would be necessary to slow them down somehow in order to allow their detection and be able to calculate their energies with any accuracy. The missing energy in the calculation was supposed to be accounted for by the emission of various types of neutrinos, in both normal and anti-neutrino stages. Rough preliminary calculations by the scientists resulted in the pessimistic estimate that it might take an estimated ten million proton versus anti-proton collisions to result in the creation of one observable decay of a W minus boson. In order to make sense out of these kind of statistical probabilities for tracking, it would require a large drift chamber having a three dimensional array of detectors and measuring 17 1/2 feet in length and 7 feet in diameter. This chamber would surround the intended proton collision impact site for the purpose of photographing the various particle tracks and measuring their track lengths. The entire chamber was itself surrounded by a powerful dipole magnet exerting a strong transverse magnetic field. Electromagnetic calorimeters which consisted of semi-cylindrical half shells of a scintillating material housed in lead "gondolas," were located one on each side of the beam and wired to electric light photo-multipliers. The calorimeters had to be of very high sensitivity and accuracy to measure the energy of the decay product leptons, whether electron or muon. The visible energy imbalance in the equations would presumably represent the energy lost in unseen neutrinos.

By July 9, 1981, the first operational test of the proton versus counter circulating anti-protons was tried. The entire staff had spent months of grueling hours of overtime work to the exclusion of almost any private life or recreation and everyone, including Carlo Rubia himself, was in a state of intense nervous apprehension. Although they had initially been confident that their scheme would work, many less optimistic skeptics with reasonably good credentials in physics had predicted the two beams would scatter and deflect each other into the walls of the enclosing stainless steel tube and disappear into the metal. The 300 billion or so of anti-protons which had just taken only 21 microseconds to make a complete circuit of the 27 kilometer synchrotron ring circumference would be totally lost[2]. To the intense relief of all the observers in the control room this calamity did not happen. Rubia rushed off to a physics conference in Lisbon, Portugal, in order to announce the good news.

Things did go wrong elsewhere, however, and the UA-1 detector was a constant worry. The very large and heavy "C" shaped hadronic calorimeters, which were made of scintillating material backed by lead and encased in an iron shell, were manufactured about two inches beyond the design specifications and did not fit well. Moreover, during the construction, it was discovered that the concrete ceiling of the underground chamber had been made two and one half feet too low, thus preventing the long arm of the maintenance crane from reaching its full extended height. This made it impossible to remove the scintillation calorimeters for repairs or alteration or adjustments to the housing so that they could fit around the beam tube properly. The solution was to excavate a pit in the existing floor in order to lower the crane machinery. There were also delays and arguments about the installation of the very extensive electrical wiring for the purpose of recording and measuring particle events in the computer data storage. Frequently nothing seemed to work properly during the first tests.

Eventually all equipment was in place and the initial test for the entire system was scheduled for April 26, 1982. A routine first preliminary task before starting up was to clean out the interior of the vacuum pipe, which ran longitudinally through the detector assembly. The customary method was to heat up the pipe electrically in order to drive off all stray oxygen atoms while, nevertheless, keeping the delicate detector apparatus cool by blowing compressed air through its own tunnel. The unintended result of this attempt was a dense cloud of red dust emanating from inside the detector. This was a serious bad signal and the cleaning operation was terminated immediately. As one can imagine, Rubia's volatile temper went off the scale. The cause was a large amount of dust and dirt ("rat turds" in the inelegant, but vivid, language of Sheldon Glashow) from a contaminated air supply, or perhaps from the inside of the ductwork. This dust was blown all over the hot vacuum tubing and into the detector's innards, thus damaging all the electronics. This disaster immediately put the UA-1 team at a great disadvantage in their competition with the French team at UA-2, who themselves were eagerly preparing to run. Excuses presented by Rubia to the CERN Commission management for a delay in the scheduled start-up became voluble and emotional. Headlines ran in the New York Times newspaper about "Breakdowns at CERN" to the secret delight of jealous American physicists. The cleanup of the dust problem took valuable time which was profitably used to develop a

2. An excellent description of the entire discovery was written by Gary Taubes in his book *Nobel Dreams*, (New York, Random House, 1986).

better and more reliable control of the high voltage Direct Currents required for the operation of the detector heart or cylindrical drift chamber, known in the shop jargon as the "C. D." This chamber was filled with a network of countless fine wires all electrostatically charged. The passage of any charged particle through the narrow spaces between would instigate a spark of light which could an be picked up by a photo-sensitive detector and the signal relayed to a computer as a registered event in the data bank. Problems of reliability had actually arisen in the regulation of these dangerously high voltages with the result that the system had to be disassembled and higher quality parts installed.

In spite of all these difficulties and with the application of feverish work by Rubia's electronic technical people, they were ready for the next start-up test of the entire system by the rescheduled date of October of 1982. During the elapsed time the French physicists at the UA-2 had been busy studying jets produced from quarks dislodged from nuclei by the proton versus anti-proton collisions. Absolutely no evidence of W bosons had been detected during routine runs by either team until November when Rubia's team reported a display which indicated a lone candidate. The tracks were reconstructed on paper and a dotted line at the end indicated where a particle would have struck an electric calorimeter inside the detector. The computerized record of this event indicated a high momentum electron, which was an auspicious sign for a W minus decay process. The only serious doubt that the experimenters had was their calculated deductions for an invisible neutrino to account for the missing energy in a conservation balance.

A request was filed with CERN management Directors to allow a two week postponement of the normal run termination date of December 6, 1982, but it was refused on the grounds that the total electric power utility load in the region would be overtaxed. It had been standard procedure to shut down the synchrotron operations during the dark winter months of January and February. Not incidentally, the scientists themselves badly needed a vacation from work. This valuable time was not actually wasted, however, for much of the manpower needed for operating both the UA-1 and UA-2 detectors was thereby released for analyzing the data obtained from around one million collision events. Roughly 150,000 of them were finally selected as significant enough to reconstruct graphically on the facility's huge frame computers. The end result was six good candidates from either of the two possible decay modes shown below:

$$W^- \rightarrow e^- + \nu_e$$
$$W^+ \rightarrow e^+ + \nu_e$$

No candidate for the Z zero boson decay had yet been found.

A general workshop session was held at Rome University from January 12 to January 14, 1983. These were followed by more collaborative seminar meetings at the CERN auditorium in Geneva before a packed room of more than four hundred people. Rubia, of course, was absolutely confident that the W boson had been discovered and was anxious to announce it to the world; but the assembled physicists, guests and staff needed to be convinced. Ultimately, they all decided that the mass of the boson was in the range of 80 GeV / c^2 and that at least six of the identified events were legitimate. A long and detailed report was hurriedly prepared and delivered to the international publication of *Physics Letters* on 23, January, 1983. Beneath the cautious title of *Experimental Observation of Isolated Large Transverse Energy Electrons with Associated Missing Energy* on the first page was a list of no less than 136 names of collaborating physicists and engineers! Big science had certainly arrived.

Section 19.7 - The Search for the Z Zero Boson

Although neither the UA-1 or UA-2 teams at CERN could report any really reliable evidence for the Z^0 boson thus far, the theorists were still not so totally discouraged as to believe such a discovery was not possible. They knew perfectly well that the expected mass of the Z Zero boson was significantly higher than that of the W. Their calculations indicated that, although the probability of finding the Z decay might be only one tenth of that for the W, it was still not impossible. Some people early in history had even been hopeful that it might decay into the proverbial Top Quark, but that turned out to be wishful thinking. Its expected mass was just a bit too high for the energy capacity of the existing machines. Nevertheless, the decay of the Z Zero into a lepton plus a second anti-lepton seemed the most probable and reliable signal for its detection, since it was dramatically simple. The Rubia staff pinned their hopes on a finer focused proton beam of higher density and improvements to the UA-l detector in order to achieve this new goal.

The next start-up of the Super Proton Synchrotron was scheduled for April 12, 1983. On June 1, only two months later, the Rubia group announced their identification of five Z Zero decay events, four of which were the predicted electron and positron pair. Only one of the events was described as being the muon plus anti-muon decay mode. On July 15 the French UA-2 group also reported verifying a Z Zero decay. The separately calculated team estimates of the mass of the Z boson varied slightly, but they agreed on a predicted value of 94 GeV. The announcements from both teams conceded that the discovery of the Z Zero was a success of a very moderate kind; and the general consensus of the scientists was that a very high energy electron versus positron collider would be more promising for verification of the Z boson.

Section 19.8 - The Reconfirmation of the Z Zero Boson

The middle of the 1980's decade launched a race among the world's largest particle accelerators to achieve higher collision energies and to push the frontier of research on quarks and bosons even further. It was further decided that high velocity collision of leptons, especially that of electrons with positrons, was the cleanest way to go in order to find the Z Zero boson and also avoid any messy confusion with other hadrons. The greatest obstacle to pursuing this method, however, was the colossal waste of electric energy input with the then popular circular ring collector and collider synchrotron machines being employed. The acceleration of electrons around a circular track required numerous powerful magnets spaced at regular intervals around the circumference of the beam tube just to restrain electrons from flying off into the metal sidewalls. Worse yet, the electrons placed in rapid motion generate a separate electromagnetic field which represents yet another energy loss to the environment. A partial compensation for these difficulties might be to add more super-conducting radio frequency cavities in the ring which could add an accelerating force; but each one of these pieces of equipment is terribly expensive and adds again to the power consumption costs. All these modifications, of course, translate into millions of dollars of capital costs and operating costs. Thus pushing the limits of high energy nuclear particle experimentation is becoming more and more expensive and making ever greater demands on public tax money. Nevertheless, the scientific goals, when explained, seem persuasive enough thus far to encourage the politicians to furnish additional funds for selected projects and to upgrade some of the largest accelerators. A competitive race ensued.

The verification of the existence of the Z Zero boson became the objective of both Stanford University in California and the French, English, and Italian consortium at Geneva, Switzerland. Both opted for the electron versus positron collision approach. The people at the older and 1.9 mile long Stanford Linear Accelerator (SLAC) decided back in 1982 to make some daring alterations to the twenty year old facility in order to convert it into what they named the Stanford Linear Collider (SLC) instead. The scientific and engineering difficulties were prodigious, although the basic idea seemed deceptively enough. The scheme was to accelerate two beams by means of bursts of radio frequency wave power from almost one hundred brand new Klysteron radio oscillator tubes with both beams separated inside the same confining and air evacuated tube, one beam consisting of negatively charged normal electrons and the other being positively charged positrons. Not only that, but the beams needed to be boosted to the required kinetic energy by repetitive cycles inside the tube. An entirely new cooled and high vacuum ring tube was thus required at the extreme end of the older linear accelerator tube. Seen from above in plan view the new elliptical ring suggested to one's imagination a glass vase for flowers or else a chemist's Erlenmeyer flask. In the initial design it was expected that each initial test burst of lepton particles would reach a kinetic energy level at the end of the accelerator machine of nearly 50 GeV. Located at the upper apex of the pear shaped ring and at the end extremity of the straight tube there was a large magnet. This device was supposed to separate the two different beams of leptons of contrary electrical charge into two divergent paths – the electrons going to the right hand side or clockwise in plan and the positrons going into the left hand entrance or counter clockwise in direction of the bottom ring. Each beam would then make a separate 180 degree circle in order to reach another very strong focusing magnet to reduce the beams down to a diameter comparable to a spider's web measured in microns. Coming from opposite directions, the two different bunches of electrons and positrons were guided into a collision point apex inside the large and new Mark II detector. The detector itself was a complicated four thousand ton iron apparatus having several concentric rings surrounding the collision point which contained successively a drift chamber, Cerenkov flash detectors, a liquid Argon calorimeter, and finally an iron shell calorimeter for tracking the muon fragments.

As one might expect, there were many technical difficulties to overcome in order to make such a complicated apparatus work satisfactorily. One of the first frustrations were the 6 foot high and 67 megawatt power rated Klysteron oscillator tubes. From the very beginning of the design process these were a calculated risk, since nothing larger than 35 megawatt radio frequency generators had been previously available on the manufacturing market. Moreover, these Klysteron vacuum tubes were notoriously difficult to manufacture and required micro-millimeter machine tolerances in dimensions.

During the starting stages of first purchasing and testing all the new equipment shipped to the accelerator site only about thirty percent of the tubes were deemed satisfactory. The remainder went back to the factory for adjustments and/or retooling.

Yet another different set of problems was how to manage to make the desired anti-electrons at the site. It was urgent to collect and compress all the electrons which were boiled off a high voltage cathode emitter into a narrow confined beam having a density of charged particles comparable to ordinary air at room temperatures at sea level elevation. The solution was the insertion of a pair of secondary rings, called damping rings, placed near the middle of the linear accelerator tunnel. Exactly correct timing in microseconds for the application of electric energy to all the numerous magnets in the entire system, which were supposed to maintain a uniform flow of electrons and positrons consistently in bunched beams, required fine tuning controlled by means of computers.

All this preparatory work took until the spring of 1987 in order to have the equipment in place. It took another two years and around a thousand skilled people working almost around the clock to achieve a successful computerized control system and the satisfactory operation of the entire machine. Finally, at sunrise on April 11,1989, the unequivocal identification of a Z zero boson by means of its decay pattern into a quark and anti-quark was registered in the Mark II detector to the satisfaction of the Director, Barrett Millikan. The entire story was a remarkable achievement accomplished for an astoundingly low cost for those times. The public announcement appeared in the New York Times newspaper on April 3.

Section 19.9 - The European Z Zero Factory

In the meantime the CERN facility located near the border of Switzerland with France had been frantically busy with the conversion of their own gigantic machine with exactly the same objective as the California team. They had already named the new and modified accelerator the "Large Electron Positron" collider or L. E. P. for short. The initial start-up date was scheduled for Bastille Day and the bi-centennial celebration of the French revolution against the monarchy. Thus the first public demonstration of the new installation took place with great pomp and ceremony on July 14, 1989, with Carlo Rubia as the new Director General and his predecessor, the German Herwig F. Schopper, attending.

The actual discovery of their first Z Zero boson had to wait until thirty days later after serious engineering operations. The counter–circulating beams of electrons and positrons, each representing 45.5 GeV of kinetic energy, were brought into collision. Eight minutes later at just five minutes before midnight on August 13, 1989, the first Z particle was detected and identified. In the ensuing four months of continuous experiments some 100,000 similar events were registered. The Z Zero mass was calculated to be 91.18 GeV. The CERN Large Electron Positron collider had clearly demonstrated that it had the capability to churn out hundreds of times more Z Zeros in a given run of the machine than their brave competitor. Much more important, however, was the independent confirmation in two separate experiments that a theoretically predicted particle actually existed.

Section 19.10 - The Higgs Boson

The verification of both the W and Z bosons was a real triumph for the theorists, as well as the experimenters. They quickly became an integral part of the Standard Model. Nevertheless, the whole idea of invisible particles helping to generate mass from pure energy and organizing the various subnuclear components in order to form some cohesive structure which exhibits the properties of stable matter is somewhat spooky; and the scientists realize it. All the intricate explanations of Non-Abelian gauge invariance and symmetry breaking do little to convince the ordinary lay person why such a mechanism would be necessary. More disconcerting to the physicists themselves, however, are some anomalies in the basic assumptions in the carefully constructed standard model. One of the most obvious contradictions, as we previously mentioned, is how the W boson in its two different varieties can exert either a positive or negative electromotive force and, even more perplexing, why both the W and Z bosons have such heavy masses at all. Photons and gluons are assumed to have no mass at all. It would be most convenient for the theoreticians if the three varieties of neutrinos had no mass at all for full compatibility with the Standard Model. That question is by no means resolved Another annoying discrepancy with the Standard Model is that at much higher energies, which is synonymous with very high temperatures, all the physicist's equations for symmetry and mass in particles work out nicely; but at the low energy extremes, such as the infra-red spectrum, they begin to scramble and yield infinities. In other words, they do not become renormalizable. Such doubt and confusion may possibly have encouraged books and newspaper articles with titles such as *The End of Science* or those people who may have some religious penchant for mysticism in their personal philosophy. One of the skeptics is Andrew Pickering, who is a professor of Sociology, among other qualifications, at the University of Illinois and the author of a book entitled *Constructing Quarks*. He suggests that the scientists might be only finding what they decided to search for and which is compatible with their mathematics, which is hardly a friendly comment.

To those earnest scientists who are leading their profession and who are constrained by their entire life work to be naturally defensive on the subject, such critical outside comments add to their discomfiture and resentment. Fortunately a mathematician in England came to their rescue. His name was Peter Ware Higgs and he was born at Newcastle in 1929. He was educated at King's College in London and afterwards received a lectureship appointment at the University of Edinburgh in 1960. He was subsequently appointed to a full Professorship in theoretical physics in 1980. He invented something now called the Higgs Field, which is a still unknown and non-zero field which can exist in the vacuum of space. He also postulated the existence of a very massive vector boson having a zero spin which would be able to supply mass to both the W or Z bosons by spontaneously breaking down existing field symmetry. Its existence is an extension of the previous logic of the Standard Model and remains, at the date this book was written, entirely conjectural.

The Higgs "mechanism," which is a strange word to use in such an abstract mathematical conjecture, is the assumption that a massless photon and two companion particles having the property of mass can change into one very massive vector boson with only one remaining particle of lesser mass such as the W or Z bosons. All this hypothetical process became an integral part of the electro-weak theory and is beyond the scope of this book. Professor Steven Weinberg, first at Harvard University and later at the University of Texas in Austin, became a champion of this entire idea in collaboration with Leonard Lederman. The reader may remember that Lederman was Director of the Fermi National Laboratory near Chicago from 1979 to 1986. Thus the proof of the existence of this postulated Higgs scalar boson in a laboratory quickly became of urgent importance in the minds of the physicists for demonstrating the authenticity of the entire Standard Model. It would also polish off our understanding of the strong and weak nuclear forces and atomic radio-activity.

The generally conceded symbol for the Higgs boson appears to be the capital letter " H." Its size might be unbelievably small, such as 10^{-29}, in spite of the great mass. The mass, however, was and is still a matter of some debate. The least that it could be is 365 GeV, but some pessimists say that it could be as much as 1000 GeV (or 1 TeV). That is admittedly a pretty wide spread of numbers, as well as a confession of ignorance perhaps: but as we enter the twenty–first century many physicists have persuaded themselves that a reasonable guess would be around 729 GeV / c^2 . This last number would make its actual verification plausible in collider machines envisioned for the near future, assuming governments are willing to spend the money. Lederman's enthusiasm for the project made him an active proponent for a new billion dollar synchrotron, which is scarcely a surprise coming from a famous experimenter. He wrote a book partly about his interesting career and partly about the super-collider for discovering the Higgs boson, which he christened the "God Particle." The casual jocularity of his title may seem a trifle irreligious to some people; but his zealous belief in the existence of the Higgs boson is genuine nonetheless and his name for it may have been the closest that he ever came to an admission of a deity.

Chapter 20
Thoughts on Quantum Weirdness

There are more things in heaven and earth, Horatio , than are dreamt of in your philosophy! – from " Hamlet " by William Shakespeare .

Section 20.1 - Introduction

Quantum Weirdness, as it is commonly termed, is part and parcel of a larger discussion of the real differences between quantum mechanics and the old fashioned concepts of classical science under which many of us still labor. It must be acknowledged to the thoughtful reader that there has been a philosophical argument raging off and on between scientists about the implications of quantum mechanics for what we complacently call "reality" for seventy five years now. Earlier in this book the author mentioned somewhat casually the Einstein-Podolsky-Rosen debate or E. P. R. problem, as it is frequently called. Other names for the same discussion are the collapse of wave function or "measurement" problem, although in later years the focus of attention shifted more to "non-locality" or "action at a distance" nomenclature. All relate to the same fundamental perplexity about the strangeness of the mathematics for quantum mechanics. It is a curious fact that Erwin Schrodinger, who was one of the first original inventors of the wave function mathematics of modern quantum mechanics, admitted to a profound personal dissatisfaction with the ambiguities it created. Nor did he keep it a secret; but wrote several dissertation papers and private letters on the subject, culminating in 1952 with a now discredited attempt to explain it by means of "wave packets" instead of particles. Werner Heisenberg even wrote a small book on the implications of the new theory on philosophy.

Much of this intellectual furor erupted in the thirteen years between Neils Bohr's original ideas on the simple hydrogen atom and electron jumps between energy levels in 1913 and Schrodinger's paper in 1952. It is not so surprising that these men, who had received excellent educations in the older classical European cultural pattern, were disturbed by the new physics. After all, quantum mechanics and Einstein's Relativity Theory had suddenly consigned a great many pet ideas of the late 1800's to the scrap basket. To many of the current crop of physics students who have matured in the materialistic ethos of urban dwellers and are the beneficiaries of all the technological end products of the industrial revolution, including the computer, the problem does not seem so important. Today the potential metaphysics of it all is frequently dismissed as superficial "gedanken" (German for "thought, notions, ideas, or supposition") speculations more suitable for science fiction novels. In fact, the outspoken adversarial spite and rancor for theorists preoccupied with any such philosophical conjectures by a few self-described "realists" or practical laboratory physicists can be startling.

The emotion sometimes generated on the subject suggests that many people in the physics field would prefer didactic facts and the certainty of experimental and observational data to hypothetical speculations. Indeed, one should not be surprised at such a reaction since this is what we all have been taught in school as the core of the "scientific method." Moreover, if one is momentarily able to divorce oneself mentally from all the cultural conditioning of our modern western industrialized civilization with its popular worship of technology, then one must objectively admit that all this talk of particles, bosons, quarks, and gluons is predominately materialistic in perception, if not intent. In fairness to the reader, this author must confess that he believes the discussion of the E. P. R. quandary to be entirely justified. More than that, the questions raised represent the early threshold of an ultimate discovery of a new and more complete comprehension of our universe. In a short article in *Physics Today* (July, 2000) by Professor N. David Mermin of Cornell University in Ithaca about using quantum weirdness as a method of cryptography for encoding messages, he ended his story with the following significant statement or thought: "...new Gedanken applications of quantum mechanics like the one I described offer fertile ground for reconsidering the knowledge-versus-reality muddle we have been thrashing about in for the past sevent–five years. It would be interesting, but maybe a bit embarrassing, if the new breed of computer scientists were the ones to straighten us out."

One heritage of the Neils Bohr days of discussion was the "wave collapse" notion brought about by laboratory measurements. This brought about a contentious debate on whether the human being operating the measuring instruments could not be a part of the cause as well. In other words, was it at all possible that the mentality and consciousness of the human being making the experiment could influence the results of the observation? The Germanic tendency toward involved philosophy is evident. The fact that such a question was seriously asked years back is quite remarkable, since it is obviously an anthropomorphic attitude. It is a question that most physicists loathe and resent.

In any event the entire issue has come out of the closet in the last seven years or so and escaped the long dominance of Neils Bohr's "Complementary Theory" or Copenhagen Interpretation and his quasi-mystical ambiguities. It has now been the subject of numerous book, newspaper and magazine articles and television shows. Not only that, quantum weirdness and particle entanglement problems have become a subject of genuine fascination for thinking people in all walks of life, to the acute distress of the professional physicists.

Section 20.2 - The E. P. R. Conundrum

Albert Einstein never during his lifetime became reconciled to the mathematics of quantum mechanics; and it's certainly not because he was stupid. Nor did he ever claim that the results of applying their equations to laboratory demonstrations were wrong. It was the inability to predict exactly what would happen in advance without a measurement process and later an apparent non-local connection between two particles emitted from the same atom or source which really bothered him. Worse yet, when two or more particles originate from a common origin in what is called an "entangled" state, then the wave function equation becomes superposed with different frequencies and the quantum mechanics mathematics yields only probabilities! Hence Einstein's tart comment that "God does not play dice." For people not well trained in physics the entire argument is apt to seem extremely subtle and arcane.

Having come from the old classical school of education, Einstein felt strongly that any ultimate laws must be causal and absolutely deterministic. Mere probabilities were not good enough for him and he had many discussions with Neils Bohr about this point. In preparation for a meeting of physicists he rallied two friends named Boris Podolsky and Nathan Rosen to write a joint paper for publication in the *Physical Review* in 1935 with the title *Can Quantum Mechanical Description Of Physical Reality Be Considered Complete?* Their conclusion was that the wave function does *not* do so. When the challenge was first written it was most probable that they were actually thinking about particles such as electrons or protons. The intended source could easily have been from the atom of a naturally radioactive element or else a beam of low energy, say 14MeV protons from an accelerator machine directed into a target of liquid hydrogen. The actual application of the mathematics was incidental. The condition for a satisfactory or "complete" theory of reality was clearly predicated:

> "If, without in any way disturbing a system, we can predict with certainty (that is, with a probability equal to unity) the value of physical quantity, then there exists an element of physical reality corresponding to this physical quantity."

From Einstein's point of view, invoking any "spooky" action at a distance between particles was absurd and consequently the values of momentum and spatial location could not have been created by the act of measurement. In his mind, for a pair of particles which originated from a close state of total momentum (such as two fermion particles of spin one-half prepared in an unstable initial state of a total spin of zero) they would have a total momentum of p_1+p_2 and relative positions of q_1-q_2 and Q_1-Q_2 for each of the two particles and by measuring one of the particles one could compute the values for the other.

The problem is that the appropriate quantum mechanics wave function equation does not do any of these things—except by imaginative inference. Since the early days of Neils Bohr the majority of physicists have been taught to consider the two particles, when captured in some atomic state which is not observable, to have no assigned properties, including spin or specific magnetic moment. The wave equation is a superposition of a quantity of different wave forms, hence only the act of actual measurement or perception represents reality. Rather dramatically, the physicist talk at this point of "the collapse of the wave function!"

Quoting Albert Einstein once again:

> "This situation, evidently, strikes many people as being uncomfortable. But it is, until now, the only way to calculate the quantum states and their transition probabilities that agrees with experiment. I am entirely convinced that the truth is situated far from the present teachings."

To his credit, Neils Bohr acknowledged this difficulty from the start and over time melded the paradox into his own personal world view philosophy. As he said:

> "...We have been forced to reckon with a free choice on the part of Nature between various possibilities to which only probability interpretations can be applied."
>
> "It is indeed recognition of this situation which makes recourse to a statistical mode of description imperative as regards to the expectations of the occurrence of individual quantum effects in one and the same experimental arrangement."

There is one minor objection to the E. P. R. criticism which should be mentioned. The paper ignores the Heisenberg "Principle of Indeterminacy," which in time became a fundamental doctrine of Quantum Mechanics. It states that one can establish the momentum of a particle reasonably accurately, but never its location at the same time. Conversely, one may measure the location in some point of time, but measurement of its momentum then is impossible.

$$\Delta p \text{ times } \Delta q \quad h/2\pi$$

However, we are not really concerned with extremely exact measurements here and the point does not really negate the issue under debate. Professor Abner Shimony of the University of Boston, Massachusetts, states the ensuing situation in nuclear science for the next fifty years succinctly:

> "More than any other person, Neils Bohr formulated a defense of the intelligibility of quantum mechanics within a year after its discovery and laid to rest the philosophical scruples of most of a generation of physicists. His principle of complementarity provided a point of view within the wave-particle duality (of light) and the Heisenberg uncertainty relations could be understood; and his answer to the E. P. R. was generally accepted by the community of physicists as a vindication of the completeness of quantum mechanical descriptions.
>
> ..."The actualization of potentialities, also known as the reduction of the wave packet—-arises from the linearity of all quantum mechanics; and it is deeply embedded in the present structure of quantum mechanics. It has often been claimed that this problem is specious, arising from a narrow or inadequate representation of the measuring process. We shall argue, however, that the problem is a fundamental anomaly which cannot be lightly dismissed.
>
> "Quantum mechanics holds that a maximal specification of a system—its state—does not assign a definite truth or falsity to each of its eventualities. When any state of a system is given there exist eventualities which do not have a definite truth or falsity, not because of ignorance on the part of some or all human beings, but because they are objectively different."

The men who invented Quantum Mechanics and revolutionized classical physics were by no means indifferent to the ontological implications of their equations. It is significant that Erwin Schrodinger introduced with some sardonic humor his "Cat in a Box" riddle which became part of the popular folklore on quantum mechanics. He pictured a live cat imprisoned in a large box containing an air supply, a small piece of radium isotope and an alpha particle detector apparatus, which, when electrically triggered, will release a poisonous cyanide gas from a vial inside the box upon the instant that the half life emits radiation. The question then becomes: At any given instant can anyone say whether the cat is alive or dead? Or are we to regard the animal as in a suspended existence neither alive nor dead? Even Schrodinger did not believe that.

Absurdity was the real intent of Heisenberg's picture, namely to emphasize the strange limitation of Quantum Mechanics. There is little doubt the cat knows when it is still alive, so we could be led to the riddle of consciousness.

Section 20.3 - Entanglement, Non-Locality and Experiments

The experimental apparatus existed for verifying all this and gradually many scientists became interested in making the actual experiments that had been suggested. All the while, Bohr continued to slug it out with Einstein and later Clauser. It should be realized that the entire E. P. R. discussion is about particles or photons which originate together from an "entangled" and unknown condition from within the same atom or molecule. Their peculiar behavior, which is mysteriously ambivalent from the positivist viewpoint, is the reason for the frequent suggestion of a need for "hidden variables."

The philosopher-scientist Abner Shimony is quite explicit on the problem.

> "Implicit in all principles (of quantum mechanics) is the famous Superposition Principle, which asserts that any two states can be combined (actually in infinitely many ways) to form states which have characteristics intermediate between those of the two which are combined.
>
> "In quantum mechanics it is possible for the states of 1 + 2 to be entangled at an instant of time t . Neither 1 by itself nor 2 by itself is in a definite state, yet the two together are in a definite state. Such a conception is incomprehensible from the standpoint of classical physics.——The change of the framework of physics from classical to quantum mechanical is clearly a fundamental transformation of the conception of nature.

> "Neither 1 nor 2 is a definite state. Another way to put the matter is to say the Wave Function equation – for the state applies the Superposition Principle to the states 1 and 2, both of which have a commonsensical state – which is entangled. The connection between entanglement and objective indefiniteness is very close."

The experiment for protons in slow velocities aimed into hydrogen atoms in liquefied state and detecting the resultant secondary protons as they ricochet by impact at approximate 45 degree angles of deflection from their initial flight line, was performed in 1976 by Lamehi-Rachti and Mittig. The detection of the charged proton coming from the collision is registered by a Stern-Gerlach magnet apparatus which also causes a deflection up or down in response to the particle spin being either clockwise (+) or counter-clockwise (-). The counter, which also serves to identify the polarization, has a target of a block of carbon which emits a secondary particle. The experimenter soon learns that after a short time delay the next proton coming from the source and going in a different direction will have exactly the opposite spin from the first detected almost every time. There is no theory in classical probability calculations that account for this behavior. It reflects the correlation of the atomic spins in the original molecule. Polarization of the second becomes automatically opposite. This decidedly weird behavior of "entangled" particles is a direct mathematical consequence of the standard quantum mechanics probability wave function Psi (–) theory and its anti-symmetric equations. Erwin Schrodinger, the inventor of the wave theory for atomic particles, pointed out this anomaly in a lecture in 1935 and has expressed how profoundly the philosophical implications disturbed him. It is well that we all should remember that Isaac Newton in 1686 was mystified by the action of gravity on celestial bodies at a distance for exactly the same reasons. Moreover, the instantaneous change, regardless of the distances of separation, implies to some people a faster than light signal or some inherent connection between the particles.

It was not very long before those people interested in this phenomenon changed their experimental apparatus over from particles to photons. There were very sensible and practical reasons for doing so. The same significant changes in a property occurred with a photon except that, instead of spin, it was the light wave polarity changing from vertical to horizontal which indicated the entanglement behavior. Furthermore, there were a multitude of mechanical options available in the use of mirrors, polarization screens, detectors, beam splitters, and rapid time switches which offered the experimenter many more opportunities to test the effect with physical changes.

The entanglement feature of two different photons coming from a common origin was vastly enhanced by the technique of passing a monochromatic beam of light through a translucent natural crystalline form of calcium known as calcite. By far the most important asset of these new experimental set-ups, however, was the ability to manipulate the two different polarization screens and each detector assembly at will for each photon. Thus measurements could be made at different times and even at large and measurably different distances (and presumably the time of transit) for each of the two photons as well. Changes in polarization or shielding a detector could be accomplished electrically almost instantaneously.

The end results of the measurements simply confirmed what the early discoverers already suspected and emphasized unmistakably the eerie synchronization between the two independent photons and an apparent nonlocality aspect.

At this point in our story a most remarkable character of independent mind and mathematical acumen entered the arena of debate and changed its entire history, much like dropping a single crystal into a supersaturated and cooled solution. His name was John Stewart Bell and he was an eminent physicist and synchrotron engineering designer with impeccable qualifications. He had been mulling over the peculiarities of quantum mechanics and its deviations from random classical probabilities ever since his introduction to the subject. In 1964 he wrote a short paper, for which he was paid a slight sum, for publication in a relatively unknown obscure *Physics* journal in Europe under the title *On the Einstein-Podolsky-Rosen Paradox*. It was largely ignored until a physicist named John F. Clauser at the Berkeley Laboratories at the University of California picked it up in 1969.

Clauser, along with Michael Horne, Abner Shimony of Boston University, and Richard Holt of the University of Western Ontario, published these ideas in 1969 in a paper entitled *Proposed Experiment to Test Local Hidden-Variable Theories*. This paper was published in the *Physical Review Letters*, the publication journal of the American Physical Society.

John Bell was born on July 28, 1928, to a humble working family in Belfast, Ireland, and attended the Belfast Technical School there, graduating at age sixteen. Lacking money, he was restricted to attending Queen's University in the same city. He graduated from there in 1949 and had already become determined to make a career in physics. He continued his studies under Sir Rudulf Pierls, who had spotted a star student. In order to support himself, Bell obtained a job at the English government's Atomic Energy Research Establishment at Harwell and quickly became their expert on both quantum mechanics and particle accelerators. While still at Harwell, he married a young girl from Glasgow, Scotland, who shared his interest in physics and proficiency in mathematics. She became his lifelong companion and professional partner. The Harwell establishment gave him a year's sabbatical to pursue more advanced studies at the University of Birmingham in 1952. This resulted in, as Bell put it, "I there became a quantum field theorist." In keeping with his new occupation he published a paper in 1954 entitled *Stability of Perturbed Orbits in the Synchrotron.*

Eventually he became bored and dissatisfied with the bureaucratic routine at Harwell so, when he was offered a job at the European CERN facility, they both decided to forgo their three year tenured contracts with the Harwell establishment and moved to Geneva, Switzerland in 1960. They have remained there ever since in adjacent offices.

It was Bell's opinion, given in his own words, that

> "There must be a mechanism whereby the setting of one measuring device can influence the reading of another instrument, however remote. Moreover, the signal involved must propagate instantaneously, so that the theory could not be Lorentz invariant [this is another way of saying that it could not be consistent with the theory of Relativity]."

It was not until 1972 that John Clauser and Stuart Freedman published the results in the *Physical Review Letters* of their first deliberate experimental attempt to test Bell's thesis of a statistical inequality which would demonstrate the truth of non-locality. Their written report was titled *Experimental Test of Local Hidden Variables Theories* and published in *Physical Review Letters* in 1972. Two quotations from this report are typically germane:

> "A binary selection process occurs for each photon at each polarizer (transmission or no-transmission). The selection does **not** depend upon the orientation of the distant polarizer.
>
> "The values we obtained (for deviations) are in clear violation of inequality. Furthermore, we observe no evidence for a deviation from the predictions of quantum mechanics, calculated from measured polarizer efficiencies and solid angles. We consider these results to be strong evidence **against** local hidden variable theories."

Very early experiments which are frequently cited as helping to instigate all this newer activity over an old debate were started in 1981. They were conducted by a team consisting of Alain Aspect, Philippe Grangier, and G. Roger at the Institute of Optics at Orsay, France. They published their results in the *Physical Review of Letters.* The first was titled *Experimental Tests of Realistic Theories via Bell's Theorem.* A second report under the title *Experimental Test Bell's Inequalities Using Time Varying Analyzers* followed in 1982. The last was an apparatus adjustment to allow the switching of polarizer directions between the two different orientations with micro-second differences in time. The time difference between the two entangled photons did not affect their co-ordination in the least; and all measurements confirmed the predictions of quantum mechanics equations.

Their third Letter settled the matter unequivocally with the title of *Experimental Evidence for Photon Anti-Correlation Effect on Beam Splitter.*

It must be admitted that all of these experiments with photons and using light rays introduces a new complication. Although the same quantum rules for electrons apply also to photons, the latter are not entirely identical with electrons. Light photons, being bosons, have properties in some respects different from electrons, which are fermions. The ability of light rays to "go around corners," as exemplified by their behavior with passage through narrow slits or tiny holes in an opaque screen, clearly implies a wave action and reminds us once again of the duality problem mentioned in Chapter 5. As Jeremy Bernstein in his book *Quantum Profiles*[1] states explicitly and quite forthrightly as "an apparent absurdity: A particle (whether photon or electron) moving between two points travels all possible paths between them simultaneously." While this previous statement may not be entirely scientifically exact because of the word "simultaneously," this feature of light photons was seized upon by Richard P. Feynman when he wrote his challenging little book QED[1] or *The Strange Story of Light and Matter.* Feynman was unquestionably a genius and characteristically an original thinker; and he spent a lot of time pondering on the significance of quantum mechanics. He emphatically decided with Einstein that light behaves as particles and, starting with that assumption and the ability of photons to travel all possible paths, he derived his own explanation of quantum mechanics and the famous "Feynman diagrams" for the inter-action of photons and electrons in matter.

A particular feature of photons needs to be emphasized. In some manner for reasons still somewhat obscure to human beings photons from an emission source will travel all possible and unblocked ways through space to reach a passive receptor—whether a photoelectric cell, sensitized photographic film, or the human eye. This becomes evident in any laboratory experiments which involve the different placement of mirrors, polarization screens, prisms, timed deflectors, and/or photoelectric detectors so frequently used. Moreover, when dealing with "entangled" photons from a common source such as a calcite crystal, if one redirects the two monochromatic rays of different polarization

---

1. Jeremy Bernstein, *Quantum Profiles.* (Princeton University Press, 1991).
2. Richard D. Feyman, *QED, The Strange Story of Light and Matter.* (Princeton University Press,

(vertical or horizontal) to a common screen they will either re-unite into an intense beam having normal dual combined polarization or else display the standard wave diffraction or interference pattern of dark and white lines depending upon how and where the operator places his mirrors or polarization screens in space. Photons appear to choose their own properties to suit the available physical circumstances presented to them. If the thinking experimenter wishes mentally to reject any notions of non-local influence or instantaneous communication between the two different photons in two different locations; then he is forced to adopt the equally absurd conclusion that the same photon must elect to go through both possible tracks at once. Should the reader find these possibilities difficult to accept as reasonable, then he is beginning to understand the justification for the term "quantum weirdness."

At any rate, Bell collected some of his written papers and thoughts on the subject at the suggestion of Simon Capelin at the Cambridge University Press; and they were published in 1987 under the title of *Speakable and Unspeakable in Quantum Mechanics*. Not only was he bravely challenging the conventional wisdom of many of his friends; but the book actually was a kind of challenge to the physics fraternity to do something about the questions and verify by experiments whether "hidden variables," pilot waves or non-local action effects really existed. It is required reading for those people who wonder about the strange probabilities of quantum mechanics as compared to the statistics of our macroscopic "real" world.

Leonard Mandel, Li-Jun Wang, and Xing Yu Zou set up an ingenious experimental arrangement at the University of Rochester, New York State, in 1991 with the intent of discovering when a photon behaves as a wave and when, instead, it behaves as a particle or bullet. In other words, what controls the duality phenomenon? In order to accomplish this they first passed the instigating light beam through a device which at the will of the human operator could divert the rays to either a left or righthand path at random.

If the beam is diverted to the left it is directed by mirrors to a calcite crystal or Down converter which emits two photons going in different directions. The left beam activates a photoelectric counter which makes a clicking noise, whereas the righthand photon passes to a detector screen. That screen then displays the usual interference pattern of dark and light bands after a number of photons pass. From the point of view of the experimenters this indicated that the photon was behaving as a wave; but what was causing the interference? Themselves?

Flopping the beam splitter over to the righthand side made the light beam pass into an apparatus having two calcite crystals. The photons from the first crystal were arranged so that the righthand photon called "signal" went directly to the detector screen, while the lefthand photons (called

"Idler") were themselves again passed through a second calcite crystal and to the photo-electric counter only. The wave interference pattern appeared on the detector screen as before. However, placing a barrier to block the path of the lefthand photon leaving from the first crystal effectively collapsed the entire behavior to simplicity and the detector screen displayed only a bright spot of light indicative of photons behaving once again as particles.

Clearly, we are talking here about not only duality only, but also the ability of photons to split and find many possible paths to the detector screen. Removing those paths down to one removes the interference, but removing the counter also removes the observer. That helps to make things simple too but unfortunately brings us back to Neils Bohr's allegation that the human observer or experimenter is also part of the problem. The explanation given by John A. Wheeler of Princeton University is: "quantum phenomena are neither waves or particles, but are intrinsically undefined until the moment that they are measured."

Speaking about the various probabilities offered by the quantum mechanics wave function equation, Jose L. Cereceda of the Telephone Company in Madrid, Spain, concluded: "It is what the experimenter can do, not what he bothers to do, that is important."

Section 20.4 - The Difficulties With Time And Distance

By this time the subject of apparently simultaneous opposite behavior of two separate photons originating from a common source by some presumed non-local action influencing their behavior had become respectable enough for discussion among the physics community. However, there is one very sticky problem to the entire idea which causes all scientists to balk at unqualified acceptance.

If the persons operating the experiment slightly alter the times at which the polarity screens or measurements are made at each separate photon detector locations then the delay in polarization change in one of the two setups should evidence a measurable lapse. Quite similarly, if the lengths of the distances of each separate detector from the common light source are deliberately made significantly different (one being larger than the other), then one would expect a required time lapse for the passage of that photon which must travel over the longer track. This necessarily follows from acceptance of Einstein's dictum that the speed of light in a vacuum is a fixed constant. Obviously the passage of a light beam through a transparent or translucent medium, such as glass or water or air or a quartz crystal reduces its velocity of transmission, but all those rates are known and have been calculated.

Unfortunately, no measurable change has been observed in both of these possible cases in the experiments described. This clearly defies all the conventional rules of relativity, since nothing is supposed to exceed the maximum velocity of light.

It happened that the reports of the experimental results made by John Clauser and party came to the attention of John A. Wheeler at Princeton University. Professor Wheeler is a world recognized authority on Special and General Relativity, including their effects on astronomical measurements and the bending of light rays by bodies of very large mass. He reacted immediately in 1993 by collecting three of his prized graduate students and directing an experiment for them to undertake. The project was accomplished by Raymond Y. Chiao, Paul G. Kwiat, and Aephraim M. Steinberg. By their own explanation they focused on three separate demonstrations of what are considered non-local effects for photon trajectories.

The first experiments were preoccupied with a known phenomenon called tunneling. They essentially prepared a "race track" where one twin photon path was unimpeded by any barrier between it and the detector, whereas the other twin encountered a barrier between the source and a detector. This barrier was actually a lamination of thin transparent layers of glass of different types each coated with a metallic layer only one micron thick. The great majority of the photons directed to this barrier were totally blocked, but of the millions of photons emitted roughly one percent appeared on the other side at the detector. To the consternation of the experimenters the photon which tunneled through the opaque glass barrier arrived at its detector **sooner** than the other twin, which had only to pass through air. Not only that, time calculations would indicate a traversing speed of around 1.7 times that of light! Tunneling is supposed to be allowed by Heisenberg's Uncertainty Principle.

The second experiment examined the question of timing more critically by deflecting each of the two photons by means of mirrors into a beam splitter before going to separate detectors. Essentially it verified what the University of Rochester team already reported. Given a choice of two alternate paths, some photons take both and interfere with each other at the detectors destructively. The explanation offered for this ambivalence is that under those circumstances the photon should be regarded as a wave packet which splits into two halves, each with one half the original energy. Presumably an ultraviolet light photon will become two infrared photons.

The final experiment with inferometers which can lengthen or shorten the path of each twin photon at will after it reaches a beam splitter was used to check non-local correlation between the two twin photons. The observations confirmed that, after leaving the inferometer, each photon knows what the other is doing non-locally and its behavior for selecting the next path, whether up or down, is co-ordinated instantaneously with that of its other twin. Not only that, if the twins are far apart so that there is no time for a signal to pass between them, the correlation still happens. Non-local communication remains a fact.

Explanations for the apparent contradiction with the conventional speed of light are numerous and very involved with time and the concept of a photon as an ambiguous entity until called upon to act. Any revision of Relativity has few believers still; but wide understanding of this strange phenomenon remains under intense discussion.

The question exists of whether the physical distance between two entangled particles really matters. As a matter of accepting the experimental results, the theoretical conjecture was that it did not. Dr. Nicolas Gisin of the University of Geneva and colleagues set out to answer this question unequivocally. Using already familiar experimental set-ups, Dr. Gisin sent the two entangled pairs of photons in opposite directions to small villages north and south of Geneva separated by a distance of seven miles instead of the past trials at no more than three hundred feet. The light signals were sent through glass optical fiber telephone cables. He used an inferometer apparatus similar to that used by the Raymond Chiao group for Wheeler and invented by Dr. James D. Franson of Johns Hopkins University. In all cases the two separated signals simultaneously selected the same possible path, up or down, upon exit to the detectors from the inferometer, thus indicating a complete correlation. Distance apart was irrelevant.

It is certainly hard to argue with Professor Abner Shimony's rather dryly stated conclusion:

> "There is a collary to entanglement....if the systems 1 and 2 are spatially well separated, then the entanglement of 1 and 2 constitutes a kind of *non-locality*...Clearly there is a tension between the theory of relativity and the causal interpretation of potentialities....The wiser course is to say that quantum mechanics presents us with a kind of causal connection which is generally different from anything which could be characterized classically."

Section 20.5 - The Opposition

In spite of the efforts of the few physicists preoccupied with quantum theory and its relation to the "real" world of ourselves it should be realized that the rank and file of the profession remained essentially skeptical and indifferent for sixty years. As far as they were concerned, Neils Bohr had already explained it as another aspect of duality in nature similar to light rays or "Complementarity"; and there was no point in worrying about it. Their attitude is quite reminiscent of the venerable adage of a good mechanic: "If it works, then don't fix it!"

Some people even regarded the problem incorrectly with the analogy of a half dollar coin carefully sawed into two half parts along its diameter by a jeweler in New York City to make one side carrying the "heads" picture and the other side the "tails." The jeweler has two friends about to travel abroad—one going to London and the other to San Francisco. He calls both into his shop and places the two half coins on the counter top. He then tells the two men to each pick up only one coin piece **after** the jeweler leaves the room for an appointment and take it with him on each of their airplane trips to the distant cities.

A few days later he telephones his friend in San Francisco and inquires whether he has the "heads" or "tail" side in his baggage. If the answer is "heads"; then the jeweler knows immediately that the other friend in London has the "tails" piece. So what's the problem?

More recently, as the results of various corroborating experiments on the evidence of entanglement were published, a kind of grudging recognition has happened. There remain some physicists who are still violently hostile to the entire subject and consider the question a complete waste of time, although one may wonder whether their public animosity is not psychologically derived from an abhorrence of any uncertainty or else a limitation imposed by the demands of their occupational career.

The more moderate, or at least quizzical, body of opinion which hesitates to embrace such radical non-locality or "action at distance" ideas is exemplified by people like Christopher A. Fuchs, a funded Fellow at the Los Alamos Laboratory in New Mexico, or Asher Peres, who is a Professor of Physics at the Technion-Israel Institute in Haifa, Israel, or Serge Haroche, who is a Professor of Physics at the Ecole Normale Superieure in Paris.

Robert Griffiths, a Professor of Physics at the Carnegie Mellon University in Pittsburgh, Pennsylvania, and Roland Omnes, author of a textbook on Quantum Mechanics and Professor of Physics at the University of Paris in Orsay, France, collaborated on an excellent article entitled *Consistent Histories and Quantum Measurements* in the magazine *Physics Today* in the August, 1999, issue. In sum, they both seem to agree that the situation may be remedied without major confrontation; but they also concede that quantum mechanics probabilities in action inside the sub-microscopic atom do not seem at first glance to be identical with those statistical probability rules which prevail in our own macroscopic world. To this writer it seems that the thrust of their arguments depend upon a concept called "consistent histories" and a process of decoherence over time which takes place during the atomic change or reaction and its subsequent measurement process where the two different probabilities eventually become reconciled in our macroscopic world. In effect, however, we do not know the exact state of a particle or photon at all points in time and cannot really determine it as having permanent or lasting properties until its final state. We neither know its polarization or position or which detector it chooses to pass through until a measurement has been performed and registered in a recorder or the mind of a human being.

Section 20.6 - Summary

That quantum mechanics is indeed strange should be conceded by everyone now, including the patient reader. If it merely stopped short with telling us that electrons in an atom jump instantaneously in discrete energy levels between orbits in sudden hops we could most probably accept that statement with equanimity. If we are told that no intermediate energy level is possible, and that, consequently, the electron has no real existence when passing between orbits, we might raise an eyebrow. When it is then stated that it is impossible to predict or measure experimentally both the position and velocity (or momentum) of any electron at the same instant in time, we might shrug our shoulders and say that this obstacle is the logical consequence of the Principle of Indeterminacy and Planck's Constant. But when we are told that two photons emitted simultaneously by a crystal of calcium carbonate into two different directions behave predictably and instantly when only one of them is disturbed, the reader's credulity must be taxed to the limit.

The writer, Jeremy Bernstein, in his amusing little book *Quantum Profiles* quotes the theoretical physicist, John Bell, at CERN as "I hesitated to think it might be wrong; but I knew it was rotten," in response to the question of whether he ever had doubts about quantum theory correctness. He reminded his questioner that Max Born stated in a written reply to Schrodinger: "I am myself inclined to renounce determination in the atomic world...." Bell continued to state: "One has to find a decent way to express whatever truth is in it."

Abner Shimony's only explanation for it is that our notions of space-time structure cannot be correct. Clearly our "window" on the Cosmos may be telling us something which we are yet unable to comprehend.

John Wheeler has expressed bafflement; but he points to the new efforts with the super-string theories as a genuine effort to find a compatible solution combining gravity, Relativity, time, electromagnetism, and the strong and weak forces into some kind of super-space concept of ten or eleven dimensions. He thinks we must revise our way of thinking of space and time. Yet, nonetheless, he allowed a candid confession: "I read all the papers on quantizing General Relativity written by wonderful people; but they certainly were not transparent to me—not in the way that one could see at a glance what was going on."

Chapter 21
The Advent of Higher Energy Colliders, the Topquark, and the Future

There are more things in heaven and earth, Horatio, than are dreamt of in your philosophy! – from "Hamlet" by William Shakespeare.

Section 21.1 – The Demise of the Super-Conducting Super Collider

The looming dominance of the European synchrotron in experimental particle research had begun to rankle the national pride of some physicists in the United States; and, moreover, the goal of confirming the Higgs boson, as well as the Top quark, afforded them a perfect justification for pushing the politicians for a bigger machine. The germ of the idea for a huge Super Collider quite possibly originated with a speech by Leon Lederman in August of 1982 at a conference at Snow Mass in Colorado. However, it took a while for a significant number of the scientific community to take the idea seriously and endorse it. The sheer magnitude of their initial design proposal, which was intended to dwarf the Fermilab and CERN both, was breathtaking. There were to be two separate particle circulator tubes or rings 54 miles (or 87 kilometers) in circumference with an expended energy rating of 20 TeV or twenty trillion electron volts each. With particles of opposite electric charge circulating in opposite directions in the rings their eventual deflection into collision at the detector would represent a total impact energy of 40 TeV. The early discussions were conducted with the Department of Energy. The magnets were to be helium cooled and made of superconducting wiring, hence the abbreviated name of S.S.C. Eventually, in 1986 the physicists involved submitted a formal proposal to Ronald Reagan, then President of the United States. The initial estimates given the Secretary of the Department of Energy for construction in 1989 were as low as $5.9 billion, and Congress obligingly appropriated the money.

As one might expect, the minute this scheme and its cost became public in the newspapers there was a frenzy of activity among members of both the Senate and House of Representatives and the governors of many states to have this bonanza located in their home states. Everyone had a reason to place the accelerator in their state, including promises of state contribution towards funding. To the credit of the supervising panel of scientists the final choice made environmental and geological sense, although they were compelled to acknowledge the political clout of the primarily Republican state of Texas in such matters. The final location was to be in the flat farmland plains near the small town of Waxahachie in Ellis County, northeastern Texas, roughly 25 miles south by west of Dallas. It was planned to bury the accelerator rings in a tunnel 15 feet 3 inches underground in what was determined to be a deep layer of calcium soft chalk rock. Moreover, the grassy and open rural land might be purchased cheaply before the real estate speculators interfered. Thus this decision had real merit of common sense in spite of the politics involved.

Lederman himself tells a very amusing story in his book[1] about the final location. Early in the negotiations he took the trouble to drive to the proposed site. At a subsequent conference with several state governors the Texas governor objected to his pronunciation of the name of the town. Lederman responded "Sir, I really tried. I went there and stopped at a roadside restaurant to ask the waitress there to tell me where I was. BURGER KING she enunciated." His sense of humor, of which Lederman has an abundance, was not appreciated by the Texas politician.

At any rate, President Reagan expressed his approval at a Cabinet meeting and the go-ahead decision was announced in October of 1988. Actual construction of the underground tunnel and a huge pit for the collision impact detector room started by 1990. It was a remarkable civil engineering achievement to accomplish the excavation of a perfect 17 1/4 mile diameter ring through the soft rock to millimeter precision tolerances. The surveying was done with laser beam transits and computerized readings, which was a recently introduced technology at that time. Not only that, but Roy F. Schwitters of the National Fermi Laboratory and his advisors had authorized the start of construction of the dipole super-conducting magnets which are needed to constrain the particle beam within the vacuum tubes. By this time the estimated cost for the project had climbed to $8.9 billion dollars.

In June of 1992 things began to fall apart. The House of Representatives began a spirited debate on the wisdom of spending that much money on an exotic machine which, from their point of view, would produce nothing tangible for gainful industrial production nor add to the paychecks of the great mass of ordinary voters.

---

1. Leon Lederman, *The God Particle.* (New York, Houghton Mifflin, 1993).

The economy of the country and the industrial nations was then stalled in a period of stagnation and rising unemployment which made the expenditure of billions on a pure theory gamble politically unpalatable. Moreover for the U.S. to attempt such a gigantic and hugely expensive project alone was extremely chauvinistic and complaints from scientists in other disciplines that it was draining money from other research were quite plausible. The real fact was that support from the great number of the general scientific community in the nation was not unanimous about the super-collider. Their arguments were not instigated purely from professional jealousy. Researchers in transistor electronics and computers, or DNA gene research, or medical research on cures for cancer and other diseases, or biochemistry for pharmaceuticals all united in a fear that the collider would detract from their budgets. Another criticism was directed against the nationalistic egomania of doing the project alone without the participation of scientists from other countries. Japan had offered to put up money providing they could participate in the administration. They were summarily refused. Another significant obstacle was the competition with the more glamorous program of the Space Agency and the rocket shuttle. That had a spectacular political potential and circus entertainment value for the popular ego far superior to that of the collider, which is not to detract from the Agency's genuine contribution to astronomy. Yet another factor was that Congress was in the middle of a gigantic funding program to encourage the construction of municipal sewage treatment and drinking water supply plants, as well as starting environmental pollution control. By 1993 the best and most honest estimates for the cost of the super-collider construction had risen to $11 billion dollars; and there were stories of management problems and competition between government agencies and university committees. All these factors mitigated against the dream of the physicists, then at exactly this point they shot themselves in the foot. It is a little told story.

The scientists had been advised by political pundits and people in Washington to commit as much money as possible to work contracts as quickly as possible on the theory that something physical in concrete and steel at the site would deter Congress from withholding subsequent funds needed to reach completion. That meant doing things in a hurry. The original design had based upon the assumption that they could contain each particle beam within a four inch (4") vacuum tube ring. Subsequent calculations and checking revealed that this number was too optimistic and that five inches (5") was the minimum dimension. That meant the revision and alteration or even scrapping of some equipment already manufactured, including the ring magnets. As Chairman from the Fermi National Laboratory of the guidance committee, Dr. Roy F. Schwitters had no choice but to plead for a contract change order of another $2 billion dollars, thus raising the total ante up to $13 billion dollars. The politicians revolted in disgust.

The House of Representatives of the United States at the end of 1993 terminated indefinitely the construction of the proposed super-collider in Texas, leaving a gigantic underground tunnel, several vertical shafts and foundations for a few buildings and detector equipment unfinished. The monetary deficit represented about $1.6 billion dollars or more. Despite the anguish and potential unemployment that this decision may have caused to a number of graduate students and disappointment to post-docs and many of the particle physicists throughout the world, the entire underground tunnel was equipped with ventilation and humidity control machinery and sealed off.

It is conceivable that the super-collider project or something similar to it may be resurrected some year in the future twenty–first century at a more prudent pace of expenditure and with international participation. After all, the United States seems able to spend $50 billion dollars for a new space station to be partially built by the Russians at a time when the Russian economy was crumbling into shambles. Some people wonder today whether that is really reasonable. As of the first year of the century, the Association of American physicists with the National Science Foundation is endeavoring to persuade our national congress to build a new twenty mile long linear accelerator at a minimal estimated cost of five billion dollars. One new aspect required to accomplish such a plan would be a marked improvement in the United States record for cooperation both monetarily and administratively with other nations in the world.

Section 21.2 – The Accelerators of the Future

With the collapse of the American plans for a gigantic super-collider the physics community was compelled to undertake a sober reassessment of the practical possibilities of high energy research in the world. Dreams that we might have begun to approach the energy furnace of collapsing stars and the super-novas of space here on this planet needed to be chastened with hard realities. Nuclear science had become part of the daily scene and no longer enjoyed the image of alchemists practicing their awesome magic. It was a great disappointment to many and cast a pall of anxiety over the physics educational institutions for job opportunities for their graduates. However, there were two salutary effects which emerged from all this pessimism. One was that the leading organizations turned their attention back towards defining what important and progressive experimental research could be done with existing facilities on a reasonable economic basis. The other was a shift in emphasis in some seventy of the universities to prepare more of their Master and Doctorate degree graduates for assimilation into teaching staffs in the schools and into general industry. A shift towards international cooperation was also stimulated.

In so far as innovations in the engineering aspects of the machines themselves are concerned, there has been some success in narrowing down or concentrating the pencil beam of particles to very small dimensions. The more exotic ideas being proposed include superconducting cavities, the use of lasers and plasma lenses, or using anti-muons and muons as colliding projectiles. All this thought is now directed towards the goal of achieving higher collision energies with less expensive equipment and/or less waste of electrical power. Modifications to existing facilities became the present activity. One early upgrade case in point was the Hadron Electron Ring Antage (HERA) machine at the newly expanded laboratory facilities at Hamburg, Germany. (Hera was a Greek goddess and wife of Zeus.) This accelerator collides electrons with protons coming from opposite directions in a newly constructed underground tunnel some four miles in circumference. The shower of particles resulting from the impact of extremely high kinetic energies takes place in the center of a huge and complicated detector which is appropriately named ZEUS.

This entirely new machine was first put into operation in the early spring of 1992; and by the autumn of the same year it was yielding valuable data. Professor Frank Sciulli of Columbia University was among those North Americans invited to attend the runs; and we are indebted to him for information. The essential idea was that the different electric polarity between the electrons and protons produces a mutual attraction which, combined with high relativistic velocities, makes it feasible to knock one of the three quarks out of the proton in what is call deep inelastic collision. The frequent result is streams of diverted electrons plus a shower of other particles, including jets of two quark hadrons. Study of the tracks, charges and momenta of these transitory particles in the detector is expected to yield clues to the puzzle of quark confinement, as well as the behavior of the weak and strong forces.

The science of sorting out actual measurements in the detector assembly has become an intricate technical specialty in itself. The detector may "see" one hundred thousand events per second during a collision event and this is obviously far beyond the capacity of any human being to observe. Thus, the data consists of a multitude of electrical signals all fed into high speed computers for storage in memory tapes to be analyzed later. Timing is crucially important and is measured in nanoseconds. The time of the entrance of the proton into the detector, the time of the actual collision "event" and the time of the arrival of a fragment to the detector screen downstream all are essential information. There are many problems to be expected. It is estimated that the three quarks contained in a very high speed proton carry only about one half the mass. All this is supposed to take place in a tube with a perfect vacuum inside, which is an unachievable goal. The collision of a proton with any stray gas molecule can produce a miniature explosion all its own and thus obscure any other activity. The analysis of the electronic data eventually requires a large team of specialists working for months.

The four essential phenomena for which the people at HERA are searching for are:

(1) Electron Luminosity
(2) Deep Inelastic Neutral Currents
(3) Deep Inelastic Charged Currents
(4) Photon Production

At one time, there was a great deal of speculation about the possibility of discovering lepton quarks or "lepto quarks." Reactions of physicists ranged from "oh, sure" to "who needs it?" Thus far no evidence has been produced for belief in this fantasy, much to everyone's relief.

It had become clear that the linear accelerator such as the one selected by Stanford University was uniquely efficient for colliding electrons with positrons; but the energy limits for the existing installation were frustrating. For a variety of reasons the physics theorists essentially agreed that a minimum 1 TeV, 10^{12} electron volts, energy level was needed to produce really new physics, including the Top quark and Higgs boson. Consequently the staff at SLAC in California starting thinking in 1993 about a new linear machine of 20 kilometers combined length, called the "Next Linear Accelerator" or NLC for short. Its Director, Burton Richter, even arranged in 1997 an agreement with Hirotaka Sugawara, the Director of the Japanese high energy laboratory near Tokyo, for mutual collaboration on the design and registered their conceptual machine with the United States Department of Energy. By 1997 a 42 meter long test prototype for the innovative equipment required had been actually built using advanced radio frequency wave generation and control. An accelerating structure only about 5' 10" is fed by three 50 megawatt Klystrons manufacturing an unbelievably short wave length of only 2.6 centimeters or a frequency of 11.4 gigahertz. All this means entering an entirely new and untried technology with an acceleration gradient of three times the old CLAC or 50 M volts per meter and having only partially anticipated problems with beam constriction and density.

Meanwhile, the people at the Fermi National Laboratory in Batavia, Illinois, under their new Director, John Peoples, shut down the twenty–five year old Tevatron accelerator on the 15, September, 1997. The event was celebrated with much newspaper publicity for the grant of $229 million dollars by the U.S. Department of Energy for a modification of its' Main Injector ring for a ten-fold increase in proposed proton versus anti-protons collisions by 1999.

The Brookhaven National Laboratory located near Upton, New York, had been suffering a decline during the years of 1996 and 1997 due to a clear lack of purpose, political squabbling in Congress, and quite adverse publicity about lax safety precautions and release of radioactive materials. A new Director named John Marburger was appointed and construction started on what was called a Relativistic Heavy Ion Collider or RIHC for short. It was intended to replace the Isabelle Project which was cancelled back in 1983. The design intent of the new machine is to collide two separate beams of massive gold ions with kinetic energies of 100 GeV each for a resulting collision energy of almost twice that amount and a huge momentum equation. The science Project Director was to be Satoshi Ozaki. The general expectation was to obtain a dense plasma cloud of quarks and gluons from this simultaneous collision of some tens of thousands particles. The first beginning evidence of success was announced in June of the year 2000.

All this seems consistent with the current American belief that bigger is better! Not at all surprisingly alarms were raised once again that such an approach towards simulated conditions of the "Big Bang" could cause an apocalyptic gravitational collapse of the whole mess into a miniature " Black Hole" into which all our planet's matter would be sucked!

Meanwhile, continuous improvements to the large CERN accelerator ring at Geneva had gradually made it evident that their machine had the greatest promise to ultimately reach the 1 TeV power range first. As we already know, it had been functioning since 1989 as a collider of electrons with positrons for the confirmation of the Z Zero boson. Following that period CERN made a major change to proton versus anti-proton collisions, as we explained in Chapter 19, with the result that it was producing W+ and W- pairs in 1996. They also had found time in the autumn of 1995 to cool down and decelerate the anti-protons enough for them to combine with other anti- protons to make briefly anti-matter hydrogen gas. Not content with all these past successes, however, in 1995 the Europeans were already planning an extensive modernization and improvement to their existing machine for the collision of protons with counter-rotating protons. Among other inadequacies the old beam constraining magnets on the large ring were not superconducting, easily overheated, and consumed prodigious amounts of electrical energy. They were also too limited in number. Equally important changes needed to be made on the radio frequency impulse accelerators to the ion beams. The CERN directors courageously decided to aim high and do the reconstruction in two different phases with hopefully scheduled operational dates of years 2004 and 2008. The project immediately became known as the "Large Hadron Collider" or "LHC" and, since the anticipated cost was astronomical, appeals were made to all nineteen country members. Japan promised $50 million towards instrumentation. Only a portion of the ultimate number of new super-conducting 14 meter long magnets will be installed for Phase 1 for a design center of mass collision energy of 10 TeV. The remaining "missing" magnets will be added in Phase 2 for an ultimate energy level of 14 TeV. Finding the Higgs Boson is the tacitly acknowledged goal.

Faced with such formidable competition in combination with national budget reductions for pure science, as opposed to that dedicated to military defense or purposes capable of commercial exploitation, the Americans swallowed their pride and decided to join in the international effort. The physicists were also confronted by the challenge that they could lose the opportunity to participate in a major high energy level scientific frontier. Truce negotiations between the CERN official and the United States began in May of 1996 with a promise of $81 million dollars from the National Science Foundation for the design and construction of new detectors. By August of 1997 the United States commitment to the planned LHC had reached $450 million dollars from the Department of Energy and $81 million from the National Science Foundation for a total of $531 million. In December of 1998, over 4 years after the voted shut-down of the super-collider, the agreement was formally signed by Federico Pena for the D.O.E. and Luciano Maiani for the CERN council in Washington, D.C. As any student of political science might predict, the leading organizer for the minority protest group against the contract was Republican Representative Joe Barton from the state of Texas.

The impact on the CERN facility operation was immediate. The Low Energy Anti-Proton Ring for their production and feeding into the primary accelerator was shut down at the end of 1996. The old Large Electron Positron collider itself is scheduled to be shut down by November 2, 2000. The bad news was that the entirely reconstructed Large Hadron Collider replacement is not expected to be operational until the year 2005. As Bernard de Wit, the delegate from the University of Utrecht put it; "there will be a period of thin physics."

Section 21.3 – The Top Quark

The intense urgency of the physicists to verify the existence of the Top quark somehow is perfectly understandable. Not only must it exist as a mate to the Bottom in order to justify their dedication to the Law of Symmetry; but it is also required to complete the three pairs of quark flavors in ascending masses which are the basis of the Standard Model. Measurement of its mass would also provide some basis for more reliable predictions for the Higgs boson, as well as justifying the general theory. Very early estimates for the mass of the Top quark (known as "Truth" in Europe) were so scattered and widely disparate as to be almost worthless. The only agreement was that it was construed to be very heavy. This fact in itself is a hint that our comprehension of the means of creation for quarks is not entirely complete yet. As far back as 1989 the eager experimenters at the Fermi National Accelerator in Illinois registered what they cautiously called "two interesting events" involving several jets. They had been smashing protons together with anti-protons at a combined energy of 1.8 TeV/c^2. They hesitated to identify the event then as the almost instantaneous decay of a Top quark, however, since they were well aware that physicists at CERN had done that in 1985 and were compelled to retract their overly optimistic statement. The expectation of the theorists was that the head-on collision of two quarks having near speed-of-light velocities would result in the creation of a Top and Anti-Top quark pair which would immediately decay into W+ and W- bosons with the simultaneous emission of an electron and muon going off in different directions. The muons are passed through a magnetized iron barrier in order to slow them down and they are identified by their subsequent decay. The bosons themselves are expected to subsequently break down into jet showers of quark hadrons, including the bottom quark. In this confused picture of the particles explosion the tracks of the electron and muon are the most conspicuous and identifiable in a detector. Two different detectors are used. One designated the "DO" is non-magnetic and uses liquid Argon as a calorimeter. The second detector is designated "CDF" and was a Tesla 1.5 magnitude type solenoid magnetic field which is intended to measure individual charged particles.

Continued experiments over a period of 18 months with the Tevatron accelerator through March of 1993 strengthened the convictions of the team of 439 international scientists enough that a 153 page report was submitted to the *Physical Review* on April 22nd of the same year. A long article on the tentative announcement was carried in the Science section of the *New York Times* newspaper on the 26, April, 1994, and a corroborative article of "cautious enthusiasm" appeared in the professional journal *Physics Today* in their June 1994 issue. By May of 1995 after continued observations and analysis of the two detector data, both Fermilab and the physics journals adopted more positive assertions of success.

The mass of the Top quark was calculated to be between 170 to 176 GeV/c^2. A good average is 174. This is close to the mass of an atom of gold.

The total shut-down of the Tevatron ring complex for alterations in September of 1997 terminated any further statistical studies of the W+ and W- bosons and the Top quark at that site for an anticipated three or more years.

21.4 – Some Theory Casualties Along the Road

Many serious people who choose physics as an avocation and occupation are apt to possess very vivid imaginations. Assuming that it were at all possible, if one actually read all the published propositions in both the professional journals and popular magazines concerning new physical laws or exotic particles, the reader might come to the suspicion that there are as many particles as there are graduate students searching for an original thesis.

(a) Possibly one of the most plausible is gravitons. Given the accepted fact that light is propagated and carried from stars through space by massless vector bosons called photons, it becomes a perfectly logical conjecture to consider the gravitational force between masses to be transmitted by some corpuscle called "gravitons." Although Newton and others were able to quantitize this invisible attractive force in a reliable equation dependent upon two relative masses and the distance between them, the method of transmitting this force remains mysterious to us. There have been many careful experiments conducted in an effort to discover and measure such a boson corpuscle. Thus far, all have resulted in failure. Nonetheless, astronomers still talk about gravity waves in a quantum field. The graviton is assumed to be massless and have a spin of 2. Gravity is one of the given postulates of our universe; but connecting it with electromagnetic, strong, and weak forces mathematically has defied all efforts thus far. The ardent advocates of the latest "super-strings" theories deny this and assert that they offer a solution; but some scientists still remain incredulous.

One reason advanced for the lack of any convincing detection of the much-discussed graviton is that gravity is an extraordinarily weak force compared to electromagnetism and the other forces operating within an atom.

In spite of the obvious importance of gravity to us humans, gravitational attraction between elementary particles is in the order of the 10^{-32} times weaker than their electromagnetic attraction.

The only alternative explanation remains that of Albert Einstein. Mass in some unknown way distorts and curves space around it and thus creates a kind of dent or dish in the fabric of space. The force of gravity then becomes only an acceleration of another mass as it falls inward toward the larger. However unreasonable that may seem to most of us, no better answer has arisen to seriously challenge it thus far. What clinches his case is that today's astronomers have been able to display on photographs the fact that massive large stars or suns actually bend light rays coming to us from a great distance behind them.

(b) Another more radical notion is the existence of magnetic monopoles. In 1992 the United States Department of Energy awarded a $295,000 grant to two scientists in the Department of Physics and Atmospheric Sciences to discover such monopole particles which may still be residual from the "Big Bang" creation. The experiment was named "MACRO" and took place in the neutrino detector laboratory underneath the Italian Appennine Mountains.

The notion of monopoles or single polarity carriers for the magnetic force is consistent with the concept of bosons and was given a mathematical justification by the Russian theorist Sasha Polyakov. It was expected that these extraordinarily heavy particles had to appear early in the cosmological explosion of the "Big Bang" of creation and their mass would be much more than a billion times that of the proton. The only trouble is that none have ever been detected experimentally, despite efforts.

At "first encounter," the non-professional reader might infer that this lack of confirmation of any such particle suggests grave doubt of its existence or even weakens the entire boson theory. However, the scientists were happily saved from such a confrontation shortly later by a now professor at MIT named Alan Guth. He generated an "inflation" picture and change of state scenario for the original cosmic explosion which would make the great preponderance of monopoles disappear from the matter which we see now. According to Guth, the probability of finding a monopole now is extremely remote; but a few determined souls remain undeterred.

(c) There remains another outrageous speculation that contradicts one of the most sacred precepts of modern physics. Articles have been written and published[2] suggesting that some particles, presumably bosons, may have the ability to exceed the speed of light. They were given the name of "tachyons."

From the author's skeptical point of view, there may be two different origins for belief in this proposition. The first is the most famous example of quantum weirdness in which two separate electrons emitted simultaneously from a calcite crystal of calcium chloride have opposite spins. Changing one magnetically causes the second electron to flip in reverse regardless of its distance away. This is the non-locality to which Albert Einstein objected so strenuously.

The other influence for toleration may possibly derive from Richard P. Feynman's invention of particle reaction diagrams using distance moved through space as the abscissa and time duration as the vertical ordinate.

Feynman, in an effort to assign some material and pragmatic picture of reality to quantum mechanics, derived his own unique explanation and derivation of the probabilities by a "sum of histories" analysis based on the assumption

(a) that a light photon or electron can leave a transmitting source and reach a receiving detector by any of numerous possible paths;

2. Nick Herbert, *Faster Than Light.* (Ontario, Canada, The New American Library, 1988).

(b) that path taking the shortest possible time in action represents that "amplitude" which is most probable.

His famous "Feynman Diagrams" for particle reactions were a later outgrowth of this idea.

In cases involving the emission and absorption of photons from electrons or with particles having different decay times it becomes possible for the diagrams to make it appear that the particles are moving faster or slower than the standard light velocity in space.

The best argument for the apparent possibility of faster than light particles, however, rests with Einstein's Special and General Relativity. It all depends whether you are looking at a physical phenomenon from the viewpoint of your desk on planet earth or whether you are mentally traveling on board the atomic particle itself. A space-time inertial system can make the difference. If one happens to be a philosophical fan for the "many worlds" explanation for the "non-locality" aspect of quantum mechanics, then tachyons may become less repugnant to you.

(d) In the past fifteen years or more it has become quite the vogue for accomplished mathematicians, starting with the Standard Model, to play with the "Theories of Everything." Einstein spent the latter part of his life at Princeton, New Jersey, attempting to reconcile gravity and electromagnetism mathematically and failed. "SUSY" or Super Symmetry, sometimes called the Kaluza-Klein (KK) theory, became one of the popular and earliest attempts for a while. This theory implied super-heavy and invisible partners to the quark and electron called "Squarks" and "Selectrons"; but by January of 2002 it was assigned to the status of wasted effort.

The mathematics of the theory required more than four dimensions. The general notion was that all the particles which we have seen thus far were originally two mated together in pairs which were identical to each other in properties except that their masses and spin were different. Thus a spin 1/2 fermion could be paired with some identical particle which is a spin zero boson. Thus the distinction between fermions and bosons becomes obliterated. No experiments thus far have identified such strange particles; but this in no way discouraged the proponents from naming the presumed combined particles as squarks, slectrons, sleptons, gluinos, and gauginos. It was also presumed that the lightest possible Super-Symmetric particle might be the "gravitino." The Higgs Boson was supposed to be the Minimal Super-Symmetric Model. No direct experimental evidence has ever been produced to verify this scheme; and it soon fell out of vogue.

SUSY was eventually modified into a much more plausible theory called "Grand Unified Theory" or G.U.T. The basic assumption of this second theory is that all three of the forces acting within the atom (strong, weak, and electromagnetic) were all combined into one at a few seconds after the "Big Bang" and before things chilled down by expansion of the universe. To be specific, this is not expected to happen unless the particle energies are somewhere between 10^{14} and 10^{15} GeV, which implies unbelievably high molecular heat energies. This concept actually was the theoretical basis for the Unified Electro-Weak Theory developed by Sheldon Glashow of Harvard University, Steven Weinberg of the University of Texas, and Abdus Salam of the International Center for Theoretical Physics, plus a few other people who were not so fortunate as to share the Nobel Prize money. The reader has already become familiar with all that story in Chapter 19. The eventual identification of the W and Z bosons in particle collision detectors was considered a stunning and resounding corroboration of the authenticity of the Electro-Weak Theory and encouragement for the G.U.T. specialists. A second recommendation for the "Grand Unified Theory" was that it offered an explanation for why the electric charges of the proton and electron were exactly equal and opposite. The ordinary reader is apt to ask "How can Nature really be all that deviously complicated?" The same doubts afflict many physicists with uneasiness; and this explains the great amount of attention paid to the "Grand United Theory" or G.U.T. The assumption is that all the basic forces – electromagnetic, weak, strong and gravity – must be inter-related from some common origin. Hence the name "Grand," which implies a noble ideal.

Section 21.5 – Super-Strings

Today there exists much enthusiastic excitement over a newer "Super-Strings" theory, which evolved from the Super-Symmetry ideas, but requires very advanced pure mathematics of a highly abstract nature. Those mathematicians who have earned a reputation for understanding the theory go on tours to different universities giving lectures on the subject to baffled and dimly comprehending graduate students.

For those of us who never had extremely advanced mathematical training or the inclination to do so, including the author, the Super-String theories will remain essentially opaque and beyond the real scope of this book. Mention of the theory is included here merely for cognizance and recognition of ambitious efforts to understand the workings of the cosmos which may deserve further investigation of the literature by some zealous readers.

Although he did not invent it, an ardent advocate of the Super-Strings theory who has made it his life work is a brilliant physicist named Edward Witten. Born in 1950 to another theoretical physicist and his wife, Witten took an erratic route to arrive at the same occupation as his father. He graduated from Brandeis College in Waltham, Massachusetts, in 1971 with a Bachelor of Arts degree in history. Upon graduation he had notions of becoming a journalist and also became active in the presidential campaign of George McGovern as the Democratic nominee. However, he became disenchanted with becoming a reporter and, possibly under the persuasion of his parents, entered the Graduate School at Princeton University to achieve his doctorate in Physics in 1976. For the next six years or so Witten did theoretical work on standard particle symmetry, quarks, and quantum chromodynamics with the color gluons. He was well aware that theories on electromagnetic forces and the strong and weak forces omitted gravity and to some extent relativity; but a paper by Schwarz of California Institute of Technology caught his attention in 1982. By that time the Super-String theory had been combined with the older Super-Symmetry ideas to propose that superstrings with perhaps ten dimensions could vibrate and transfer energy to mass-carrying particles, as well as producing the forces acting upon them. It was generally conceded that ten dimensions was six more than the average physicist could cope with comfortably; but the possibility was extended that at least four of the surplus dimensions, and possibly six, could somehow be "rolled up" or compacted to much smaller quantities more compatible with Planck's Constant and the dimensions of an atom. The most important point made, however, was that the strings involved gravity and that gravity and the topology of space had to become an integral part of the solution. This instantly aroused Witten's interest. Later in the same year Schwarz and Michael B. Green of Queen Mary College in London published a joint paper describing how such anomalies might be eliminated. This absolutely almost immediately converted Witten to a dedicated disciple of the Super–String Theory; and he decided to make it his life work in spite of his conviction that it would take the next fifty years for it to evolve to any final fruition. By this time Witten had moved to the protection of the Institute for Advanced Study at Princeton. He published no less than nineteen papers on the subject in 1985. From Witten's point of view the theory encompassed not only the formation of material particles with mass, but involved the non-Euclidean geometry of space and time as well. He was not the only enthusiast. By 1997 and 1998, Gordon L. Kane, a professor at the University of Michigan in Ann Harbor, Michigan, began writing articles describing how the authenticity of the String Theory could be verified with synchrotron colliders such as the Large Hadron Collider by looking for extraordinary super-massive particles and the more prosaic approach of carefully studying the mass differences between families of quarks or between electrons and tauons.

Astro-physicists also became fascinated. In November of 1997 Juan Maldacena of Harvard University had mailed from Los Alamos a paper which became the topic of intense discussion in the international conference on the Strings subject at the Institute of Theoretical Physics for the University of California in June of 1998 at Santa Barbara. Maldacena had actually been drawn into the controversy by his work on the entropy of astronomical "black holes." Witten had already suggested ways in which the six extra dimensions might have been compacted; and, in his persistent pursuit of the gravity component, he was obliged to delve into topology as it related to possible space curvature distortions. This study of various geometric surfaces which enclose volumes was already an integral part of black holes and the "event horizon" work by astronomers. Maldacena brought up the subject of "duality;" but he was not talking about photons and light waves. The puzzle was that there were the already established gauge theories for bosons and fermions which did not involve gravitational fields and there were the newer Super-String theories which did. The compatibility between the two became a matter of intense discussion. Witten seized upon this duality as an explanation for the strange confinement of the three quarks inside a proton or neutron. Maldacena came up with a topology solution for the debate which he named "D-branes" and which, when pictured in three dimensional representation, roughly simulates a conical hole in a flat surface which merges into long cylindrical throats or tube. At some point in the throat of this space curvature he claimed that the gauge field and other forces in Nature become decoupled from gravity. To the amateur, the picture looks for all the world like the illustrations used for black hole singularities. Given this analogy, it is no surprise that some people suggested that we have two different universes back-to-back, one a shadow universe to our own in a different space-time continuum. This notion has a whimsical attraction to the writer, if for no other reason that it might explain where black holes and swallowed stars go beyond our own universe. No less eminent an astrophysicist as Sir Martin Rees, Astronomer Royal of England, would have little trouble with such a quaint idea, since he has already admitted to be tolerant to the "many universe" propositions made for quantum mechanics. All these notions and more have been beautifully explained with great craftsmanship for the lay person of keen intelligence by Brian Greene in his remarkable book *The Elegant Universe*[3]. To the great surprise of many people, the book became almost an instant best seller, proving that some people are keenly interested in physics.

The patient reader at this point may find himself mentally asking "How much of all this can I really believe?" The only answer is the somewhat unsatisfactory statement given by Isaac Newton of London in 1686.

"In experimental philosophy we are to look upon propositions inferred by general induction from phenomena as accurately or very nearly true,

---

3. Brian Greene, *The Elegant Universe.* (New York, W.W. Norton, 1999).

notwithstanding any hypotheses that may be imagined, til such time as other phenomena occur by which they may be made more accurate, or liable to exceptions."

Section 21.6 – The Future

If indeed we are to expect an interim period of "thin physics" until the year 2008 when the rebuilt "Large Hadron Collider" comes into operation in Geneva, this could become an opportunity to rethink our achievements and carefully appraise the questions which deserve further research within the practical economic restraints imposed by society. The writer has little patience with the popular speculations about the "End of Science" or "End of Physics." The curiosity of human beings is such that they are going to continue to search for answers to the question "Why?." The fact that they have persisted over the centuries is rather a wonder in itself, since most other animals, insofar as we can tell, merely accept their environment and adapt to it. To the majority of physicists themselves quantum mechanics and nuclear science is a kind of picture window opportunity for them to view the workings of both our world and the entire cosmos. The trouble is that the glazing in the " window frame" is partially frosted over and much of the view is obscure in detail. "For we know in part and we prophesy in part…for now we see through a glass darkly," as the New Testament tells us.[4] There is much learning yet to be done and much that remains to be fully comprehended. For those young men active in the field the area of the unknown in our acquired knowledge beckon for discovery.

One typical example today in the science which combines parsimony in the equipment cost budget and with creative ingenuity in the experimental concept is that of a group of three individuals who obtained their graduate degrees in three different universities and are presently professors in physics in three different colleges! This collaborative group consists of Alex R. Dzierba, a professor at the University of Indiana, Curtiss A. Meyer, a professor at the Carnegie Mellon University, and Eric S. Swanson, a professor at the University of Pittsburgh. They all had been thoroughly educated and indoctrinated with the theory of quarks, gluon colors, electro-dynamic fields, and quantum chromodynamics or Q. C. D. with which the noble reader has become superficially familiar. With incontestable logic the group all decided that there seemed no good reason why gluons could not combine with other gluons to form a color-neutral particle christened a "glueball." Worse than that, quarks and anti-quarks could combine to form hybrid and unstable

4. Corinthians First Epistle, Chapter 13, Verses 9 and 12.

mesons as well. All this theoretical mess became known as " Q.C.D. exotics." One justification for this unconventional thinking was that quarks are relatively light in calculated mass and, by themselves, are inadequate to create a baryon. That is, the calculated mass of three selected quarks is not enough to account for the known mass in a combination of the three to form a typical proton. The entrance of the glueballs could solve this problem.

To their delight an experiment actually made earlier back in 1985 at CERN, Switzerland, by Claude Amsler and Frank Close appeared to corroborate this notion. A beam of anti-protons was directed into protons in the form of liquid hydrogen inside of a new design of detector called the "Crystal Barrel." This consisted of a cylindrical barrel formed by 4000 electrically charged fine wires in parallel surrounded on the outside by 1380 cesium iodide scintillation detectors. The results were interpreted to indicate the possibility of a heavy meson designated f_j (1700). This particular experiment was terminated in 1996 by the shut-down of the CERN accelerator and anti-proton ring in Geneva for improvements.

The Pittsburgh group first mentioned hoped to use a laser beam of high energy gamma ray photons, which were created by directing a strongly focused stream of electrons into a thin diamond wafer, to collide with an unspecified target inside a "crystal barrel" detector of calorimeters. The electrons had previously been directed away from the photons by a large magnet before impacting the small target. Not the least of their collective ingenuity was the use of salvaged equipment from other laboratories. They managed to obtain a super-conducting magnet, originally costing about twelve million dollars thirty years ago, from the Stanford Linear Accelerator in California for only three million. The eleven ton Cerenkov counter and lead glass detector is expected to come from the Brookhaven National Laboratory in Long Island. All this equipment is being assembled at the Jefferson National Accelerator facility at Newport News, Virginia. The readers of this book may need to wait a year or more before they read of any results in the newspapers.

In a lecture which he was giving to undergraduate students in physics at Brown University in Providence in 1999, Professor Ulrich on the subject of the Electro-Weak Theory asked the provocative question: " Where do we go from here?" One answer is that a majority of particle physicists from all countries in the world now look to the rebuilt CERN Synchrotron in Geneva as their best hope. Their expectations are not merely for a final verification and proof of the Weak Neutral Currents Theory and a confirmation of the reality of the hypothetical Higgs Boson. At stake beyond that is the entire question of bosons and their creation, their role, their behavior, their physical characteristics, and whence they came. The statement that we have reached "The End of Physics," standing by itself, is ridiculous. What is meant really is that we are approaching the practical limits of the applications of theory and surmised laws of the cosmos which can be replicated in laboratories on the earth for sensory evidence and demonstration to people; but we are not quite at that point yet! The energy transfers of the stars and galaxies are beyond us.

Metaphysical questions abound and refuse to disappear. The reader who is not professionally involved in serious science as an occupation is happily free to indulge in his or her own fancies. Not withstanding the accepted professional explanations described in this book, there still exist some embarrassing inadequacies of a practical nature in the Standard Model. After years of careful experimentation the scientists still do not really know why there are only six quarks which have the distinctly different masses which we can measure; nor can they convincingly explain why the W and Z bosons have the heavy masses which the laboratory measurements seem to indicate. What makes bosons anyway? What makes the proton so permanent? The exact masses of the three neutrinos are up for debate still; nor can we explain why the light photon, among all the heavier bosons, remains massless. We are unable to explain why the probability that any given electron will absorb or emit a photon remains at an exact constant of one in 137,03597 times. Not only that, the implicit assumption that the mass of the tiny electron remains an unchanged constant, thus making the absorption or emission of a photon in any atom purely a function of the energy in the centrifugal force as the electron hops between orbits of different diameters, remains a mystery. Why does it not disgorge or capture back a photon by changing its mass? Who made the rule? What actually comprises the dark and unseen mass which astronomers and astro-physicists both insist must exist in the outside perimeter of nebulae and throughout the entire universe? The hope is that the Superstrings Theory and various "Theories of Everything" may eventually provide an answer in the decades ahead. The joy for the amateur scientist or curious dilettante is that he or she is entirely free to form their own opinions, construct their own fantasies, venture their own guesses, and argue with any objectionable and didactic postulates at will without restraint or fear of censure by peers. Read whatever you desire and can find in literature in the subject, including the current attempts to explain the mathematicians' efforts to wrap up relativity, mass, time, and space in one tidy package of equations. Have fun doing it. There are a lot of fascinating ideas out there.

END

BIBLIOGRAPHY

Adair, Robert. *The Great Design: Particles and Fields of Creation.* (New York: Oxford Univ. Press, 1987.)

Albert, David Z. *Quantum Mechanics and ExperiEnce.* (CambriDge, Mass.: Harvard Univ. Press, 1992.)

Alexandrov, Yu A. *Fundamental Properties of the Neutron.* (Oxford, England: Clarendon Press of Oxfotrd Univ. Press, 1992.)

Baggot, James. 192. *The Meaning of the Quantum Theory.* (New York, USA: Oxford Univ. Press, 1992.)

Bahcall, John and Ostriker, Jeremiah P., eds. *Unsolved Problems in Astrophysics.* (Princeton, N.J.: Princeton Univ. Press. 1997)

Ballentine, L.E., ed. *Foundations of Quantum Mechanics Since the Bell Inequalities.* (College Park, Md.: American Association of Physics TeAchers, 1988.)

Barrow, John D. *Pi in the Sky.* (Oxford, England: Clarendon Press, 1992.)

Bell, John S. *Speakable and Unspeakable in Quantum Mechanics.* (New York: Cambridge Univ. Press, 1987.)

Bergman, Peter. *Introduction to the Theory of Relativity.* (New York: Dover Publications; 1976, first published 1942.)

Bernstein, Jeremy. *Quantum Profiles.* (Princeton, N.J.: Princeton Univ. Press, 1991.)

Bernstein, Jeremy. *Cranks, Quarks and the Cosmos.* (New York: Basic Books Division of Harper Collins Publishers, 1993.)

Bethe, Hans A. *The Road from Los Alamos.* (New York: Touchstone Books of Simon & Schuster Publishing Co., 1991.)

Blin–Stoyle, R.J. *Nuclear and Particle Physics.* (London: Chapman and Hall., 1991.)

Bohm, David. *Quantum Theory.* (New York: Prentice Hall; reprinted by Dover Books Publications in 1987.)

Bohm, David. *Three Roads to Quantum Gravity.* (New York: Walker Publishing and Co., 2000.)

Bohm, David. *Wholeness and the Implicate Order.* (London: Routledge and Kegan Paul, Ltd., 1980.)

Bohm, David and Hiley, Basil J. *The Undivided Universe: An Ontological Interpretation of Quantum.* (London: Routledge, 1993.)

Born, Max. *Einstein's Theory of Relativity.* (New York: Dover Publications, 1962.)

Born, Max. *Atomic Physics.* (New York: Cambridge Univ. Press, 1969.)

Cahn, Robert N. and Goldhaber, Gerson. *The Experimental Foundations of Particle Physics.* (New York: Cambridge Univ. Press, 1989.)

Calder, Nigel. *Einstein's Universe.* (New York: Penguin Books, 1979.)

Capra, Fritjof. *The Tao of Physics.* (Boulder, Colo.: reprinted by Bantam Books of New York. Originally by Shambhala Publications in 1976, at Boulder, Colo., 1977.)

Cheng, T.P. and Ling–Fong, Li, eds. *Gauge Invariance.* (College Park, Md.: American Association of Physics Teachers, 1990.)

Close, Frank; Marten, Michael; and Sutton, Christine. *The Particle Explosion.* (New York: Oxford Univ. Press, 1987.)

Cooper, Necia G., and West, Geoffrey B., eds. *Particle Physics: A Los Alamos Primer.* (New York: Cambridge Univ. Press, 1988.)

Coughlan, G.D., and Dodd, J.E. *The Ideas of Particle Physics: An Introduction for Scientists,* 2nd ed. (Great Britain and New York: Cambridge Univ. Press, 1984.)

Davies, Paul *The EDge of Infinity.* (New York: Simon & Schuster, 1981.)

Davies, Paul. *God and the New Physics.* (New York: Simon & Schuster, 1983.)

Davies, Paul. *The Cosmic Code.* (New York: Simon & Schuster, 1988.)

Dirac, P.A.M. *The Principles of Quantum Mechanics.* (London: Oxford Univ. Press 1930; 4th edition printed in New York by Oxford Science Publications in 1958.)

Eddington, Arthur S. *The Nature of the Physical World.* (New York: The Macmillan Co.; London: CaMbridge UnIv. Press, 1928.)

Einstein, Albert. *The Meaning of Relativity.* (Princeton, N.J.: Princeton Univ. Press, 1950.

European Physical Journal. Sec. C, vol. 3. Edited by Haidt, D. and Zerwas, P.M.) (Published by Springer Verlag of New York and Heidelberg, Germany, 1998.)

Feynman, Richard P. *QED: The Strange Theory of Light and Matter.* (Princeton, N.J.: Princeton Univ. Press, 1989.)

Feynman, Richard P. *The Character of Physical Law.* (Cambridge, Mass, MIT Press).

Feynman, Richard P. *Six Easy Pieces.* (Edited by Leighton, Robert B. and Sands, Matthew. Reading, Mass.: Helix Books of Addison–Wetley Publishing Co., 1994.)

Feynman, Richard P. *The Meaning of It All.* (Reading, Mass.: Helix Books of Addison–Wesley Co., 1998.)

Feynman, Richard P; Leighton, Robert B; and Sands, Matthew; *The Feynman Lectures on Physics Vols 1&3.* (Reading, Mass, Addison-Wesley Publishing Co, 1963).

Feynman, Richard P. and Weinberg, Steven. *Elementary Particles and the Laws of Physics.* (New York: Cambridge UnIv. Press, 1987.)

Fritzsch, Harald. *Quarks: The Stuff of Matter.* (New York: Basic Books Division of Harper Collins Publishers, 1983.)

Gaisser, Thomas K. *Cosmic Rays and Particle Physics.* (CambriDge and London: Cambridge Univ. Press, 1990.)

Gamors, George. *Thirty Years that Shook Physics* (New York, Dover Publications, 1966).

Gamow, George. *One, Two, Three—Infinity.* (New York: Dover Publications 1988; originally published by Viking Press in 1961,)

Gell–Mann, Murray. *The Quark and the Jaguar.* (New York: W. H. Freeman Co., 1994.)

Glashow, Sheldon. *Interactions.* (New York: Warner Books, 1988.)

Gleick, James. *Chaos.* (New York: Penguin Books, 1987.)

Gleick, James. *Genius: The Life and Science of Richard Feynman.* (New York: Pantheon Books, 1992.)

Green, Brian. *The Elegant Universe.* (New York: W.W. Norton and Co., 1999.)

Greenberg, O.W., ed. *Quarks: Selected Reprints.* (College Park, Md.: American Association of Physics Teachers, 1986.)

Gribbin, John. *In Search of Schrodinger's Cat: Quantum Physics and Reality.* (Toronto and New York: Bantam Books, 1984.)

Gribbin, John. *Schrodinger's Kittens and the Search for Reality.* (Boston: Little Brown & Co., 1995.)

Gribbin, John. *The Search for Super–Strings, Symmetry and Theory of Everything.* (Originally Penguin Books, republished by John and Mary Gribbin, 1998.)

Grotz, K., and Lapdor, H.V. *The Weak Interaction in Nuclear, Particle Physics and Astrophysics.* (Bristol, England: Adam Hilger. Imprint of IOP Publishing, Ltd., 1990.)

Haken, H., and Wolf, H.C. *Atomic and Quantum Physics.* (Heidelberg, Germany: Springer–Verlag, 1987.)

Han, M.Y. *The Probable Universe.* (Blue Ridge Summit, Pa.: Tab Books, 1993.)

Heilbron, J.L., and Seidel, Robert W. *Lawrence and His Laboratory, vol. 1.* (Berkeley, California: Univ. of California Press, 1989.)

Heisenberg, Werner. *The Physical Principles of the Quntum Theory.* (New York: Dover Publications, 1930.)

Heisenberg, Werner. *Physics and Beyond.* (New York and London: Harper & Row, 1971.)

Heisenberg, Werner. *Philosophical Problems of Quantum Mechanics* (Pantheon Books 1952, Reprinted by Ox Bow Press, Conn 1979).

Heisenberg, Werner. *Encounters with Einstein,* (Princeton, N.J.: Princeton Univ. Press, 1983.)

Heitler, W. *The Quantum Theory of Radiation.* (New York: Dover Publications, 1984.)

Herbert, Nick. *Quantum Reality.* (Garden City, N.Y.: Anchor Press of Doubleday, 1985.)

Herbert, Nick. *Faster than Light: Superluminal Loopholes in Physics.* (Ontario, Canada: Penguin Books, 1988.)

Hey, Tony, and Walters, Patrick. *The Quantum Universe.* (New York: Cambridge Univ. Press, 1987.)

Hiller, John R.; Johnston, Ian D.; and Styer, Daniel F. *Quantum Mechanics Simulations.* (New York: John Wiley & Sons, 1995.)

Hinton, Charles H., *Speculation on the Fourth Dimension.* Edited by Rucker, Rudolf. (New York: Dover Publications, 1980.)

Hoffman, Banesh. *The Strange Story of the Quantum.* (New York: Dover Publications; first published in 1947, 1959.)

Holton, Gerald, and Van der Waerden, B. L., eds. *Sources of Quantum Mechanics.* (New York: Dover Publications, 1967.)

Honig, William M. *The Quantum and Beyond.* (Perth, Australia: Swan River Press, 1986.)

Jeans, Sir James. *The Mysterious Universe.* (New York: The Macmillan Co., 1930.)

Jeans, Sir James. *Physics and Philosophy.* (New York: Dover Publications 1981; originally published in England: Cambridge Univ. Press, 1943.)

Johnson, George. *Strange Beauty* (A biography of Murray Gell–Mann): (New York City by Borzai Book Division of Alfred A. Knopf, Inc., 1999.)

Kafatos, Minos and Miculitsianos, A., eds. *Supernova 1987A in the Large Magellanic Cloud.* (New York: Cambridge Univ. Press, 1988.)

Kane, Gordon. *The Particle Garden.* (Reading, Mass.: Helix Books of Addison–Wesley Co,, 1995.)

Lederman, Leon. *The God Particle.* (New York: Houghton Mifflin Co., 1993.)

Lightman, Alan. *Dance for Two* Essays. (New York: Pantheon Books, 1996.)

Lindley, David. *The End of Physics.* (New York: Basic Books Division of Harper Collins Publishers, 1993.)

Lindley, David. *Where Does the Weirdness Go?* New York: Basic Books Division of Harper Collins Publishers, 1996)

Lockwood, Michael. *Mind, Brain and the Quantum.* (Oxford, England: Blackwell Press, 1989.)

March, Robert H. *Physics for Poets.* (Chicago: Contemporary Books, 1978.)

Maddox, John. *What Remains to be Discovered.* (New York City; The Free Press Division of Simon and Schuster, Inc., 1998.)

Modinos, A. *Quantum Theory of Matter.* (New York: John Wiley & Sons, 1996.)

Morris, Richard. *The Edges of Science.* (New York: Prentice Hall Press, 1990.)

Newton, Sir Isaac. *Principia* [Mathematical Principles], vols. 1 & 2. (Berkeley, Los Angeles: Univ. of California Press; oriignal first printing in 1934, reprinted 1962.)

Overbye, Dennis. *Lonely Hearts of the Cosmos.* (New York: Harper Collins, 1991.)

Pagels, Henry R. *The Cosmic Code: Quantum Physics as the Language of Nature* (New York, Simon & Schuster, 1982).

Pagels, Heinz R. *Perfect Symmetry.* (New York: Bantan Books of Simon & Schuster, 1986.)

Pais, Abraham. *Subtle is the Lord: The Science and Life of Albert Einstein.* (London: Oxford Univ. Press, 1982.)

Pais, Abraham. *Inward Bound: Of Matter and Forces in the Physical World.* (New York: Clarendon Press of Oxford Univ. Press, 1986.)

Parker, Barry. *Einstein's Dream.* (New York: Plenum Press, 1986.)

Parker, Barry. *Creation.* (New York: Plenum Press, 1988.)

Penrose, Roger. *The Emperor's New Mind.* (London: Oxford Univ. Press, 1989.)

Pettofrezzo, Anthony J. *Matrices and Transformations.* (New York: Dover Publications, 1966.)

Planck, Max. *A Survey of Physical Theory.* (New York and Mineola, N.Y.: reprint of translation by Dover Publications 1993; originally published by Methuen & Co. of London in 1925.)

Polkinghorne, J.C. *The Quantum World.* (Princeton, N.J.: Princeton Univ. Press, 1985.)

Polkinhorne, John. *Rochester Roundabout: The Story of High Energy Physics.* (Essex, England: Longman Singapore Publishers, 1989.)

Rae, Alastair. *Quantum Physics: Illusion or Reality.* (New York: Cambridge Univ. Press, 1986.)

Regents of the University of California and the U.S. Department of Energy Particle Data Group. *"Review of Particle Physics." The European Physical Journal.* (Heidelberg, Germany: Springer–Verlag, 1998.)

Rifkin, Jeremy. *Entropy: A New World View.* (New York: Bantam Books of the Viking Press, 1980.)

Riordan, Michael. *The Hunting of the Quark.* (New York: Simon & Schuster, 1987.)

Riordan, Michael and Schramm, David N. *The Shadows of Creation.* (New York: W.H. Freeman and Co., 1991.)

Rohrlich, Fritz. *From Paradox to Reality.* (New York: Cambridge Univ. Press, 1987.)

Schrodinger, Erwin. *The Interpretation of Quantum Mechanics.* (Edited by Michael Bitbol in 1995. Ox Bow Press of Woodbridge, Conn., 1955.)

Schrodinger, Erwin, *My View of the World.* (Hamburg, Germany, Paul Z. Vertag. Reprinted by On Bow Press, Woodbridge, Conn. 1961.)

Schwarz, Cindy. *A Tour of the Subatomic Zoo.* (New York: American Institute of Physics, 1992.)

Schwenk, Henry C., and Shannon, Robert H. *Nuclear Power Engineering.* (New York: McGraw–Hill Book Co., 1957.)

Shankar, Ramamurti. *Principles of Quantum Mechanics.* (New York and London: Plenum Press, 1980.)

Shimony, Abner. *Search for a Naturalistic World View. Vols. 1 & 2.* (New York and London: Cambridge Univ. Press, 1993.)

Silk, Joseph. *The Big Bang.* (New York: W. H. Freeman Co.; first published in 1980, reprinted 1989.)

Skinner, Ray. *Relativity for Scientists and Engineers.* (New York: Dover Publications, 1982.)

Smyth, Henry D. *Atomic Energy for Military Purposes.* (Princeton, N.J.: Princeton Univ. Press, 1945.)

Stranathan, J.D. *The Particles of Modern Physics.* (Philadelphia: The Blakiston Co., 1945.)

Sutton, Chrstine. *Spaceship Neutrino.* (Cambridge, England: Cambrdige Univ. Press, 1992.)

Taubes, Gary. *Nobel Dreams: Power, Deceit and the Ultimate Experiment.* (New York: Random House, 1986.)

Tolman, Richard. *Relativity, Thermodynamics and Cosmology.* (New York: reprint by Dover Publications 1987; originally published in 1934 by Oxford Univ. Press in England.)

Tomonaga, Sin–itro. *The Story of Spin.* (N.p.: Univ. of Chicago Press, USA, 1997.)

Trefil, James. *From Atoms to Quarks.* (New York: Scribner's Sons, 1980.)

Weinberg, Steven. *Dreams of A Final Theory: A Scientist's Rearch for the Ultimate Laws of Nature.* (New York: Vintage Books Division of Random House, 1994.)

Weisskopf, Victor F. *The Privilege of Being a Physicist.* (New York: W.H. Freeman Co., 1989.)

Whitehead, Alfred North. *An Inquiry Concerning the Principles of Natural Knowledge.* (New York: Dover Publications, 1982.)

Wilber, Kenneth, ed. *Quantum Questions.* (Boulder, Colo.: New Science Library of Shambhala, 1969.)

Wilczek, Frank, and Devine, Betsey. *Longing for His Harmonies.* (New York: Hortan & Co., 1988.)

Williams, W.S.C. *Nuclear and Particle Physics.* (New York: Oxford Univ. Press, 1991.)

Wolf, Fred Alan. *Parallel Universes.* (New York: A Touchstone Book of Simon & Schuster Publishing Co. 1988.)

Wolf, Fred Alan. *Taking the Quantum Leap.* (New York: Harper & Row, 1989.)

Zee, A. *An Old Man's Toy: Gravity in Einstein's Universe.* (New York: The Macmillan Publishing Co., 1989.)

Zukav, Gary. *The Dancing Wu Li Masters: Overview of the New Physics.* (New York: Bantam Books of the Viking Press, 1979.)